MOLECULAR THERMODYNAMICS: A STATISTICAL APPROACH

MOLECULAR THERMODYNAMICS: A STATISTICAL APPROACH

JAMES W. WHALEN
The University of Texas at E1 Paso

A Wiley-Interscience Publication
JOHN WILEY & SONS, INC.
New York • Chichester • Brisbane • Toronto • Singapore

In recognition of the importance of preserving what has been written, it is a policy of John Wiley & Sons, Inc., to have books of enduring value published in the United States printed on acid-free paper, and we exert our best efforts to that end.

Library of Congress Cataloging in Publication Data:

Whalen, James W.
Molecular thermodynamics : a statistical approach / James W. Whalen.
p. cm.
"A Wiley-Interscience publication."
Includes index.

ISBN 0-471-51478-0
1. Thermodynamics. I. Title.
QD504.W47 1991 90-39767
CIP

Printed in the United States of America

10 9 8 7 6 5 4 3 2 1

PREFACE

Simply stated, the objectives of thermodynamics would be to construct a framework and the related detail that incorporates the observable features of all known substances and their mixtures, in all physical states, their transitions between states, and all chemical changes that they can undergo. These objectives are very broad.

The subject has been explored for more than a hundred years. The subject detail would fill several volumes and is expanding. Even a listing of its application would fill many more volumes. There is an introduction to the subject in every textbook of physical chemistry and there are numerous, excellent, detailed treatises available to satisfy all levels of interest. From its introduction by Gibbs and Boltzmann to this date, contributors to the subject have included many of the most respected scientists of all time. Thermodynamics is not a neglected subject that requires refinement. Another book would seem to require some justification.

The study of thermodynamics normally follows the historical development in which a minimum number of unambiguous observables are incorporated into a maximally general framework. The selection of observables that will define a structure of complete generality can, to a student, seem arbitrary and remote from its intended use. Also, since the number of observables and the processes to which they can apply are large, the intricate detail required to select and accommodate meaningful observables requires some ingenuity to construct and considerable discipline to follow. There is no way to avoid assimilation of the detail in application. All that a different approach to the subject can hope to accomplish is a larger understanding of the framework and its applicability. The by-product can be an increased inclination to master the detail.

The approach which I have incorporated into this text is based on statistical molecular argument that the student seeking an introduction to the subject may find more congruent with present day background than the traditional arguments of continuum mechanics. In following this line of development, topics do surface in a different sequence than is encountered in the traditional development. This may pose a difficulty, but I hope not an insurmountable one, for those familiar with the general subject. The number of worked examples and end of chapter problems may, to those accustomed to a more traditional presentation, seem minimal. This is deliberate since it is imagined that extended effort will be expended on understanding the structure. There is a large number of problems in the cited sources for the subject matter of this text for those who wish to acquire more applications experience. I have drawn heavily on those sources and wish to acknowledge all of the authors as well as many others not specifically cited in end of chapter references.

I wish to thank Dr. M. I. Davis, Dr. K. W. Hipps and Dr. L. C. Porter for looking at the manuscript and for their helpful suggestions. I also wish to thank Dr. A. Höpfner of the Institut für Physicalische Chemie, Universität Heidelberg, and Dr. J. Sauer of the Institut für Organische Chemie, Universität Regensberg, for their hospitality during early stages of the writing. The book could not have been completed without the assistance and encouragement offered by my wife.

JAMES W. WHALEN

El Paso, Texas
December, 1990

CONTENTS

MOLECULAR THERMODYNAMICS: A STATISTICAL APPROACH

INTRODUCTION

Chemical thermodynamics is a formal framework for the consideration of material and energy exchange processes. The framework consists of four fundamental statements (referred to as the zeroth, first, second, and third laws of thermodynamics) and the body of implication (largely mathematical) that follows when the application of the laws is subject to boundary conditions (constraints on the process circumstances). The four laws are phenomenological in the sense that they are a minimum set of statements that encompasses existing experience. The body of the subject lies in the first and second laws; the zeroth and third laws can be regarded as fundamental to the structure but their impact is rather operational. Many, seemingly different, first and second law statements will be encountered. Such statements differ, not in truth, but in generality and focus.

A commonly encountered combined first and second law statement asserts that the energy of the universe is constant and its entropy[1] is increasing. Another such statement denies the possibility of perpetual motion of either the first or second kinds. For our use a statement like the first has complete generality but zero focus. A statement like the second, which has its origin in concerns relating to heat engines, is too sharply focused. Neither kind of statement will serve usefully as a basis for developing operational chemical thermodynamics.

In the classical development of chemical thermodynamics we would begin with a combined first and second law statement such as "in an isolated system

[1] Entropy is a feature of the energy that relates to the state of disorder.

energy is conserved (constant) and an increase in entropy accompanies any spontaneous process." The isolated system is taken to be a bulk system of any chemical complexity and that part of the surroundings that can interact with the system through either energy or material exchange. Such exchanges should be reflected by observable features of either the system or surroundings or both. The identification of meaningful observables and their incorporation into an internally self-consistent structure are the results that we call classical thermodynamics. The bulk system is a material collection identified by mass and chemical composition. Its internal, microscopic detail is not of concern in the formulation.

Microscopic detail (i.e., molecular behavior) is normally introduced as an extension to the classical development. It is, however, possible to develop the thermodynamic structure from step by step statistical argument. This approach would seem to have the advantage that certain thermodynamic parameters, such as entropy, that are difficult to conceptualize in terms of bulk material processes arise naturally as statistical terms in the molecular argument. If we adopt this course in the following discussions and in identifying the structures developing from them and if we retain the traditional language of the classical subject, our development will overlay that which follows from the classical route.

Our discussions will begin with the motions of atoms and molecules and the manner in which these motions relate to particle energies. We will then be interested in the average energy and the distribution of energies in large collections of particles. It will be necessary to introduce statistical constructs that produce meaningful averages in many particle systems when subject to stated constraints and to consider the significance of such terms. The statistical approach, supported by conventional mathematical formalism, will replicate the classical thermodynamic structure and provide the equilibrium conditions for the distribution of particles between phases and among species. We will have introduced the subject on the basis of statistical rather than phenomenological argument, providing at the same time the framework in which many energetically related quantities for processes of concern can be obtained by direct calculation.

Such calculations do not replace experimental measurements carried out on macroscopic systems. The correctness of the involved statistical treatment and the limitations of the simplifying assumptions involved must be measured against experiment. Mathematical complexity and/or ignorance of detailed structure will ultimately limit our ability to calculate but the identification of important statistical parameters and their relationship to the system observables should lend larger understanding to the subject, even when the molecular argument has become too involved to follow.

It will be assumed throughout this treatment that the student's background is adequate as regards fundamental mathematical and physical concepts, particularly when supplemented by the excellent referenced texts. However, recognizing that the variety in style and symbol usage may introduce difficulties, it will be convenient to provide certain background material when extended analytic detail in a particular notation is required. So that the basic thread of argument can be more easily followed, this material and much of the development detail will be provided in appendices. Since our arguments will begin with molecule energies, Appendix A on Energy should be reviewed and supplemented by the cited references if necessary. A very brief summary of those energetic concepts that we require immediately is as follows.

Newton's laws of motion are phrased in terms of forces. Energy is a second-order implication. If the force (a vector[2] quantity) acting on a body is defined by the particle motion, $\mathbf{f} = (d/dt)m\mathbf{v}$ and $\mathbf{f}\cdot\mathbf{v} = (d/dt)(\frac{1}{2}mv^2)$ where $\mathbf{v}$ is velocity (a vector[2] quantity) and m is mass, we refer to the term $\frac{1}{2}mv^2$ as the kinetic energy. If the force is defined by the negative gradient of a function of particle position (r), $\mathbf{f} = -\nabla U(\mathbf{r})$, we refer to the function $U(\mathbf{r})$ as the potential energy. Molecules and atoms, individually and in large collections, will always have kinetic energy and collections of them, even in the absense of external fields, may have potential energies deriving from internal fields.

In the limit of small particles moving in small dimensions a more general (quantum) mechanics requires that only discrete (quantized) values of the energies be permitted. All electronic and nuclear motions and the internal motions in diatomic and polyatomic molecules fall within this constraint. In most cases, center of mass motions (translation and rotation of the molecule) do not require the quantum treatment. When quantum considerations apply it is required that the allowed energy states be identified. In most center of mass motions the quantum treatment degrades to an energy continuum. But since our statistical approach will require the assignment of energy values to particles (atoms and molecules) in our systems of interest, it will normally be convenient to phrase our arguments in terms of discrete energy values (as would be associated with quantum considerations) as opposed to energy intervals (as would be associated with continuum considerations). We can then simplify the results as appropriate.

A normal source of energy level information for electronic, vibrational, and rotational energy levels will be spectroscopic data that represent allowed transitions from one energy state to another. Discrete energy packages (photons) associated with electromagnetic energy fields can be absorbed (or emitted) serving as probes which identify transitions and therefore the energy level spacing, e.g., for absorption of photons

$$(\epsilon_n - \epsilon_0) = \Delta\epsilon_{0\to n} = \epsilon_{\text{photon}}$$

where ϵ_n is some excited state and ϵ_0 is the ground (lowest energy) state. Transitions between excited states are, of course possible

$$(\epsilon_{n'} - \epsilon_n) = \Delta\epsilon_{n\to n'} = \epsilon_{\text{photon}}$$

[2] See Appendix B.

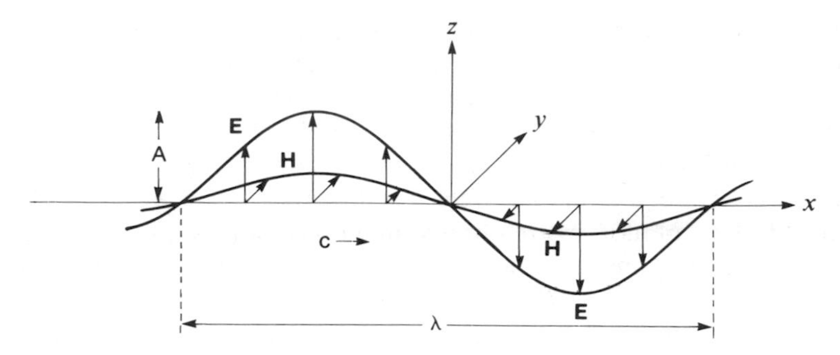

Figure I.1. Illustration of an electromagnetic wave with wavelength λ and amplitude A propogating along the X axis. The electric field vector **E** points along the z axis; the magnetic field vector **H** points along the Y axis.

as are transitions from higher to lower states accompanied by the emission of photons having the discrete energy associated with the particular transition. Photon energies are given by the relationship

$$\epsilon_{\text{photon}} = h\nu$$

where h is a fundamental constant (Planck's constant) arising from quantum argument and ν is a frequency associated with the waveform of the electromagnetic wave. Transitions, of course, tell us nothing about the absolute energy of the ground state.

The electromagnetic wave can be represented as perpendicular, in-phase electric and magnetic field oscillations as indicated in Figure I-1. The waveform is characterized by an amplitude A that relates to its intensity and by wavelength λ and velocity c that relate to its energy. The velocity associated with the waveform for any particular wavelength is dependent on the medium through which the field is propagated. In vacuum c is $3 \times 10^{10}\,\text{cm sec}^{-1}$ for all λ. Frequency, $\nu\,\text{sec}^{-1}$, is defined as c/λ. Photon energies are often characterized by wave number, $\tilde{\nu}\,\text{cm}^{-1}$, where $\tilde{\nu}$ is defined as $1/\lambda$. Then

$$\epsilon_{\text{photon}} = hc\tilde{\nu}.$$

Figure I-2 relates the electromagnetic energy spectrum in terms of commonly identified regions to the molecule motions of interest and to electronic excitation. Translational energy levels are too closely spaced to be resolved by photon energies. Various quantum expressions that include molecule features relate to the spectroscopically identified energy levels through the constant h and sets of quantum constraints that are integer numbers running from 0 or 1 to unbounded higher values. The energy level values for modes of particle motion with which we will be concerned are as follows[3]:

[3] See Appendix A on Energy for detail and extension.

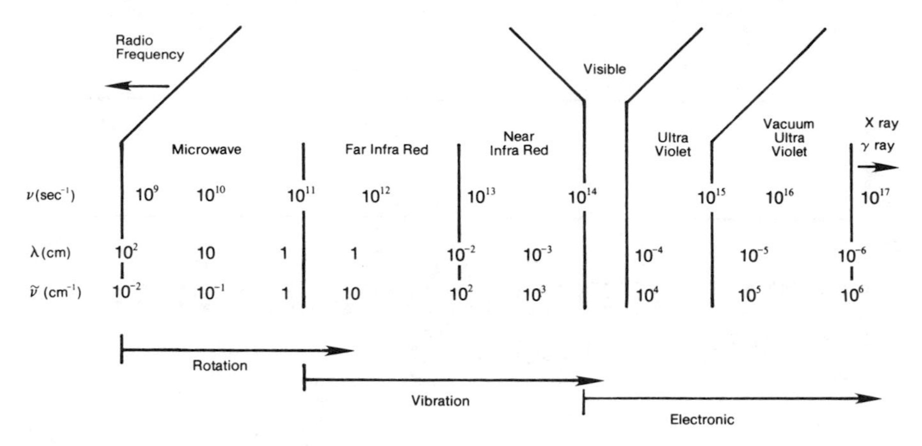

Figure I.2. The electromagnetic spectrum indicating regions that are energetically equivalent to molecular motions.

translation:

(in x, y, z coordinates) $\quad \epsilon_i = \dfrac{(n_x^2 + n_y^2 + n_z^2)h^2}{8m\mathrm{V}^{2/3}} \qquad n_x, n_y, n_z = 1, 2, 3, \ldots, \infty$

rotation:

(about the mass center) $\quad \epsilon_i = \dfrac{j(j+1)h^2}{8\pi^2 I} \qquad j = 0, 1, 2, \ldots, \infty$

vibration: $\quad \epsilon_i = (v + \tfrac{1}{2})h\nu_e \qquad v = 0, 1, 2, \ldots, \infty$

where V is volume, I is moment of inertia, and ν_e is the fundamental vibrational frequency associated with a particular vibrational mode. The $n_x \cdots j \cdots v$ are quantum numbers that particularize allowed energy levels. General expressions for electronic energy levels cannot be expressed in compact form although the quantum constraints are quite clear.[4] Ground state energies for rotation and vibration are zero and $\frac{1}{2}h\nu_e$, respectively. For translation the ground state energy is that associated with the particle effectively at rest and the electronic ground state energy for atoms is taken to be zero for chemical argument.

[4] See Appendix A on Energy for detail and extension.

1

THE STATISTICAL APPROACH TO THERMODYNAMICS

The fundamental assumption of the statistical approach to thermodynamics is that the behavior of materials in bulk can be described by appropriate averages obtained from summed activities of the constituent particles. Materials in bulk will contain some multiple of Avogadro's number (order 10^{24})[1] of particles (atoms, molecules, or their constituent nuclei and electrons). We will refer to such collections as macroscopic systems where we use the term system to refer to any collection of material particles enclosed by a boundary. Anything not included in a defined system will be considered as the surroundings. Any macroscopic system will have a microscopic description, i.e., a description in terms of its individual particles. The microscopic description we will address initially is in terms of particle energies.

The total internal energy of a macroscopic system is the sum of the kinetic and potential energies associated with the various active modes of motion of the individual particles together with the sum of the potential energies resulting from interactions between the particles. We anticipate that there is a distribution of kinetic energies among the individual particles and among their several modes of motion and that the kinetic energy of a single particle is constantly changing as a result of energy exchange occurring during collision or in the coupling of internal motions.

We characterize this exchange process as resulting from very weak interactions to distinguish the process from those particle interactions originating in

[1] From this point "order" and "order of" will be represented by the symbol (O).

various collective forces acting between particles. These collective forces, which are referred to as van der Waals forces, vary with particle structure. They are effective when the interparticle distances of separation are a few molecular diameters or less. In very low-density (dilute) particle systems the potential energy contribution to the total energy may be negligible. Although the kinetic energies of individual particles are continuously redistributed as a result of very weak interactions and the potential energy of an individual particle with reference to other particles will vary with particle motion, the total internal energy will remain constant if the system is prohibited from exchanging energy or particles with other systems or with the surroundings. We summarize this statement as "the internal energy of an isolated system is conserved (constant)."

On the basis of the above remarks we assume that we can construct a macroscopic system as a collection of N [(O) 10^{24}] identical particles. We can imagine selecting these particles from a larger collection having random kinetic energies ϵ_κ where the subscript κ (a general running index) identifies any of many possible values of ϵ. We can enclose the particle collection by fixed boundaries and isolate it from the surroundings. We can make the volume sufficiently large that the average particle separation will be large compared to the distance over which van der Waals forces are effective so that, even with structured particles, we need consider only the kinetic contribution to the total energy. We will have constructed an isolated system of N particles occupying volume V, and having energy $E(=\sum N_\kappa \epsilon_\kappa)$. We are not immediately interested in the value of E but we are assured that it will remain constant under the imposed constraints although the original ϵ_κ values will certainly change. If we further insist that the particles have no structure so that the kinetic energy is entirely translational, we can formulate a simple expression for the quantized energy levels that are available to particles in the system.

The quantized energy levels for particles of mass m occupying volume V as we have noted are given by

$$\epsilon_i = \frac{(n_x^2 + n_y^2 + n_z^2)h^2}{8mV^{2/3}} \tag{1.1}$$

where h is Planck's constant and the quantum number set n_x, n_y, n_z relates to the three degrees of translational motion. The translational energy levels of particles of given mass depend only on the volume to which they are confined. We presume that the translational motions are independent, i.e., no single n_x or n_y or n_z value depends on any other value. Clearly a particular energy level ϵ_i can be realized by many different values of n_x, n_y, n_z and by permutations of the values among n_x, n_y, n_z. Each distinguishably different expression we refer to as a quantum state. The total number of quantum states corresponding to a particular energy level can be quite large. We shall refer to this number of quantum states as the degeneracy of the energy level and designate an energy level and its associated degeneracy as ϵ_i and g_i, respectively.

We anticipate that, although the n_x, n_y, n_z are unbounded by quantum considerations, there will be some limit to the number of energy levels in the sense of availability that will permit many of the N particles to occupy each of the available energy levels associated with the volume that they occupy; let the number of particles in energy level ϵ_1 be N_1, in energy level ϵ_2 be N_2, etc. Generally, the number of particles in energy level ϵ_i is N_i. The distribution number set $N_1, \ldots, N_i$ is a feature of the microscopic description of the macroscopic system having N particles $(=\sum N_i)$ and energy $E(=\sum N_i \epsilon_i)$ while occupying volume V. We regard the parenthetical conditions as constraints on the distribution numbers, recognizing that there are many different distribution number sets that satisfy these constraints. We will find it convenient to call each unique distribution number set a complexion of the system. Each complexion is, then, a particular microscopic description of the system.

It would be our assumption that the exchange of energy on particle collision would produce a constantly changing complexion representing, over any long period of time, all complexions that are compatible with the two constraints and that all such complexions are equally probable. Any single complexion can be replicated in a very large number of ways by interchanging particles among the energy levels while retaining the particular distribution numbers $N_1, \ldots, N_i$. A statistical expression that proves adequate for our immediate need (but see Appendix C on Statistics) provides for interchange of particles among energy levels correcting for sequential permutations of particles within an energy level assignment. We write

$$t = N! \prod^{i} \frac{g_i^{N_i}}{N_i!} \tag{1.2a}$$

where t (term) is the number of ways in which a single unique distribution can be achieved. There is one such term for each complexion and the number of ways, Ω, in which all (κ) unique complexions can be achieved is

$$\Omega = \sum^{\kappa} \left(N! \prod^{i} \frac{g_i^{N_i}}{N_i!} \right)_{\kappa}. \tag{1.3a}$$

We obtain Eq. (1.2a) by the following argument. Number N particles sequentially. Assign the first N_1 particles to energy level ϵ_1, the next N_2 to energy level ϵ_2, etc. to the assignment of N_i particles to energy level ϵ_i. The N_1 particles assigned to energy level ϵ_1 can each be given g_1 assignments, i.e., there are $g_1^{N_1}$ different assignments for the N_1 particles. The $g_1^{N_1}$ different assignments in energy level ϵ_1, combine with all such assignments in every other energy level leading to product over all $g_i^{N_i}$. Retain the $N_1, \ldots, N_i$ distribution numbers but make all possible interchanges of particles between the assignments of particles to the $N_1, \ldots, N_i$ values. There are N! such sequencing assignments. Among these there are arrangements in which only the sequencing within an energy level is changed, e.g., the arrangement of *klm* ··· is changed to *lmk* ··· or to *mlk* ···. These do not lead to recognizably different complexions. We must divide by each $N_i!$ to obtain the number of identifiable unique complexions. This leads to product over all $N_i!$.

In the sequencing that lead to Eqs. (1.2a) and (1.3a) we have imagined that each particle is uniquely identifiable. Generally speaking, molecules of a single species are not uniquely identifiable and those interchanges represented by the N! factor do not always lead to distinguishably different complexions. Our expression for indistinguishable particles should be

$$t = \prod^{i} \frac{g_i^{N_i}}{N_i!} \tag{1.2b}$$

and

$$\Omega = \sum^{K} \left(\prod^{i} \frac{g_i^{N_i}}{N_i} \right). \tag{1.3b}$$

We must identify Ω in further use as Ω_I (indistinguishable particles) or Ω_D (distinguishable particles).

The distinction is important since, in gas phase systems Eqs. (1.2b) and (1.3b) would be the valid expressions. We will, however, encounter particles assigned to a site where the site is distinguishable by virtue of its fixed location. Then Eqs. (1.2a) and (1.3a) would be the valid expressions. Operationally we can select the appropriate expression on the basis of whether the particles are localized (e.g., in a crystal structure) or nonlocalized (e.g., in a gas phase).

All of the above statistical expressions have an extremely important feature that can be illustrated in a very simple way. Consider, as a special numerical example, the case of no degeneracy; then Eq. (1.2a) becomes

$$t = \frac{N!}{\prod N_i!}. \tag{1.2c}$$

Now, assign just 20 particles to 5 such nondegenerate energy levels.[2] For a 4-4-4-4-4 distribution, t is (O) 10^{11}. For a 10-7-1-1-1 distribution, t is (O) 10^8. We conclude that, although this is an admittedly large perturbation, the value of t can change drastically with change in the distribution numbers. But, more importantly, as the number of particles and the number of energy levels increases, the distinction between values for various distribution numbers escalates rapidly. For 200 particles distributed equally among 10 energy states t is (O) 10^{191} and to obtain the thousand-fold less probable distribution that we obtained for the 20 particle, 5 level case we require a term $200!/(25!)^5(15!)^5$, clearly not so extreme but, still, a large perturbation. However, if the 200 particles are distributed

[2]Our conclusions will apply to Eqs. (1.2b) and (1.3b) also but numerical values for degeneracy cannot be addressed in any simple manner.

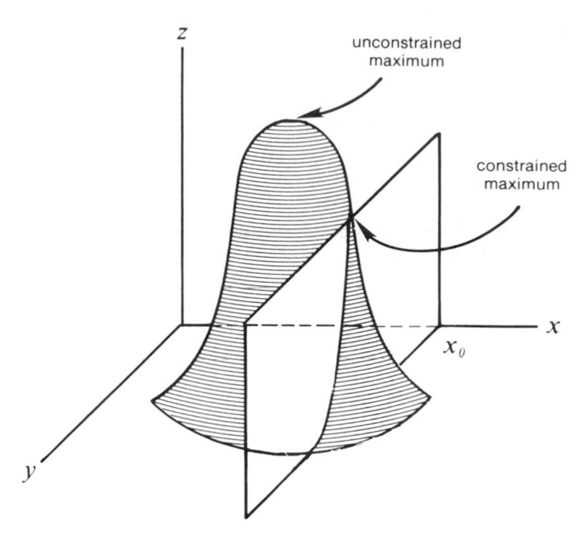

Figure 1.1. Illustration of an unconstrained maximum in the function $f(x, y, z)$ and the maximum under a constraint $x \geqq x_0$.

evenly among 20 energy states t is (O) 10^{244} and to obtain the thousand-fold less probable distribution we require a term $200!/(12!)^{10}(8!)^{10}$, i.e., we need shift only 20 particles. With very large N (10^{24} qualifies) the magnitude of t peaks so sharply for the even distribution that we can replace the sum of terms in Eq. (1.2a) by the maximum valued term.

The maximum valued term represents a particular unique complexion, i.e., a particular set of distribution numbers $N_1, \ldots, N_i$. The sum of the N_i must total N and the sum $N_i \epsilon_i$ must total E. These requirements are constraints on the distribution numbers. Such constraints do not alter the conclusion that the statistical term sum can be represented by its maximum value but that maximum value will not coincide with the equal distribution leading to the maximum term in the absence of constraints. This is crudely illustrated in Figure 1.1.

The problem reduces to the well known mathematical solution for a conditional maximization.[3] We require the unique set of distribution numbers $(N_1^*, \ldots, N_i^*)$ that maximizes the Eq. (1.3a) (or 1.3b) expression subject to the constraints $\sum N_i^* = N$, $\sum N_i^* \epsilon_i = E$. To outline the solution[4] we

1. recognize that N is a large number and address the maximization of ln t, rather than t, simplifying the arithmetic of factorials since, if N is large, $\ln N! = N \ln N - N$ (the Stirling approximation)[5] and

$$\ln t = N \ln N + \sum^{i} N_i \ln g_i - \sum^{i} N_i \ln N_i \tag{1.4}$$

[3] See Appendix B.
[4] See Appendix D.
[5] See Appendix B.

2. recognize that, for $t = t_{max}$,

$$\delta \ln t = 0$$

subject to

$$\sum^{i} \delta N_i = 0 = \sum^{i} \epsilon_i \delta N_i$$

3. introduce Lagrangian undetermined multipliers[5] $-\alpha$, applied to the constraint on constant N, and $-\beta$, applied to the constraint on constant E, and follow the technique of their use[6] to produce the distribution number set $N_1^*, \ldots, N_i^*$ associated with $t_{max} = \Omega_D$ obtaining,

$$N_i^* = g_i e^{-\alpha} e^{-\beta\epsilon_i} \tag{1.5}$$

which we will refer to as the Boltzmann distribution.

We would note that, since $N \ln N - N$ is constant for all distribution number sets, the same expression would be obtained for the Eq. (1.3b) term, i.e., Eq. (1.5) is general, $(N_i^*)_I = (N_i^*)_D = N_i^*$.

The undetermined multiplier α can be immediately identified in terms of a basic statistical term which we will call the partition function and designate as z,

$$\sum^{i} N_i^* = N = e^{-\alpha} \sum^{i} g_i e^{-\beta\epsilon_i}$$

$$e^{-\alpha} = \frac{N}{\sum^{i} g_i e^{-\beta\epsilon_i}} = \frac{N}{z}$$

with z defined as

$$z = \sum^{i} g_i e^{-\beta\epsilon_i}. \tag{1.6}$$

Then

$$\alpha = \ln \frac{z}{N}. \tag{1.7}$$

[6]The maximum when constraints apply is obtained by adding the unconstrained expression for a maximum and all constraint equations

$$\delta \ln t - \alpha \sum \delta N_i - \beta \sum \epsilon_i \delta N_i = 0$$

$$\sum^{i} (\ln g_i - \ln N_i - \alpha - \beta\epsilon_i)\delta N_i = 0$$

and applying normal maximization procedures recognizing that, for arbitrary variations δN_i, each term in the bracketed sum must be 0.

Also we recognize that the probability of a randomly selected particle having energy ϵ_i is just the fraction of particles having energy ϵ_i

$$p_i^* = \frac{N_i^*}{N} = \frac{g_i e^{-\beta\epsilon_i}}{z}. \tag{1.8}$$

The undetermined multiplier β is then identified as

$$e^{-\beta\epsilon_i} = \frac{p_i^* z}{g_i}; \qquad \beta = \frac{1}{\epsilon_i} \ln \frac{g_i}{p_i^* z}. \tag{1.9}$$

We anticipate that more physically meaningful identities will attach to both quantities as the subject develops.

The distribution number set (N_i^*) is a characteristic of the system under consideration. The set of particles initially having random energies totaling E, or even a set of particles initially all having the same energy $\epsilon = E/N$, retains the value E when isolated, but ultimately arrives at a particular distribution of particles among energy levels where the distribution is characterized by a single valued term of statistical origin, β. Another set of randomly chosen particles with their associated energies would have produced a different system energy, say E', and our arguments would produce a different set of distribution numbers $(N_i^*)'$ and different values β', α' for the same energy levels $\epsilon_1, \ldots, \epsilon_i$. If the same set of particles were confined to a different volume the energy levels [Eq. (1.1)] and distribution numbers [Eq. (1.5)] would be different.

It is of importance to note that, when a distribution number set is achieved in an isolated system (E, V, and N are constant) and regardless of the dynamic exchange of particles among energy levels, the system appears unchanging with time in any sampling process that we initiate. We will refer to this steady state as the equilibrium state and note that the equilibrium state is characterized by $t^* = t_{max} = \Omega$, i.e., the equilibrium state reflects the most probable distribution of particles among energy levels because that distribution can occur in an overwhelming number of ways and not because of any a priori statistical favor. Our initial statistical statement assigned equal probability to all distributions.

We address the problem of further identification of the undetermined multiplier β as follows. The energy total and average energy per particle can be expressed in terms of the partition function z, i.e.,

$$E = \sum^i N_i^* \epsilon_i = N \sum^i \left(\frac{g_i e^{-\beta\epsilon_i}}{z} \right) \epsilon_i$$

$$= \frac{N}{z} \sum^i \epsilon_i g_i e^{-\beta\epsilon_i} \tag{1.10a}$$

$$\bar{\epsilon} = \frac{E}{N} = \frac{\sum^i g_i \epsilon_i e^{-\beta\epsilon_i}}{z}. \tag{1.10b}$$

We note that

$$\left(\frac{\partial \ln z}{\partial \beta}\right)_V = \frac{1}{z}\frac{\partial \sum^{i} g_i e^{-\beta\epsilon_i}}{\partial \beta} = -\frac{\sum^{i} g_i \epsilon_i e^{-\beta\epsilon_i}}{z} \tag{1.11a}$$

or

$$-\left(\frac{\partial \ln z}{\partial \beta}\right)_V = \frac{E}{N} = \bar{\epsilon}. \tag{1.11b}$$

The undetermined multiplier α has already been related to the partition function so that a closer examination of this term seems in order.

The partition function and the distribution numbers will be of largest importance in our further discussions. It will be a notational and conceptual convenience to recognize that wherever the term $g_i\epsilon_i$ appears, we can refer to quantum states rather than energy levels, i.e., the partition function expression

$$\begin{aligned} z = \sum^{i} g_i e^{-\beta\epsilon_i} &= g_1 e^{-\beta\epsilon_1} + g_2 e^{-\beta\epsilon_2} + \cdots g_i e^{-\beta\epsilon_i} + \cdots \\ &= e^{-\beta\epsilon_1} + e^{-\beta\epsilon_1} + \cdots (\text{totaling } g_1 \text{ terms}) \\ &\quad + e^{-\beta\epsilon_2} + \cdots (\text{totaling } g_2 \text{ terms}) \\ &\quad + \cdots e^{-\beta\epsilon_i} + \cdots (\text{totaling } g_i \text{ terms}) \end{aligned}$$

can be written

$$z = \sum^{j} e^{-\beta\epsilon_j} \tag{1.12}$$

and

$$N_i^* = N \frac{g_i e^{-\beta\epsilon_i}}{\sum^{i} g_i e^{-\beta\epsilon_i}}$$

can be written

$$N_j^* = \frac{e^{-\beta\epsilon_j}}{\sum e^{-\beta\epsilon_j}} \tag{1.13}$$

where, as noted earlier, each ϵ_j is a unique quantum state, i.e., an energy value characterized by a unique set of quantum numbers, and N_j is the population of that state. We will usually employ the energy state notation for the partition function and for the distribution numbers while retaining the energy level notation for Ω. When both appear in the same expression the energy level notation will be used.

Equation (1.1) demonstrates that a term in the partition function sum is determined by Planck's constant ((O) 10^{-26} erg sec [7]), mass, volume, and the magnitude of the quantum numbers n_x, n_y, n_z. The smallest energy step is associated with unit change in one quantum number (say n_x) in the constant volume system of identical particles,

$$\Delta\epsilon = [\epsilon_{(n_x+1)^2} - \epsilon_{(n_x)^2}] = \frac{h^2}{8m}\left(\frac{2n_x+1}{V^{2/3}}\right). \tag{1.14}$$

For particles of mass (O) 10^{-22} g, independently occupying a volume of several liters, $\Delta\epsilon$ is (O) $n \times 10^{-32}$ ergs. Even if n_x is large, this is a very small quantity, i.e., the steps are so small that we can regard the energy as continuous, replacing the sum by an integral. For a single dimension of motion

$$\sum^{j} e^{-\beta\epsilon_{j,x}} = \int_1^{\infty} e^{-(\beta h^2/8mV^{2/3})n_x^2}\,dn_x. \tag{1.15}$$

Recognizing in this integral form something very close to the definite integral

$$\int_0^{\infty} e^{-a^2x^2}\,dx = \frac{\pi^{1/2}}{2a} \tag{1.16}$$

we extend the lower limit of integration to zero (with negligible error), setting

$$a = \left(\frac{\beta h^2}{8mV^{2/3}}\right)^{1/2}$$

to obtain

$$\sum^{j} e^{-\beta\epsilon_{j,x}} = \left(\frac{2\pi m V^{2/3}}{h^2\beta}\right)^{1/2}. \tag{1.17a}$$

Each dimension of motion will produce the same term so, for three-dimensional motion,

$$z = \left(\frac{2\pi m}{h^2\beta}\right)^{3/2} V. \tag{1.17b}$$

From Eq. (1.12), it follows that

$$\bar{\epsilon} = \left(\frac{\partial \ln z}{\partial\beta}\right)_V = -\frac{\partial}{\partial\beta}\left[\frac{3}{2}\ln\frac{2\pi m V^{2/3}}{h^2} - \frac{3}{2}\ln\beta\right]_V \tag{1.18a}$$

[7] A detailed discussion of units is contained in Chapter 3.

$$\bar{\epsilon} = \frac{3}{2}\frac{1}{\beta} \tag{1.18b}$$

$$E = -N\left(\frac{\partial \ln z}{\partial \beta}\right)_V = \frac{3N}{2\beta}. \tag{1.18c}$$

The association of $\ln z$ with $\bar{\epsilon}$ and $\ln z^N$ with E suggests that z is a particle partition function while z^N is a system partition function.

This suggestion can be pursued through the following set of considerations. In terms of the Eq. (1.6) distribution numbers N_i^*, Eq. (1.5), as it follows from Eq. (1.3a), becomes

$$\ln \Omega_D = N \ln N - N + \sum^{i} N_i^* \ln g_i - \sum^{i} N_i^* \ln N_i^* + \sum^{i} N_i^* \tag{1.19}$$

where each $N_i^* = N p_i^*$. Then

$$\ln \Omega_D = N \ln N - N + N\sum^{i} p_i^* \ln g_i - N\sum^{i} p_i^* \ln N p_i^* + N\sum^{i} p_i^* \tag{1.20}$$

and, since the probabilities must sum to unity,

$$\ln \Omega_D = N\left[\sum^{i} p_i^* \ln g_i - \sum^{i} p_i^* \ln p_i^*\right]. \tag{1.21}$$

The $\ln p_i^*$ are related to the partition function by the logarithmic form of Eq. (1.8),

$$\ln p_i^* = \ln g_i - \beta\epsilon_i - \ln z.$$

If we multiply each such term by $N p_i^*$ and sum over all i, we obtain

$$N\sum^{i} p_i^* \ln p_i^* = N\sum^{i} p_i^* \ln g_i - \beta N\sum^{i} p_i^* \epsilon_i - N \ln z \sum^{i} p_i^* \tag{1.22}$$

so that

$$\beta E + \ln z^N = N\left[\sum^{i} p_i^* \ln g_i - \sum^{i} p_i^* \ln p_i^*\right] = \ln \Omega_D \tag{1.23}$$

where, again, z^N is associated with the system energy E but, in this case also with the particular distinguishable particle statistical term Ω_D.

Had we begun with Eq. (1.3b), Eq. (1.21) would have read

$$\ln \Omega_I = \sum^{i} p_i^* \ln g_i - \sum^{i} p_i \ln p_i - N \ln N + N$$

and to obtain this expression from Eq. (1.22) we should have to subtract $\ln N!$

from each side giving

$$\beta E + \ln \frac{z^N}{N!} = N\left[\sum^{i} p_i^* \ln g_i - \sum^{i} p_i^* \ln p_i^* - \ln N! \right] = \ln \Omega_I \qquad (1.24)$$

where the term $z^N/N!$ is the partition function term associated with E and the indistinguishable particle statistical term.

Clearly, when we are applying the partition function to a collection of particles it will be necessary to recognize whether the particles are to be treated as distinguishable or indistinguishable (localized or nonlocalized). We, therefore, define two system partition functions, $\mathscr{Z}$,

$$\mathscr{Z}_D = z^N = \left(\sum^{i} g_i e^{-\beta \epsilon_i} \right)^N = \left(\sum^{j} e^{-\beta \epsilon_j} \right)^N \qquad (1.25)$$

$$\mathscr{Z}_I = \frac{z^N}{N!} = \frac{\left(\sum^{i} g_i e^{-\beta \epsilon_i} \right)^N}{N!} = \frac{\left(\sum^{j} e^{-\beta \epsilon_j} \right)^N}{N!}. \qquad (1.26)$$

In Eqs. (1.23) and (1.24) we have replaced $\ln \Omega$ with the much more tractable term $(\beta E + \ln \mathscr{Z})$ to which we assign the symbol σ. The maximization of σ is a measure of equilibrium in the isolated system just as the maximization of $\ln \Omega$ is an equilibrium measure. We must, of course, identify σ with respect to localized vs nonlocalized particle systems just as we identify Ω and $\mathscr{Z}$

$$\sigma_D = \beta E + \ln z^N \qquad (1.27)$$

$$\sigma_I = \beta E + \ln \frac{z^N}{N!}. \qquad (1.28)$$

We should recall at this point, however, that the Eq. (1.5) distribution number set was independent of the distinguishability question. Since the total energy is always $\sum N_j^* \epsilon_j$ in the independent particle collection it follows that total energy is also independent of the distinguishability question. We can, then, regard Eq. (1.18c) as general, writing

$$E = -\left(\frac{\partial \ln \mathscr{Z}_I}{\partial \beta} \right)_{V,N} = -\left(\frac{\partial \ln \mathscr{Z}_D}{\partial \beta} \right)_{V,N} = -\left(\frac{N \partial \ln z}{\partial \beta} \right)_V. \qquad (1.29)$$

This also follows from the observation that the N term is independent of β when N is a specified constant.

The conclusions reached to this point are in part general and in part restricted to the particular carefully identified particle set to which we have confined most of our attention. The statistical approach that we have outlined is general and the two undetermined multipliers are a general feature of that approach. The general relationships are those which identify statistical terms [Eqs. (1.2b), (1.3b), (1.4), (1.25), (1.26)], the distribution of particles among energy states [Eqs. (1.5), (1.13)], the partition function [Eqs. (1.6), (1.12), (1.25), (1.26)], and important interrelationships between these quantities and the energy of an isolated system [Eqs. (1.12), (1.27), (1.28)]. Details such as Eqs. (1.1) and (1.14) through (1.18c) clearly relate to the dilute collection of structureless, identical, indistinguishable, independent particles. These particular qualifications, as the reader will have perceived, were chosen with their relaxation to include more complex collections in mind. Dilute specifies a very low-density gaseous phase. Relaxation of this constraint leads to dense gases and ultimately condensed (solid or liquid) phases. Structureless, confines attention to translational energy only, excluding internal energies associated with other modes of motion and with van der Waals interactions. Identical particles confines attention to a single species and a single set of quantum states that are available to all particles. Indistinguishability is a characteristic of the gas phase in which a single particle species has neither identity nor localized position. The energy of an independent particle set is the sum of individual particle energies. The energy of a single independent particle in the set is not dependent on any other particle energy.

To maintain a proper perspective on the utility of our results we should note that a system approximating our carefully identified particle set is a well-insulated, closed container of large volume containing a known mass of helium gas. All helium atoms are identical. In the gas phase helium atoms are indistinguishable. Although the atoms have structures, the energy levels other than translation are those of excited electronic states and helium ions plus electrons. Normally we can consider these to be unavailable states. Interactions between helium atoms are quite weak and in a low-density system the potential energy of interaction is completely trivial.

Also this, or any other real gas system, does not have to be constructed by the prescription of selecting individual molecules with random energies. We could take from our surroundings any bulk sample of a pure substance in known amount, expand it to occupy as a gas the entirety of a large volume V, and place it in isolation. Its energy total E in isolation would be the energy carried into isolation with it. If, as a result of the expansion or other transitory effects prior to isolation, the distribution numbers are not the equilibrium N_j^* we have only to wait, with confidence that they will be at some later time and will remain so.

If the atoms or molecules of this particular substance at the specified E and V do not approximate our structureless, independent particle system our arguments would still appear to be generally applicable although requiring further development. Other energy levels can be included in our summations to accommodate the additional modes of motion of structured particles providing that we can develop appropriate expressions. If there is a potential energy of interaction that will depend on particle position in the field of all other particles, our increasing familiarity with kinetic argument would lead us to believe that the potential energy contribution will be constant ($= U$) in the isolated system and that the kinetic energy sum over individual particles will not be the total energy E_Σ,[8] but $E_\Sigma - U$ with the statistical argument unchanged $\sum N_j^* \epsilon_j + U = E_\Sigma$ although the ϵ_j may be altered and U may be difficult to quantify. If the system contains several species, we anticipate that each species will have a set of energy states that is closed to other species. These states may be the normal (single species) states or they may be perturbed states which could be difficult to treat.

The skeptical reader may observe at this point that extensions to structured and interacting systems and to their mixtures must be limited to rather simple substances and, even then, an ability to calculate E, σ, and $\mathscr{Z}$ in terms of β, V, and N does not provide a complete description in terms of observable features. This is, of course, true. It is through the consideration of energy exchange processes by which we can enlarge the statistical concepts and provide the observable features. We defer, for the moment, any extension to structured and interacting systems to apply our rudimentary understanding to such exchange processes which we define as follows.

Consider two identical particle isolated systems which have different E, V, and N values. Each of them is characterized by the available energy levels, the distribution numbers, and the partition function. The difference between distribution numbers, energy levels, and partition functions is as well defined as the terms themselves. We can equally well regard the two systems as being the same system in two different isolated thermodynamic states (state 1 and state 2) and the differences in the statistical terms as being characteristic of the change in thermodynamic state. We can well imagine that E, V, and N changes may accompany the relaxation of isolation constraints on any system and that controlled, programmed relaxation of isolation constraints will permit us to accomplish any desired change such as the transition from state 1 to state 2. This is a process and we anticipate the identification of process variables that are characteristic of the process and that relate to differences in the statistical properties of the two states. Such observables should confirm statistical calculations for simple system processes and provide a framework for dealing with theoretically intractable systems.

In the discussions that follow it will be convenient to refer to the fictitious particle system that has formed the present framework for argument (structureless, indistinguishable, independent, nonlocalized) as "the ideal gas."

SYMBOLS: INTRODUCTION AND CHAPTER 1

f = force, $\mathbf{f}$ = force vector

$v, \mathbf{v}$ = speed, velocity vector

[8] We will often use the summation sign subscript to represent the total value of a term that has additive components

∇ = Laplacian operator
U = potential energy
$r, \mathbf{r}$ = radius, radius vector
$\epsilon_0, \epsilon_i, \epsilon_n$ = energy levels
h = Plancks constant
ν = frequency
λ = wavelength
$\mathbf{E}$ = electric field vector
A = amplitude of wave motion
$\mathbf{H}$ = magnetic field vector
c = velocity of light
$\tilde{\nu}$ = wavenumber
n_x, n_y, n_z = translational quantum numbers
j = rotational quantum numbers
v = vibrational quantum numbers
N = number of particles
ϵ_K = particle kinetic energy
$N_1 \cdots N_i \cdots N_K$ = number of particles having energy $\epsilon_1, \ldots, \epsilon_i, \ldots, \epsilon_K$
m = particle mass
V = volume
E = total kinetic energy in a system of particles
t = term (in a statistical expression)
$\Omega, \Omega_I, \Omega_D$ = sum of all terms, sum for indistinguishable particles, sum for distinguishable particles
g_i = degeneracy of an energy level
$-\alpha, -\beta$ = undetermined multipliers
p_i = fraction of particles in an energy level *i*, probability that a random particle has energy ϵ_i
ϵ_j, N_j, p_j = energy, particle number, probability $\cdots$ for an energy state *j*
$\epsilon_j^*, N_j^*, p_j^*$ = designation applying to systems that are in statistical (thermodynamic) equilibrium
z = a particular sum over energy states available to a particle, called the partition function
e = Naperian logarithim base
$\mathscr{Z}, \mathscr{Z}_I, \mathscr{Z}_D$ = the partition function for a particle system, of indistinguishable particles, of distinguishable particles
$\sigma, \sigma_I, \sigma_D$ = designation of $\ln \Omega$, $\ln \Omega_I$, $\ln \Omega_D$

2

ENERGY EXCHANGE I: THERMAL AND MECHANICAL PROCESSES

The closing remarks of Chapter 1 direct our attention to the identification of process parameters that will relate to the changes in E, V, N and the associated statistical terms when the thermodynamic system undergoes a state to state transition. The process parameters should be observable features of the process or directly relate to observable features. Presently we understand that the isolated system macroscopic features E, V, N dictate the microscopic features $\beta, \mathscr{Z}, \Omega$ (or its analog σ) through the distribution number set $N_1^*, \ldots, N_j^*$. Among these terms only V and N are observables so that although we can characterize the ideal gas rather completely in principle,[1]

$$\mathscr{Z}_{\mathrm{I}} = \frac{z^{\mathrm{N}}}{\mathrm{N}!} = \left(\frac{2\pi m}{\beta h^2}\right)^{3\mathrm{N}/2}\left(\frac{\mathrm{V}e}{\mathrm{N}}\right)^{\mathrm{N}} \tag{2.1}$$

$$\mathrm{N}_j^* = \mathrm{N}\frac{e^{-\beta\epsilon_j}}{\sum^j e^{-\beta\epsilon_j}} = \frac{\mathrm{N}e^{-\beta\epsilon_j}}{z} \tag{2.2}$$

$$\mathrm{E} = \sum^j \epsilon_j \mathrm{N}_j^* = \frac{\mathrm{N}\sum^j \epsilon_j e^{-\beta\epsilon_j}}{z} \tag{2.3}$$

[1] We will not continue to identify every usage of the statistical terms with the subscript I or D to simplify notation.

$$\sigma_{\mathrm{I}} = \ln \Omega_{\mathrm{I}} = -(\ln \mathscr{Z}_{\mathrm{I}} + \beta \mathrm{E}), \tag{2.4}$$

in practice we must identify either E or β in terms of some system observable.

Also, the constraints under which the statistical relationships were developed require that the relationships apply only to equilibrium circumstances in an isolated (ideal gas) system. We imagine that we can remove the ideal gas constraint by extending our calculations. The remaining constraints derive from general statistical features. We cannot follow the detail of finite processes involving the ideal gas system or more complex systems. We can consider only their initial and final isolated equilibrium states, relating the process to changes in the equilibrium properties. However, the process concept can be broadened in an extremely powerful manner.

First we note that during any energy exchange process involving both system and surroundings, the isolated system must be our carefully defined particle system and that part of the environment that can possibly engage in any interaction with this system of interest. We will refer to our system of interest as system $\mathscr{A}$ (or systems $\mathscr{A}, \mathscr{B}, \mathscr{C}$, etc., if we wish to include other systems of thermodynamic concern or subdivide the original system) and to the interacting environment as system $\mathscr{E}$. All statistical arguments relevant to isolated systems must apply to the supersystem $\mathscr{A} \cdots \mathscr{E}$. In particular, for equilibrium in such a supersystem, the combinatorial term, e.g., $\Omega_{\mathscr{A}\mathscr{E}} = \Omega_{\mathscr{A}}\Omega_{\mathscr{E}}$ must attain its maximum value by whatever redistribution of particles among energy states is required.[2] We can presently evaluate the maximum value, $\Omega_{\mathscr{A}}$, for only one part of that isolated system, "the ideal gas," when we are given its isolation E, V, and N values. We suspect, in view of the complexity of the environment as a thermodynamic system, that the maximum value of $\Omega_{\mathscr{E}}$ will not be avaliable or even interesting except in process terms. We are certain that the maximum value of $\Omega_{\mathscr{A}}$ will reflect the equilibrium state of system $\mathscr{A}$, independently from system $\mathscr{E}$ when there are isolation barriers confining system $\mathscr{A}$. In this circumstance system $\mathscr{A}$ will have fixed values of E, V, and N.

We construct such an isolated supersystem consisting of isolated system $\mathscr{A}$ and an isolated portion of the environment. If we now relax internal isolation barriers to permit system $\mathscr{A}$ and $\mathscr{E}$ to exchange energy and/or material, we imagine that the isolated supersystem will attain some new, overall state at equilibrium and that E, V, and N for system $\mathscr{A}$ will change. Any such process can be terminated by reimposing the isolated system barriers and the final equilibrium state for system $\mathscr{A}$ will be determined by the instantaneous system values of E, V, and N at the time the exchange was terminated.

Any overall process can be carried out to produce the same change in state (initial and final values of E, V, and N) by several different routes. One of the many process routes between two equilibrium states could consist of a number

[2] Since a combinatorial term Ω represents the most probable state of a system, Eq. (1.21), we should regard $\Omega_{\mathscr{A}}$ and $\Omega_{\mathscr{E}}$ as probabilities that combine, in the standard manner, as products. See Appendix C.

of infinitesimal sequential processes and, in the event that at no point in this differential process do the system properties differ more than infinitesimally from equilibrium properties, we refer to a reversible process. In microscopic terms, although both available energy levels and distribution numbers may change substantially in the course of the process, interruption of the process and isolation of system $\mathscr{A}$ at any intermediate point in the reversible differential process would produce no discernible change with time in either distribution numbers or energy levels for system $\mathscr{A}$ with the instantaneous E, V, and N values at the moment of interruption. Since system $\mathscr{A}$ is in a continuous equilibrium state we conclude that it is possible, for processes that are truly reversible or for hypothetical reversible processes that we conceptualize, to apply our statistical arguments using the continuously changing thermodynamic variables to express changes in $\mathrm{E}_{\mathscr{A}}$, $\mathscr{Z}_{\mathscr{A}}$, and $\sigma_{\mathscr{A}}$ from their initial values to their values at any point in the process.

As applied to the system energy,

$$\mathrm{E} = \sum^{j} \mathrm{N}_j \epsilon_j = \mathrm{N} \sum^{j} \wp_j \epsilon_j, \tag{2.5}$$

any change in energy can only be accomplished by change in N, by change in the population of energy states or by change in the energy states,

$$\Delta \mathrm{E} = \Delta \sum^{j} \mathrm{N} \wp_j \epsilon_j. \tag{2.6}$$

The energy change cannot be dependent in any way on the process route between specified initial and final values of $\wp_j$, ϵ_j, and N.[3] If changes in $\wp_j$, ϵ_j and N are accomplished as a series of infinitesimal steps with the particle system in a state of continuous statistical equilibrium, the change in E during any infinitesimal process step is[4]

$$d\mathrm{E} = \mathrm{N}\left(\sum^{j} \wp_j^* d\epsilon_j^* + \sum^{j} \epsilon_j^* d\wp_j^* \right) + \left(\sum^{j} \epsilon_j^* \wp_j^* \right) d\mathrm{N}. \tag{2.7}$$

If an infinitesimal quantity of energy is added to the system it can be accommodated only by change in occupancy or in energy state values. If particles are added to (subtracted from) the system under the condition that $\sum^j \wp_j^* \epsilon_j^* (= \bar{\epsilon}^*)$ for the added (subtracted) particles is $\bar{\epsilon}$ for the system, then E changes in direct proportion to dN with $\wp_j^*$ and ϵ_j^* remaining constant. If $\bar{\epsilon}^*$ for the added (subtracted) particles is not $\bar{\epsilon}^*$ for the system, then an additional

[3] E is only one of the thermodynamic variables that will emerge for which changes are independent of the process route. For all such (state) variables F, $\int_1^2 dF = F_2 - F_1$. Terms dF are referred to as perfect differentials. See Appendix B.

[4] In any reversible process the continuously changing $\wp_j$ and ϵ_j must always be equilibrium values and all such quantities must be designated as such by the * notation. Quantities such as E, Ω, σ, or $\mathscr{Z}$, do not require such designation since they have meaning only as equilibrium values.

increment ($\pm\delta$E) of energy has been added to (subtracted from) the system that must be accommodated by changes in either $\wp_j^*$ or ϵ_j^* or both. We can imagine that all such energy changes can be accomplished if the isolated system of interest is permitted to interact with the environment across boundaries that are energy or material permeable and that the energy or material transfer can be made infinitely slow.

A boundary that can be made permeable at will to N-type particles can selectively accommodate changes in N. When molecules cannot be exchanged across the system boundaries, $d\mathrm{N} = 0$ and we will refer to the system as a closed system. If change in N is a permitted feature of any process we will refer to an open system.

Normally, we will not wish to mix our carefully designed particle system with the environment. Open system arguments will be of most utility when we wish to subdivide a system of interest making internal particle exchange possible but maintaining closed boundaries, at least conceptually, with respect to the environment. Therefore, in many cases, including the arguments that follow, it will be a simplifying convenience to deal with the closed system expression

$$d\mathrm{E} = \mathrm{N}\left(\sum^{j} \wp_j^* d\epsilon_j^* + \sum^{j} \epsilon_j^* d\wp_j^*\right) \tag{2.8}$$

to assess interactions with the surroundings. Conclusions reached on the closed system basis will in no way be prejudiced by that restriction and the results obtained can be generalized to the open system by insertion of and attention to that term. In Eqs. (2.7) and (2.8) the N, $\wp_j^*$, ϵ_j^*, and $\bar{\epsilon}^*$ appearing as multipliers are instantaneous values of those quantities during infinitesimal changes in N, $\wp_j^*$ and ϵ_j^*.

We have noted [via Eq. (1.1)] that changes in energy states for the ideal gas are effected by change in system volume. We would expect this to be a general feature, recognizing that interacting particle (nonideal) systems would, in addition, exhibit changes in potential energy as average particle–particle distances change and that at very small average particle–particle distances the ϵ_j values could be influenced. If change in energy level spacing is always a first-order volume feature of gaseous systems and the only volume feature of the ideal gas system, displacement of a system boundary to decrease the volume will expand energy levels. If the population of each level is unchanged, the system energy will be increased. If the boundary displacement is such as to increase the system volume, parallel argument leads to the conclusion that the system energy has decreased.

In mechanics the application of a force to effect a displacement is an energy form called work.[5] The differential element of work is

$$dw = f\,dx \tag{2.9a}$$

[5] See Appendix A and Appendix B.

and work is the line integral over the force-displacement path

$$W = \int_c f dx. \tag{2.10a}$$

In the case at hand it is easy to see that an external force applied over a displaceable system boundary of area A, which accomplishes an infinitesimal linear displacement of the entire boundary, accomplishes an infinitesimal volume change and defines a work element proportional to the change in volume

$$dw = \frac{f_{ex}}{\mathsf{A}}(\mathsf{A}dx) = \frac{f_{ex}}{\mathsf{A}} d\mathrm{V}. \tag{2.9b}$$

Energy expended externally in the form of work has been transferred to the system in the form of change in energy level spacing. The magnitude of the external energy expended is the work accomplished and the sign can be positive or negative depending on the direction of boundary displacement, i.e., increase or decrease in volume. In order that the work term sign reflect the system energy change (increase in energy with decrease in volume) we write

$$W = -\int_c \frac{f_{ex}}{\mathsf{A}} d\mathrm{V}. \tag{2.10b}$$

The concept of work as introduced in Appendix A on Energy derives from forces dependent on the location of particles in various fields. In all such cases, molecules in the field of other molecules, charged particles in electric fields, particles with magnetic moment in magnetic fields, etc., the change in total energy can be represented by the work element required to effect a change in particle location, i.e., to effect a change in potential energy. Our mechanical work process does not appear to be related to potential energy. There are no potential energies of interaction in the ideal gas. Energy exchange across the displaceable boundary is stored in (or taken from) the expanded (contracted) energy level set. We can, however, regard the increase in kinetic energy associated with volume decrease as an increase in system work capacity in that energy added to the system would be (in principle) recoverable as work performed by the system in returning to the original volume by exactly reversing the force-displacement path,

$$\int_{c\,\mathrm{V}_1}^{\mathrm{V}_2} \frac{f_{ex}}{\mathsf{A}} d\mathrm{V} = -\int_{c\,\mathrm{V}_2}^{\mathrm{V}_1} \frac{f_{ex}}{\mathsf{A}} d\mathrm{V}.$$

In systems more complex than the ideal gas, work terms other than that represented by Eq. (2.10b) would be of importance. For present purposes the displaceable boundary and the mechanical work associated with its displacement are sufficient constraints. Other work contributions can be accommodated by introducing other boundary constraints as problems require.

When energy is transferred to or from our closed system of interest by processes other than those that alter energy states, it can be accommodated only by a redistribution of particles among the energy state set, i.e., by change in the $\not{p}_j^*$. We refer to such changes as thermal energy transfer and to the energy exchanged as heat (Q). Energy exchanged as heat is represented by the line integral along the particlar heat transfer route followed in the thermal exchange process

$$Q = \int_c dq \tag{2.11}$$

where Q is taken to be positive if energy (heat) is added to the system of interest and negative if energy (heat) is taken from that system. We conceptualize a boundary across which thermal energy can be transferred and refer to it as a diathermal boundary.[6] Its counterpart, across which thermal energy cannot be transferred, is an adiabatic boundary.

In terms of the boundary constraints that we have constructed for controlled energy exchange processes, our isolated system of particles is a closed, rigidly bounded, adiabatic system. Selective relaxation of the isolated system constraints permits energy exchange processes to occur across open and/or displaceable and/or diathermal boundaries. All energy changes of present concern accomplished by interaction of the closed system with other systems or the surroundings are accommodated by heat transfer (Q) and work (W) processes,

$$\Delta E = E_2 - E_1 = (Q + W)_N = \left[{}_c\int_1^2 dq - {}_c\int_{V_1}^{V_2} \frac{f_{ex}}{A} dV \right]_N \tag{2.12}$$

$$\Delta E = \Delta \left\{ N \left[\left(\sum^j \not{p}_j^* \epsilon_j^* \right)_2 + \left(\sum^j \not{p}_j^* \epsilon_j^* \right)_1 \right]_N \right\}. \tag{2.13}$$

Equation (2.12) deserves some discussion. In the classical development of our subject it is known as the first law of thermodynamics and is obtained from phenomenological argument. The terms Q and W are finite process terms, dq and dw $[=(f_{ex}/A)dV]$ are increments of heat and work that, taken over the process path, provide the Q and W values. The notation $(\ )_N$ has been used to signify that N is constant throughout the closed system process. The values E_2, E_1, V_2, V_1, and the terms $(\sum^j \not{p}_j^* \epsilon_j^*)_1$ and $(\sum^j \not{p}_j^* \epsilon_j^*)_2$ are associated with the

[6]The mechanism of thermal energy exchange across the diathermal boundary is not important to our arguments but we would recognize that systems having different average energies can exchange energy by particle collision with their boundary of separation through the same "very weak interactions" that maintain the ever-changing but overall constant distribution number set in our isolated system.

initial (1) and final (2) equilibrium states for our system when it is isolated from the surroundings. If the heat and work increments conform to the requirements of reversibility, the p_j^* and ϵ_j^* are equilibrium values at every stage of the process and the differential elements of heat and work are associated with these statistical terms,

$$(dq^*)_N = N\sum^{j} \epsilon_j^* \, dp_j^* \tag{2.14}$$

$$(dw^*)_N = N\sum^{j} p_j^* \, d\epsilon_j^*. \tag{2.15}$$

The reversible path is one of many routes between states E_1, V_1, N_1 and E_2, V_2, N_1. Each particular alternative, nonreversible route will have associated with it unique Q and W values and the incremental dq and dw values related to that route cannot be associated with the equilibrium statistical terms.

In developing the arguments for reversible processes we have introduced a distinction between such conceptual routes between equilibrium states and the spontaneous (irreversible) routes that would be followed if we suddenly relaxed and then reimposed the isolation constraints. To quantify the opening remarks regarding maximization of the combinatorial term Ω, a finite process can occur if Ω (or $\ln \Omega$) for the isolated supersystem can increase

$$(\ln \Omega_{\mathscr{A}\mathscr{E}})_2 - (\ln \Omega_{\mathscr{A}\mathscr{E}})_1 = \Delta(\ln \Omega_{\mathscr{A}} + \ln \Omega_{\mathscr{E}}) > 0. \tag{2.16}$$

The process will continue in the direction of increasing $\Omega_{\mathscr{A}\mathscr{E}}$ until the maximum value is reached at which point

$$d(\ln \Omega_{\mathscr{A}} + \ln \Omega_{\mathscr{E}}) = 0. \tag{2.17}$$

The spontaneous process has directionality and a driving force and the driving force is

$$d(\ln \Omega_{\mathscr{A}} + \ln \Omega_{\mathscr{E}}) > 0 \tag{2.18}$$

for every incremental step of the finite process. For any process occurring in the isolated supersystem we can write

$$d(\ln \Omega_{\mathscr{A}} + \ln \Omega_{\mathscr{E}}) = d(\sigma_{\mathscr{A}} + \sigma_{\mathscr{E}}) \geqslant 0 \tag{2.19}$$

for every process element, describing the finite process conditions (>0) and the equilibrium condition ($=0$). Along a reversible route between the initial and final states the zero equality must (virtually) hold for every infinitesimal step. To make this observation useful it is required that we relate changes in $\ln \Omega$ or σ to observable features of the process.

To identify such features we should concentrate on individual process elements. We can concentrate on thermal energy exchange by imposing closed, rigid, diathermal boundaries between the system and the enviroment. Energy will be exchanged across the diathermal boundary if $\ln \Omega_{\mathscr{A}\mathscr{E}}$ can increase and the condition under which this can occur is that $\beta_{\mathscr{A}} \neq \beta_{\mathscr{E}}$. Thermal equilibrium prevails under the condition that $\beta_{\mathscr{A}} = \beta_{\mathscr{E}}$.

We demonstrate these requirements by noting that regardless of our ability to construct detailed statistical expressions for the environment, both it and the most general system of thermodynamic interest will be described by combinatorial terms that involve distribution numbers.[7] Also the variation in $\ln \Omega$ for a closed system is independent of localization or nonlocalization so generally,

$$(d \ln \Omega)_{V,N} = d\left(\sum^{i} N_i^* \ln g_i - \sum N_i^* \ln N_i^* + \sum N_i^*\right) \tag{2.20}$$

$$= \sum^{i} \ln g_i dN_i^* - \sum^{i} \ln N_i^* dN_i^*$$

with

$$\ln N_i^* = \ln g_i + \ln N - \ln z - \beta\epsilon_i$$

so that

$$(d \ln \Omega)_{V,N} = -\ln N \sum^{i} dN_i^* + \ln z \sum^{i} dN_i^* + \sum^{i} \beta\epsilon_i dN_i^* = \beta dE \tag{2.21}$$

for either closed system since $\sum dN_i^* = 0$. In the isolated supersystem, $E_{\mathscr{A}\mathscr{E}} = E_{\mathscr{A}} + E_{\mathscr{E}} = \text{constant}$ so $dE_{\mathscr{A}} = -dE_{\mathscr{E}}$ and

$$\beta_{\mathscr{A}} dE_{\mathscr{A}} + \beta_{\mathscr{E}} dE_{\mathscr{E}} = d \ln \Omega_{\mathscr{A}\mathscr{E}} \geqslant 0$$

$$dE_{\mathscr{A}}(\beta_{\mathscr{A}} - \beta_{\mathscr{E}}) \geqslant 0 \tag{2.22}$$

and the equilibrium condition ($=0$) is fulfilled for an infinitesimal variation $dE_{\mathscr{A}}$ only when $\beta_{\mathscr{A}} = \beta_{\mathscr{E}}$. In the event that we wish to conduct a reversible thermal energy transfer across a closed, rigid diathermal boundary, $\beta_{\mathscr{A}} - \beta_{\mathscr{E}} = d\beta$ cannot differ more than infinitesimally from zero at any stage in the process. The process direction for this case is such that $dE_{\mathscr{A}} > 0$ when $\beta_{\mathscr{A}} > \beta_{\mathscr{E}}$ and a finite quantity of thermal energy will be transferred to or from system $\mathscr{A}$ when $\beta_{\mathscr{A}}$ differs from $\beta_{\mathscr{E}}$ by a finite amount.

In the event that the closed, diathermal boundary is displaceable, work can accompany the thermal energy transfer process. Our conditions for reversibility require that equilibrium prevails throughout the process so that if dq is a

[7] Concern about the environment can be removed by making it an effectively infinite reservoir of our ideal gas analog helium. Being effectively infinite, any interaction with our system of concern leaves the environment unchanged.

reversible element dw is also,[8]

$$(dE)_N = dq^* + dw^* = dq^* + N\sum^j \wp_j^* d\epsilon_j^*. \tag{2.23}$$

A quite important result follows from Eq. (2.23) since it is possible to show[9] that

$$d \ln \Omega = d\sigma = \beta\left(dE - N\sum^j \wp_j^* d\epsilon_j^* \right) \tag{2.24}$$

which we rephrase as

$$(dE)_N = \frac{1}{\beta} d\sigma + N\sum^j \wp_j^* d\epsilon_j^* \tag{2.25}$$

relating [via Eq. (2.23)] $dq^*(= N\sum^j \epsilon_j^* d\wp_j^*)$ to $d\sigma$ and to the important combinatorial term $d \ln \Omega$

$$d\sigma = \beta dq^* = d \ln \Omega. \tag{2.26}$$

This is general and true whether or not the reversible work term is present, although clearly the dq^* value will depend on the presence or absence of that

[8] Although the notation q_{rev} and w_{rev} will be commonly encountered, q^* and w^* are preferable for our purposes where the * designations refer to the equilibrium state that prevails during reversible processes.

[9] For $\sigma = \ln \mathscr{Z} + \beta E, d\sigma = d \ln \mathscr{Z} + E d\beta + \beta dE$ and, where $\mathscr{Z} = \mathscr{Z}(\beta, \epsilon_j, N)$ (see Appendix B on Mathematical Considerations),

$$d \ln \mathscr{Z} = \left(\frac{\partial \ln \mathscr{Z}}{\partial \beta}\right)_{\epsilon_j, N} d\beta + \sum^j \left(\frac{\partial \ln \mathscr{Z}}{\partial \epsilon_j}\right)_{\beta, \epsilon_k, N} d\epsilon_j + \left(\frac{\partial \ln \mathscr{Z}}{\partial N}\right)_{\beta, \epsilon_j} dN.$$

The closed system argument eliminates the final term and the first term [via. Eq. (1.18c)] is $-E d\beta$ so

$$d \ln \mathscr{Z} = -E d\beta + \sum^j \left(\frac{\partial \ln \mathscr{Z}}{\partial \epsilon_j}\right)_{\beta, \epsilon_k, N} d\epsilon_j^*.$$

Each additive component of the second term is

$$\left(\frac{\partial \ln \partial \mathscr{Z}}{\partial \epsilon_j^*}\right)_{\beta, \epsilon_k, N} = \frac{N}{z}\frac{\partial z}{\partial \epsilon_j^*} = \frac{N}{z}\left(\frac{\partial e^{-\beta \epsilon_j^*}}{\partial \epsilon_j}\right)_{\beta, \epsilon_k, N} = -\frac{\beta N e^{-\beta \epsilon_j^*}}{z} = -\beta N \wp_j^*.$$

so that

$$d \ln \mathscr{Z} = -E d\beta - \beta N \sum^j \wp_j^* d\epsilon_j^*$$

and Eq. (2.24) follows from the $d\sigma$ expression.

term and its particular value if present. If w is not reversible, q is not reversible and Eq. (2.25) is not valid.

There are further large implications associated with Eq. (2.26) since any thermal exchange process involving the system and the environment must be a reversible process in the environment that can be as large as we choose to include and therefore unchanged by any interaction with our system of interest: $(dq_{\mathscr{E}}) = (dq_{\mathscr{E}})^*$. If the supersystem is isolated, $dq_{\mathscr{E}} = -dq_{\mathscr{A}}$ so that $d\sigma_{\mathscr{E}}$ is always determined by $-dq_{\mathscr{A}}$. Then

$$d\sigma_{\mathscr{A}} - \beta dq_{\mathscr{A}} \geqslant 0 \tag{2.27}$$

and, if the infinitesimal heat quantity is transferred under the condition that β is commonly valued and constant and if, in addition, the system volume is constant, the work term in Eq. (2.8) is absent and $d\mathrm{E}_{\mathscr{A}} = dq_{\mathscr{A}}$ so

$$(d\sigma_{\mathscr{A}} - \beta d\mathrm{E}_{\mathscr{A}})_{\beta,\mathrm{V}} \geqslant 0. \tag{2.28}$$

This result is of very large importance since, from Eqs. (1.27) and (1.28)

$$(d\sigma_{\mathscr{A}} - \beta d\mathrm{E}_{\mathscr{A}})_{\beta,\mathrm{V}} = (d\ln \mathscr{Z})_{\beta,\mathrm{V}} \tag{2.29}$$

which we rephrase as

$$\left[d\left(\frac{\sigma}{\beta} - \mathrm{E}_{\mathscr{A}}\right)\right]_{\mathrm{V},\beta} = \frac{1}{\beta}(d\ln \mathscr{Z}_{\mathscr{A}})_{\mathrm{V},\beta} = -(d\mathrm{A})_{\mathrm{V},\beta} \tag{2.30}$$

defining the function A as

$$\left[\mathrm{E} - \frac{\sigma}{\beta}\right]_{\mathrm{V},\beta} = \mathrm{A} \tag{2.31}$$

which we will call the Helmholz free energy following classical practice.

The Eq. (2.19) relationship that includes the σ (or $\ln \Omega$) change in the surroundings as a criterion for the directionality of spontaneous processes is now rephrased in terms of a thermodynamic function for the system of interest

$$(d\mathrm{A})_{\beta,\mathrm{V}} \leqslant 0. \tag{2.32}$$

It also follows from Eq. (2.27) that, for the system of interest,

$$d\sigma_{\mathscr{A}} \geqslant \beta dq_{\mathscr{A}} \tag{2.33}$$

the equality prevailing when $dq_{\mathscr{A}}$ is a reversible heat element $dq^*_{\mathscr{A}}$ for which Eq. (2.26) applies.

From Eqs. (2.11), (2.31), and (2.33) it follows that $d\mathrm{A} = dw^*$ where dw^* represents all reversible work elements. Further identification of observables should derive from a close examination of the work process. For present concerns we have confined our attention to mechanical work. To analyze the work process and identify process observables associated with it we must first identify the mechanical variables. Mechanical work is the product of a force and a displacement. Displacement of a system boundary, when it is subjected to unequal internal and external forces and free to move, changes a system dimension and therefore the system volume. Volume is one mechanical variable; we consider the problem of assessing the internal forces that are pertinent.

Consider a single particle in a closed, rectangular system of volume V which is equal to the product of three linear dimensions (a, b, c). If the particle is in a particular energy state it has a particular kinetic energy. If the volume is large, quantum effects can be ignored and we can consider this energy value in the classical sense as equal to $\frac{1}{2}m\mathrm{v}^2$ where v is a vector velocity that can be resolved into velocity components (v) perpendicular to the three parallel wall pairs. The frequency of collision with a given wall depends on the velocity component perpendicular to that wall and on the distance traveled between collisions, $v_x/2a$, for the x axis parallel wall pair separated by distance a. On each collision the particle suffers a change of momentum mv_x to $-mv_x$ so that the total change in momentum (p_x) per unit time is

$$\frac{dp_x}{dt} = 2mv_x\frac{v_x}{2a} = \frac{mv_x^2}{a}. \tag{2.34}$$

By Newton's second law $\mathbf{f} = d\boldsymbol{P}/dt$; the change in momentum with time defines force. In this circumstance force per unit area is

$$\frac{f_x}{bc} = \frac{mv_x^2}{abc} = \frac{mv_x^2}{\mathrm{V}} \tag{2.35}$$

where b, c represent the remaining dimensions of the bounded volume V. Consideration of other walls would produce a similar expression. If there are N independent particles in the container their unique contributions will be additive and we define pressure as the cumulative force per unit area exerted on the bounding wall due to particle collisions with that wall,

$$\mathrm{P} = \frac{m}{\mathrm{V}}\sum_{\mathrm{K}=1}^{\mathrm{N}} v_{x,\mathrm{K}}^2. \tag{2.36}$$

Since there will be a distribution of velocity components (following the distribution of particles among the energy states available, even if those states are so closely spaced as to represent an energy continuum) we construct the average velocity component value

$$\overline{v_x^2} = \frac{\sum^{\mathrm{K}} v_{x,\mathrm{K}}^2}{\mathrm{N}}$$

so that

$$P = m\frac{N\overline{v_x^2}}{V}. \tag{2.37}$$

We would assume that the average velocity component would be equal in all directions (otherwise the gas would concentrate in some volume domain) so that, from

$$\overline{v^2} = \overline{v_x^2} + \overline{v_y^2} + \overline{v_z^2}$$

and

$$\overline{v_x^2} = \overline{v_y^2} = \overline{v_z^2} = \tfrac{1}{3}\overline{v^2}$$

or with

$$\bar{\epsilon} = \tfrac{1}{2}m\overline{v^2}$$

and Eq. (1.18c),

$$PV = \tfrac{2}{3}N\bar{\epsilon} = \tfrac{2}{3}E_t = \frac{N}{\beta}. \tag{2.38}$$

These simple relationships apply to the ideal gas but the concept is completely general. Any confined particle gaseous system will exert a force on the confining walls. The forces exerted by structured particles can reflect their internal motions only to the extent, if any, that the internal motions affect translation since only translation contributes to the pressure. Gas phase translation and internal motion are completely independent and structure has no effect in this regard. If the structure leads to particle–particle interactions, particle motion and therefore the pressure will reflect their interactions, leading to more involved expressions for the PV vs E relationship. For gases in general, the pressure exerted by the confined particle system is a statistical reflection of the translational energy. In the isolated system the pressure is an equilibrium property. Any isolated system is as completely characterized by any three of the quantities P, V, β, and N as by E, V, and N. We will refer to the particular P, V, β, N relationship for any system as the equation of state for that system.

On the basis of Eq. (2.22) any two systems in equilibrium across a diathermal boundary will have the same β value. With β now identified by the PV/N value for the ideal gas, it becomes an observable. The β value for any system of interest is equal to PV/N for an ideal gas system when the two are in thermal equilibrium across a closed, rigid, diathermal boundary. The PV/N term can certainly be classified as observable; N and V are specifications of the system and the internal (system pressure) is measured by the external force per unit area (a measurable mass acted on by the gravitational field) required to maintain V at its prescribed value in a closed, nonrigidly bounded ideal gas system. As previously noted, the real gas helium can serve as an ideal gas to a very, very close approximation in a dilute system.

For historical and practical reasons, it will be convenient to define a scale in terms of $1/\beta$ which increases with the PV term. Such a scale is referred to as temperature and is defined by

$$\frac{PV}{N} = \frac{1}{\beta} = kT \tag{2.39}$$

where k is a scaling factor known as Boltzmann's constant, which brings the PV/N value into congruence with an arbitrary historical scale. The gross features of this scale are that it extends from the kinetic energy zero for the ideal gas [$PV/N = \frac{2}{3}(E/N) = kT$] to unbounded higher values in units of degrees Kelvin (K).

We are, of course, not restricted to the ideal gas or to a gas behaving ideally for the actual laboratory measurement of β or T for any system of interest. The physical properties of many substances, e.g., the volume of liquid mercury or the resistance of a platinum wire, when in diathermal contact with the ideal gas, or a gas behaving ideally, can be demonstrated to change in a sufficiently linear manner with PV/N (at least over certain ranges) that clever exploitation of such behavior can produce laboratory thermometers.[10] We can readily measure T, just as we can measure P, even though they are of purely statistical origin. We reserve treatment of units and numerical values for a general discussion in the following chapter.

A further useful feature of Eq. (2.38), defining a system pressure in terms of internal forces acting on the bounding walls of the system, is that we can now address the work terms [Eqs. (2.9 and (2.10)] through system parameters. We develop the arguments as follows. If the boundary conditions characteristic of isolation are relaxed to provide interaction with the surroundings through closed, adiabatic, displaceable boundaries, the system dimensions are subject to change through boundary displacement unless external forces on the bounding walls are exactly equal to the internal forces (pressure times area) acting on those walls. In kinetic-molecular terms, if the external force per unit piston area (P_{ex}) exerted by the surroundings is constant and greater than the internal force exerted by particle collison with unit piston area (P) the boundary (piston) moves to diminish one of the system dimensions (say a) resulting in an increased frequency of particle collison with all walls, an increase in force exerted internally on all walls, and therefore an increase in system pressure P. If uninhibited, the process will continue until internal and external forces are exactly balanced. Although the bounding wall is still free to move it remains stationary: mechanical equilibrium prevails. The system will remain in the state

[10] This assertion, deriving from Eq. (2.22), is equivalent to the classical statement of the zeroth law of thermodynamics: "If two bodies are each in thermal equilibrium with a third body, they will be in thermal equilibrium with each other."

characterized by E_2 (or $P_{ex} = P_2$), V_2, and N whether or not the rigid boundary constraint is reimposed. We conclude that our closed, adiabatic system and its surroundings are in mechanical equilibrium across a displaceable boundary when there is no pressure difference across the boundary just as system and surroundings are in thermal equilibrium when there is no difference in β (now related to T) across the diathermal boundary. The energy exchange process in microscopic terms has taken the system from one equilbrium state (the original isolated state) to another equilibrium state (a different isolated state). Both equilibrium states are characterized by an energy level set and by the particle distribution among the quantum states that maximize the statistical term Ω.

In mechanical terms the energy of the system has been increased in direct proportion to the work performed on the system,

$$\Delta E = W = -\int_{c\,1}^{2} \frac{f_{ex}}{A}(A dx) = -\int_{c\,1}^{2} P_{ex} dV. \tag{2.40}$$

The path of integration for the state 1 to state 2 transition follows the instantaneous external pressure and the system volume changes values throughout the process. If the external pressure is constant throughout, as in the above example, the integral reduces to $-P_{ex}\Delta V$.

Clearly the condition of continuous mechanical equilibrium is not met during an arbitrary constant pressure work process and the process as described above is irreversible. It is, of course, possible in principle to continuously adjust P_{ex} so that it differs only infinitesimally at any point in the process from the system pressure P. Provided that the adjustments are also carried out sufficiently slowly that the distribution of particles among the instantaneously available quantum states is always that distribution conforming to the maximum value of Ω for the instantaneous E, V, N values at that point in the process, the work process is reversible and

$$dE = dw^* = -P dV \tag{2.41}$$

where we recognize the virtual equality of P_{ex} and P at every stage of the process.

We summarize the previously derived general expressions in terms of the new variable T recognizing that, for $\beta = 1/kT$, $d\beta = -(1/kT^2)dT$ and,

$$p_j^* = \frac{e^{-\epsilon_j/kT}}{z}; \qquad z = \sum^{j} e^{-\epsilon_j/kT} \tag{2.42}$$

$$E = -\left(\frac{\partial \ln \mathscr{Z}}{\partial \beta}\right)_{V,N} = kT^2\left(\frac{\partial \ln \mathscr{Z}}{\partial T}\right)_{V,N}. \tag{2.43}$$

Generally, there exists for all substances an equation of state of the form

$$f(P, V, T, N) = 0. \tag{2.44}$$

From Eq. (1.24g)

$$\sigma = \ln \mathscr{Z} + \beta \mathrm{E} = \ln \mathscr{Z} + \frac{\mathrm{E}}{k\mathrm{T}} \tag{2.45}$$

we define

$$\mathrm{S} = k\sigma = k \ln \mathscr{Z} + \frac{\mathrm{E}}{\mathrm{T}} = k \ln \Omega \tag{2.46}$$

referring to S, so defined, as the entropy. From Eqs. (2.14) and (2.26) we write,

$$dq^* = \mathrm{T}d\mathrm{S} = \mathrm{N}\sum^{j} \epsilon_j^* d\wp_j^* = k\mathrm{T}d\ln\Omega. \tag{2.47}$$

This is entirely general.[11] For a closed, constant volume system $(dq^*)_{\mathrm{V,N}}$ must be used. For systems bounded by other constraints dq^* will reflect those constraints.

In terms of the new thermodynamic variable entropy, the Helmholz free energy becomes

$$\mathrm{A} = -k\mathrm{T}\ln\mathscr{Z} = \mathrm{E} - \mathrm{TS}. \tag{2.48}$$

The equilibrium condition requires that

$$(d\mathrm{A})_{\mathrm{T,V}} = -k\mathrm{T}d\ln\mathscr{Z} = 0.$$

The circumstance under which a spontaneous process can occur is that

$$(d\mathrm{S})_{\mathrm{E,V}} = (d\ln\Omega)_{\mathrm{E,V}} > 0 \tag{2.49}$$

in any isolated system for every process step or that

$$(d\mathrm{A})_{\mathrm{T,V}} = -k\mathrm{T}(d\ln\mathscr{Z})_{\mathrm{T,V}} < 0 \tag{2.50}$$

for every process step in the system of thermodynamic interest. Equations (2.49) and (2.50) and their extensions are known as the Second Law of Thermodynamics in the classical development.

From Eq. (2.39) we note that, for the ideal gas,

$$\mathrm{A} = -k\mathrm{T}\left[\mathrm{N}\ln\left(\frac{2\pi mk\mathrm{T}}{h^2}\right)^{3/2}\frac{e}{\mathrm{N}} + \mathrm{N}\ln\mathrm{V}\right] \tag{2.51}$$

[11] Since S and T are statistical terms that are defined only for the equilibrium state, it is intuitively reasonable that, among all dq process terms, only dq^* is directly related to S since only for dq^* is the system in statistical equilibrium throughout the process.

so that

$$\left(\frac{\partial A}{\partial V}\right)_{T,N} = -\frac{N\ell T}{V} = -P. \tag{2.52}$$

Although this result is obtained from the ideal gas partition function, it is generally true for all systems and we can regard pressure as so defined.[12]

We also note that the results embodied in Eqs. (2.12), (2.40), and (2.47) can be summarized as

$$(dE)_N = (TdS - PdV)_N \tag{2.53}$$

which we will regard as a combined, closed system, classical first and second law statement.

With Eq. (2.41) we have the ideal framework for the approach to variation in N within a system of thermodynamic concern. The system partition function $\mathscr{Z}$ depends on T, V, and N, therefore the Helmholz free energy A is also dependent on these quantities. Consider identical particle systems $\mathscr{A}$ and $\mathscr{B}$ externally isolated but internally in contact across a displaceable, diathermal but nonpermeable barrier. When internal thermal and mechanical equilibrium previal each system is at uniform and constant T and V. Imagine now that the $\mathscr{A}$–$\mathscr{B}$ system is placed in a thermostated environment at the $T_{\mathscr{A}} = T_{\mathscr{B}}$ temperature and the system–environment boundaries are made rigid and diathermal while the internal boundary is made material permeable. If any spontaneous process can occur it will do so by a decrease in the system free energy, i.e., for the $\mathscr{A}$–$\mathscr{B}$ system

$$(dA)_{T,V} < 0.$$

Since, in this case, T and V remain constant throughout the process, it is required that

$$-\ell T\left(\frac{\partial \ln \mathscr{Z}}{\partial N}\right)_{T,V} < 0$$

i.e., that for a nonlocalized (gaseous) system

$$-\frac{\partial}{\partial N}(N\ell T \ln z - N\ell T \ln N + N\ell T)_{T,V} = -\ell T \ln \frac{z}{N} < 0. \tag{2.54}$$

Recalling [Eq. (1.7)] that α, the undetermined multiplier attached to the constraint on distribution numbers (and therefore N), was given by $\ln z/N$, we

[12] As we shall see in later development, generally $PV = N\ell T[1 + f(V)]$. The additional terms $N\ell Tf(V)/V$ will originate in the potential energy component of the partition function.

introduce the notation

$$\mu = -kT \ln \frac{z}{N} = -kT\alpha \tag{2.55}$$

defining μ as the driving force for spontaneous material transfer across a material permeable boundary at constant T and V for the system. We will refer to this driving force as the chemical potential.[13] We anticipate that, since chemical potential, together with observables T and P, are the driving forces for all of those spontaneous energy exchanges that we are currently recognizing, we shall have ample opportunity to develop further implications.

To summarize and consolidate the conclusions of this chapter, the following remarks may be helpful. The undetermined multiplier β, originally introduced to accommodate the maximization constraint on constant energy in the isolated system, was related to the Kelvin scale of absolute temperature ($\beta = 1/kT$). On the basis of this accomplishment, the term entropy $S(= k \ln \Omega)$ replaced the term $\sigma(= \ln \Omega)$ in our arguments. The undetermined multiplier α, originally introduced to accommodate the maximization constraint on constant N, in the isolated system, was related to a term which we designated as the chemical potential ($\alpha = -\mu/kT$). Pressure, defined in terms of force per unit area exerted on the bounding walls by the confined particle system, was quantified by kinetic argument and shown to be related, for the ideal gas, to the energy density of the system $[P = \frac{2}{3}(E/V)]$. The concept of the equation of state, $f(P, V, T, N) = 0$, was introduced and the form $PV = NkT$ developed for the ideal gas. Boundaries were conceptualized across which thermal (heat) energy, mechanical (work) energy, and material (particles) could be exchanged with other systems or the surroundings. We found that gradients in T or P across appropriate boundaries with the surroundings or gradients in μ across (normally) internal open boundaries result in energy or material exchanges that constitute a process.

The concept of a reversible process was introduced as a sequence of infinitesimal steps the sum of which constitute a finite process. There are never more than infinitesimal gradients in T, P, or μ across the interaction boundaries. During a reversible process, the thermodynamic system is in a continuous state of (virtual) equilibrium with any system with which it is interacting. During a reversible process, i.e., at any infinitesimal step of such process, fixed energy states prevail in the thermodynamic system of interest and the equilibrium condition defined the distribution of particles among them.

Where the interaction boundaries are with the environment, the thermodynamic system and its immediate surroundings can be considered to be an isolated supersystem to which statistical argument must apply. The statistical argument of most concern is that any process occurring spontaneously in the isolated supersystem must result in an increase in the combinatorial term Ω

[13] The concept is not confined to internal particle exchanges as close examination of the argument will demonstrate; it is just that we will normally be concerned with internal changes in N.

or its analog entropy. Generally, the statement $dS \geqslant 0$ describes both the equilibrium circumstance and the possibility that a higher maximum exists for the isolated system. This result was related to a new thermodynamic parameter, which we called the Helmholz free energy, and defined as $A = E - TS$. The isolated supersystem criterion was replaced by the minimization of free energy for the thermodynamic system of interest under constant T and V constraints.

As a further aid in assimilating the contents of this chapter, the most important new results are identified, excluding important equations developed in Chapter 1 in which only the $\beta = 1/kT$, substitution has been made.

Microscopic:

$$dE = N\left(\sum^{j} p_j^* d\epsilon_j + \sum^{j} \epsilon_j d p_j^*\right) + \left(\sum^{j} p_j \epsilon_j\right) dN \tag{2.7}$$

$$dq^* = N \sum^{j} \epsilon_j d p_j^* \tag{2.14}$$

$$dw^* = N \sum^{j} p_j^* d\epsilon_j \tag{2.15}$$

$$A = -kT \ln \mathscr{Z}; \qquad dA = -kT d \ln \mathscr{Z} \tag{2.48, 2.50}$$

$$\mu = -kT \ln \frac{z}{N} \tag{2.55}$$

$$S = k \ln \Omega = k \ln \mathscr{Z} + \frac{E}{T} \tag{2.46}$$

Macroscopic:

$$P = -\left(\frac{\partial A}{\partial V}\right)_{T,N} \tag{2.52}$$

$$(\Delta E)_N = Q + W \tag{2.12}$$

$$PV = NkT \tag{2.39}$$

$$A = E - TS \tag{2.31}$$

$$dq^* = TdS \tag{2.21}$$

$$dw^* = -PdV \tag{2.41}$$

$$dE = TdS - PdV \tag{2.53}$$

Excepting Eq. (2.55), which applies to a system of indistinguishable particles, i.e., gas phase, and Eq. (2.39), which represents the ideal gas, the above expressions are general and qualified only by our ability to formulate the partition function and its related terms.

SYMBOLS: CHAPTER 2

New Symbols

$\mathscr{A}, \mathscr{B}, \mathscr{C}$ = designation of a particular system of particles
$\mathscr{E}$ = designation of a portion of the environment surrounding a system
dw = element of mechanical work
W = sum of all work elements in a process
A = area
dq = element of thermal energy (heat)
Q = sum of all heat elements in a process
A = Helmholz free energy
$\boldsymbol{P}, p_x$ = momentum vector, momentum component (x direction)
P = pressure
$\bar{v}_x$ = average x-component of velocity
$\overline{v^2}$ = mean square velocity
T = temperature (Kelvin scale), identification of β
k = Boltzmann's constant
K = Kelvin scale for temperature
S = entropy (replacing σ as identifying $\ln \Omega$)
μ = chemical potential, identification of α

Carryover Symbols from Chapter 1

$\epsilon_i, \epsilon_i^*, \epsilon_j, \epsilon_j^*, p_j, p_j^*$, E, V, N, Ω, α, β, p, $\mathscr{L}$, z, σ, f, **f**

3

UNITS AND CALCULATIONS

It will be useful to digress from the line of statistical argument at this point to quantify terms that, to now, are only symbols representing features of the system. For this purpose it will be helpful to assemble those statistical results that are operational. The following equations are not closely identified here since all have been discussed to the limit of our present understanding in the preceding two chapters.

$$p_j^* = e^{-\epsilon_j/kT}/z \tag{3.1}$$

$$E = \frac{\sum^i \epsilon_j e^{-\epsilon_j/kT}}{z} = kT^2\left(\frac{\partial \ln \mathscr{Z}}{\partial T}\right)_{V,N} \tag{3.2}$$

$$S = k \ln \mathscr{Z} + \frac{E}{T} \tag{3.3}$$

$$A = -kT \ln \mathscr{Z} \tag{3.4}$$

$$\mu = -kT \ln \frac{z}{N} \tag{3.5}$$

Until we have developed the capability to deal with structured particles, the system is gaseous, the only energy is translational, the system partition function is that for indistinguishable particles,

$$\mathscr{Z}_{\mathrm{I}} = \frac{z^N}{N!}; \qquad z = z_t = \left(\frac{2\pi m k T}{h^2}\right)^{3/2} V, \tag{3.6}$$

and the available energy states are

$$\epsilon_{j,n} = \frac{n^2 h^2}{8mV^{2/3}}. \tag{3.7}$$

When any term component, other than the pure numbers, in the above expressions is quantified by having a numerical value assigned to it, the assignment must be accompanied by a unit specification (dimension) and this unit specification must be consistent with all other units appearing in the expression. For example, we have noted in Chapter 2 that a thermodynamic temperature scale can be defined in terms of the ideal gas PV product with $PV/N = kT$. To introduce dimensional arguments, it will be useful to start with this relationship.

Clearly, with $PV = \frac{2}{3}E$, the PV product has dimensions of energy and NkT must have those same dimensions. Also, with $\bar{\epsilon} = \frac{3}{2}\beta$, β must have dimensions energy^{-1}. Where β is then identified with T^{-1} and is assigned arbitrary units of degrees Kelvin (K), k (Boltzmann's constant) must have units energy K^{-1} molecule^{-1}. Energy was defined in terms of a force distance product. We must develop fundamental units for force and establish a more complete analysis of the temperature scale before the above remarks and their extensions can have meaning. We begin with force and energy.

Mass, length, and time are sufficient fundamental quantities to completely define those forces and energies with which we have immediate concern. The unit electrical charge will be required in some of our considerations involving structured particles. Although the scientific community subscribes to and encourages the usage of SI (meter, kilogram, second) units, virtually all thermodynamic work in chemistry done before 1970 uses the cgs (centimeter, gram, second) system. We shall use units convenient to the circumstance at hand with the explicit intent to encourage the student to acquire a ready facility for conversion. We concentrate here on mechanical and thermal units on the basis of present need. Electrical units are discussed in some detail in Appendix D on Intermolecular Forces.

Force is the product of mass (m) and acceleration ($d\mathbf{v}/dt$) where $\mathbf{v}$ is velocity ($d\mathscr{L}/dt$).[1] Time is t and $\mathscr{L}$ is a linear dimension. In momentum ($\mathbf{p} = m\mathbf{v}$) terms, force is rate of change of momentum $[m(d^2\mathscr{L}/dt^2)]$. Energy is the product of a force and a linear dimension $[m\mathscr{L}(d^2\mathscr{L}/dt^2)]$. Force and energy units depend on our choice of systems. In the cgs system we define the dyne as the force

[1] See Appendix B for vector notation and rules for combining vectors.

required to give a 1 g mass an acceleration of 1 cm sec^{-2}. In the SI system read 1 kg mass and 1 m sec^{-2} to define the Newton. Then

$$1 \text{ dyne} = 1 \text{ g cm sec}^{-2} = 10^{-5} \text{ kg m sec}^{-2} = 10^{-5} N.$$

The respective energy units are the erg and the joule

$$1 \text{ erg} = 1 \text{ g cm}^2 \text{ sec}^{-2} = 10^{-7} \text{ kg m}^2 \text{ sec}^{-2} = 10^{-7} \text{ J}$$

$$1 \text{ erg} = 1 \text{ dyne cm} = 10^{-7} N \text{ m} = 10^{-7} \text{ J}.$$

Planck's constant h appears as a quantum scaling factor in all quantized energy expressions. When translational energy levels are obtained from the Eq. (3.7) quantum expression, Planck's constant must have units (energy mass volume$^{2/3}$)$^{1/2}$ = (g cm^2 sec^{-2} g cm^2)$^{1/2}$ = erg sec or (kg m^2 sec^{-2} kg m^2)$^{1/2}$ = J sec. The value of h, established from energy level spacings for internal motions as given by spectroscopic data is 6.626×10^{-27} erg sec (6.626×10^{-34} J sec).

In our arguments to this point we have employed the number of particles in a thermodynamic system as a measure of material quantity. This is a convenience in statistical expression that we shall continue to employ. In relating our results to laboratory systems we will ultimately utilize the relative molar mass (known in the SI system as RMM) but referring to it as the mole weight (M) and defining M as the laboratory weight in grams of Avogadro's number ($N_A = 6.022 \times 10^{23}$) of particles. Each particle of mass m is $M \times 10^{-23}/6.0221$ g ($M \times 10^{-26}/6.0221$ kg) when m appears in a statistical expression. As the subject develops we will require concentration units to describe mixtures and these will be the familiar, conventional units traditional to the chemical laboratory; molarity, molality, and mole fraction with the mole defined as Avogadro's number of particles.

Some of our arguments have developed in terms of volume and concentrations will often require volume units. In the cgs system the volume unit is centimeter^3 and in the SI system meter^3. In the laboratory it is the liter, which is defined as 1 cubic decimeter (dm^3). Therefore, 1 liter = 1 dm^3 = 10^{-3} m^3 = 10^3 cm^3 = 10^3 ml. The angstrom (Å = 10^{-8} cm = 10^{-10} m) is widely used for molecular dimensions. A comparable SI unit is the nanometer (1 nm = 10^{-9} m = 10^{-7} cm = 10 Å). Density (ρ) expressed in g cm^{-3} is readily associated with experience whereas kg m^3 is difficult; 1 g cm^3 = 10^{-9} kg m^{-3}.

The present inclination of chemists to avoid the SI pressure unit derives from the common laboratory association of pressure with the column of mercury supported by the standard atmosphere. The standard atmosphere is defined as 760.00 cm Hg. One millimeter of Hg as a pressure unit is referred to as 1 Torr. The standard atmosphere is reconciled with the definition of pressure in terms of force exerted by confined particles in their momentum exchange with unit cross section of bounding wall by requiring that pressure have dimensions of force distance^{-2} or, in terms of fundamental units, mass length^{-1} time^{-2}. These unseemly units can be associated with the practical process by thinking of a gas confined to a volume that has one movable wall. The gas pressure at any prescribed volume is measured by the external force on the movable wall required to maintain the volume at the prescribed value. The external force can be a mass acted on by the gravitational field. The mass associated with the standard atmosphere is, per cm^2, the mass of a column of mercury 76.000 cm in height and 1 cm^2 in cross section, i.e., 76.000 cm^3. Taking the density of mercury as 13.594 g cm^{-3} and the gravitational constant as 980.00 cm sec^{-2}, we obtain 1.0133×10^6 g cm^{-1} sec^{-2} atm^{-1}, or 1.0133×10^6 dyne cm^{-2} atm^{-1}. This is equivalent to 1.0133×10^5 $N\,m^{-2}$ atm^{-1}. In the SI system 1 $N\,m^{-2}$ is known as a Pascal. There is another force $length^{-2}$ unit that is known as the bar and defined to be 10^5 N m^{-2} so that the bar is approximately 1 atmosphere (1/1.0133 = 0.9868 atm bar^{-1}).

Except for certain very low temperature measurements, the ideal gas thermometer is not a practical measuring device. We employ reference points scaled to ideal gas behavior. Any single component system in two-phase (at constant pressure) or three-phase equilibrium can serve as such reference as evidenced by the constancy and reproducibility of the ideal gas PV product when the multiphase system and the ideal gas are in equilibrium across a diathermal boundary.[2]

Our temperature scale divides the energy interval between the two-phase water systems solid–liquid and liquid–gas, both under one atmosphere pressure, into 100 equal parts, referring to each such increment as a Centigrade degree. The Centigrade temperature scale is constructed on the basis of assigning value zero to the solid–liquid reference point. Extending this scale upward provides an unlimited upper boundary. Extending the Centigrade scale down to the kinetic energy zero by extrapolation of the PV product from low values to that point establishes an ideal gas scale of temperature. The Kelvin scale (K) adopts the Centigrade degree interval beginning the scale at the kinetic energy zero. The constant k (Boltzmann's constant) is the scaling factor that brings the ideal gas PV/N value into congruence with T = 273.16 K for the ideal gas in equilibrium with the three-phase water system. The value of k is 1.381×10^{-16} erg $molecule^{-1}$ K^{-1} (1.381×10^{-23} J $molecule^{-1}$ K^{-1}). When $N = N_A$, the $N_A k$ product has the value 8.314×10^7 erg $mole^{-11}$ K^{-1} (8.314 J $mole^{-1}$ K^{-1}).

Because of its relevance to gas phase calculations ($PV/T = Nk$), the $N_A k$ product is denoted as R and referred to as the gas constant; a molar quantity. Then, $PV = nRT$ where $n = N/N_A$. The PV product, an energy term since volume × force/area = force × distance, is most conveniently expressed as liter atmospheres in which case $N_A k = R$ must have dimensions liter atmospheres $mole^{-1}$ K^{-1}. With the atmosphere defined as 1.0133×10^6 dyne cm^{-2}, the dimension 1 atm is 1.0133×10^9 dyne cm or 1.0133×10^2 J. Then R is 8.314/101.33 = 0.082051 atm $mole^{-1}$ K^{-1}.

[2]We imagine that developing thermodynamic argument will provide an explanation for this empirical observation.

With these descriptions of consistent units we should now be able to obtain numbers for statistical terms that we have identified to this point. We can, for example, quantify the partition function, distribution numbers, energy, entropy, free energy, and chemical potential for the ideal gas if we assign a mass to the particles. It will be more interesting, however, to work with real systems. We have noted that, if dilute, the real gas helium should behave like the ideal gas in spite of its structure. Let us assume that the dilute real gas argon, which is still structurally simple, will also behave ideally. The mass of the helium atom is $4.00/6.02 \times 10^{23} = 6.64 \times 10^{-24}$ g and the mass of the argon atom is $13.95/6.02 \times 10^{23} = 2.32 \times 10^{-23}$ g. If we take N to be N_A, occupying the ideal gas volume of $22{,}400\,\text{cm}^3$ at 273.16 K and 1 atmosphere pressure, which we will refer to as standard (STP) conditions, we have all the numbers required to evaluate the partition functions.

Until we become more familiar with the calculations, it will be well to ensure that the units that we anticipate using do indeed produce the dimensionless number required for z. We have

$$z = \left(\frac{2\pi m k \mathrm{T}}{h^2}\right)^{3/2} \mathrm{V} \quad \left[\frac{\left(\dfrac{\text{g}}{\text{molecule}}\right)\left(\dfrac{\text{erg}}{\text{molecule K}}\right)\text{K}}{\left(\dfrac{\text{erg sec}}{\text{molecule}}\right)^2}\right]^{3/2} \text{cm}^3$$

$$= \left(\frac{\text{g}}{\text{erg sec}^2}\right)^{3/2} \text{cm}^3 = \left[\frac{\text{g}}{\left(\dfrac{\text{g cm}^2}{\text{sec}^2}\right)\text{sec}^2}\right]^{3/2} \text{cm}^3 = \frac{\text{cm}^3}{\text{cm}^3}$$

and we are free to proceed numerically without concern for unit consistency.

Since it is $\ln z$ that will be most useful in our calculations, we will address our first calculations to that term,

$$\ln z = \frac{3}{2} \ln\left(\frac{2\pi k}{h^2}\right) + \frac{3}{2} \ln \frac{M}{\mathrm{N_A}} + \frac{3}{2} \ln \mathrm{T} + \ln \mathrm{V}. \tag{3.8}$$

Collecting constant terms first,

$$\frac{3}{2} \ln \frac{2\pi k}{\mathrm{N_A} h^2} = \frac{3}{2} \ln \frac{(2)(3.14)(1.38 \times 10^{-16})}{(6.02 \times 10^{23})(6.63 \times 10^{-27})^2} = \frac{3}{2} \ln (0.0328 \times 10^{15})$$

$$= \tfrac{3}{2}[\ln 0.0328 + \ln 10^{15}] = \tfrac{3}{2}[-3.42 + 2.303(15)] = 46.69,$$

$$\ln z = 46.69 + \tfrac{3}{2} \ln M + \tfrac{3}{2} \ln \mathrm{T} + \ln \mathrm{V}. \tag{3.9}$$

The expression is now adapted to general use with substitution of particular M, T, V values. If we are interested in helium gas under standard temperature and pressure conditions:

$$\tfrac{3}{2}\ln M = \tfrac{3}{2}\ln 4.00 = 2.08$$
$$\tfrac{3}{2}\ln \mathrm{T} = \tfrac{3}{2}\ln 273.2 = 8.42$$
$$\ln \mathrm{V} = \ln(2.24 \times 10^4) = 10.02$$
$$\ln z_{\mathrm{He}} = 46.69 + 2.08 + 8.42 + 10.02 = 67.21.$$

To obtain $\ln z_{\mathrm{Ar}}$ under the same conditions of V and T we calculate

$$\tfrac{3}{2}\ln M_{\mathrm{Ar}} = \tfrac{3}{2}\ln 39.95 = 5.50$$

and substitute this value for the M_{He} value to obtain

$$\ln z_{\mathrm{Ar}} = 67.21 - 2.08 + 5.50 = 70.03.$$

In either case we obtain the effect of change in temperature or volume by changing the appropriate term in the $\ln z$ expression. This is accomplished most efficiently if we take out the original values while inserting the new values. As regards T, we simply add $\tfrac{3}{2}\ln \mathrm{T}_2/\mathrm{T}_1$, where T_2 is a temperature of interest and T_1 is the original 273.16 K. If we assume that the change in T is accomplished at constant volume, this is the entire change in $\ln z$. If the change in T is accomplished at constant pressure, or if volume is changed at constant temperature a term $\ln \mathrm{V}_2/\mathrm{V}_1$ must be added to the partition function. For an ideal gas at constant pressure $\ln(\mathrm{V}_2/\mathrm{V}_1) = \ln(\mathrm{T}_2/\mathrm{T}_1)$ and the new partition function is given by removal of the volume term and addition of $\tfrac{5}{2}\ln(\mathrm{T}_2/\mathrm{T}_1)$. The change with either temperature or volume is completely independent of the mass, and therefore the identity of the particles so long as the ideal gas relationship is obeyed. The substitution of $\ln(\mathrm{T}_2/\mathrm{T}_1)$ for $\ln(\mathrm{V}_2/\mathrm{V}_1)$ is just an arithmetical convenience to produce the numerically equivalent change in $\ln z$. We cannot avoid dealing with the proper values of T and V when we are interested in the $\ln z$ value itself.

We can make the general observation at this point that $\ln z$ is (0) 67 for helium under standard conditions of temperature and pressure and does not change greatly with either temperature, mass, or volume (about ± 3 units for 10-fold changes in T or M and about ± 2 units for 10-fold changes in V). Over a very wide range of conditions, then, $\ln z$ is (0) 67 and z itself is (0) 10^{29}. This very large number as a translational partition function value should not be too surprising since it largely reflects the energy level degeneracies which can be enormous. Where $z = \sum^i g_i e^{-\epsilon_i/k\mathrm{T}}$, for low energies the degeneracies are the only important feature of the sum since $e^{-\epsilon_i/k\mathrm{T}} \to 1$ as $\epsilon_i/k\mathrm{T} \to 0$ and all such terms contribute g_i to the partition function. The $\epsilon_i/k\mathrm{T}$ are not required to be very small for this to occur, e.g., $e^{-0.01}$ is 0.990. At the other end of energy values the sum eventually terminates [e^{-20} is (0) 10^{-9}] even though the degeneracies are becoming much larger.

We should carry these considerations with regard to z and $\ln z$ one step further since it is $\mathscr{Z}$ and $\ln \mathscr{Z}$, which are of interest in molecule collections. The relationship is complicated by the nonlocalized character of gas phase translational expressions. We have

$$\ln \mathscr{Z} = \mathrm{N} \ln \frac{z e}{\mathrm{N}}$$

and for order of magnitude considerations we take the value of z for helium gas under standard conditions (1.54×10^{29}) to obtain the molar ($\mathrm{N} = \mathrm{N_A}$) partition function

$$\ln \mathscr{Z}_{\mathrm{He}} = 6.02 \times 10^{23} \ln \frac{(1.54 \times 10^{29})(2.72)}{(6.02 \times 10^{23})} = 6.93 \times 10^{24}. \tag{3.10}$$

We note that $\ln \mathscr{Z}$ is a number of (0) N even though we were required to construct the term on the basis of dividing z^N by N!; $\mathscr{Z}$ itself is (0) $10^{10^{24}}$. The number is inconceivably large. Fortunately we are not required to utilize it in our general arguments but it is good to know that division by N! did not produce a numerical catastrophe.

The $\ln \mathscr{Z}$ and $\ln z$ values lead immediately to the evaluation of entropy, Helmholz free energy, and chemical potential for the ideal gas or for any dilute monatomic gas. First we note that $\mathrm{E} = \frac{3}{2}\mathrm{N}k\mathrm{T}$ for such systems [Eq. (1.18c); $\mathrm{E} = \frac{3}{2}\mathrm{N}/\beta$] and that

$$\mathrm{S} = k \ln \mathscr{Z} + \frac{\mathrm{E}}{\mathrm{T}} = k \ln \mathscr{Z} + \tfrac{3}{2}\mathrm{N}k$$

$$\mathscr{Z} = \frac{z^N}{\mathrm{N}!}$$

$$\mathrm{S} = \mathrm{N}k\left[\ln \frac{z}{\mathrm{N}} + \frac{5}{2}\right] \tag{3.11}$$

$$\mathrm{A} = \mathrm{E} - \mathrm{TS} = -k\mathrm{T} \ln \mathscr{Z} \tag{3.12}$$

$$\mu = -k\mathrm{T} \ln \frac{z}{\mathrm{N}}. \tag{3.13}$$

The $\frac{3}{2}\mathrm{N}k\mathrm{T}$ value that we calculate for the monatomic, structured, ideal gas is the translational energy ($\mathrm{E_t}$), i.e., the energy above what the atom has at rest and in its ground electronic state ($\mathrm{E_0}$). So as not to lose sight of this consideration, which will become important when electronic excitation is a possible energetic feature of our considerations, we note that the total energy should always be represented as $\mathrm{E} - \mathrm{E_0}$ although we will often prefer the notational convenience of E. A complete discussion of the nonzero value for E at $\mathrm{T} = 0$ cannot be developed in the ideal gas framework so we note at this point only that we should qualify any calculations as referring to translational components of the thermodynamic parameter of interest.

From Eq. (3.11) we obtain a general expression for the entropy of the ideal gas (or dilute monatomic gases) of any molar mass M occupying a volume V at temperature T,

$$\mathrm{S} = \mathrm{N}k\left[\ln\frac{2\pi k}{\mathrm{N}h^2} + \ln e^{5/2} + \ln M + \frac{3}{2}\ln \mathrm{T} + \ln\frac{\mathrm{V}}{\mathrm{N}}\right]. \tag{3.14}$$

It will often be more convenient to express the volume feature in terms of pressure, which, for ideal gases gives, after collecting terms,

$$\mathrm{S} = \mathrm{N}k[22.33 + \tfrac{3}{2}\ln M + \tfrac{5}{2}\ln \mathrm{T} - \ln \mathrm{P}]. \tag{3.15}$$

This relationship for monatomic gases behaving ideally is known as the Sakur–Tetrode equation. It is required that the dimensions of $k\mathrm{T}/P$ reduce to $\mathrm{cm}^3\,\mathrm{molecule}^{-1}$. When pressure is expressed in $\mathrm{dyne\,cm}^{-2}$ units, this will be accomplished

$$\frac{(\mathrm{erg\,molecule}^{-1}\,\mathrm{K}^{-1})\mathrm{K}}{\mathrm{dyne\,cm}^{-2}} = \frac{\mathrm{dyne\,cm\,molecule}^{-1}}{\mathrm{dyne\,cm}^{-2}} = \mathrm{cm}^3\,\mathrm{molecule}^{-1}.$$

To complete the exercise with a particular example we obtain the molar[3] ($N = N_{\mathrm{A}}$) entropy ($\tilde{S}$) of helium gas under standard conditions $(\tilde{\mathrm{S}}^0)_{\mathrm{He}}$

$$\begin{aligned}(\tilde{S}^0)_{\mathrm{He}} &= 8.314 \times 10^7[22.33 + \tfrac{3}{2}\ln 4.00 + \tfrac{5}{2}\ln 273.2 - \ln 1.013 \times 10^6]\\ &= 204.8 \times 10^7\,\mathrm{erg\,K^{-1}\,mol^{-1}} = 204.8\,\mathrm{J\,K^{-1}\,mol^{-1}}.\end{aligned}$$

For the more massive argon atom we replace $\frac{3}{2}\ln 4 (= 2.08)$ with $\frac{3}{2}\ln 39.95 (= 5.53)$ to obtain

$$(\tilde{\mathrm{S}}^0)_{\mathrm{Ar}} = 204.8 - 2.08 + 5.53 = 208.3\,\mathrm{J\,K^{-1}\,mole^{-1}}.$$

The Helmholz free energy for the monatomic ideal gas also follows directly as

$$\mathrm{A} = -k\mathrm{T}\ln\mathscr{Z} = -\mathrm{N}k\mathrm{T}\ln\frac{ze}{\mathrm{N}}$$

and, for helium under standard conditions, we obtain the numerical value most

[3] Use of the tilde symbol over a thermodynamic expression will designate a molar quantity. This practice is not common but will reduce our notational problems.

easily from the already accomplished (molar) calculation for $\ln \mathscr{Z}_{He}$ [Eq. (3.10))

$$\tilde{A}^\circ_{He} = -(1.38 \times 10^{-23})(273.2)(6.93 \times 10^{24}) = -26.12 \times 10^3 \text{ J mole}^{-1}.$$

For argon a similar calculation provides

$$\tilde{A}^\circ_{Ar} = -29.63 \times 10^3 \text{ J mole}^{-1}.$$

Since the Helmholz free energy has been defined in terms of energy and entropy (both functions of temperature), the energy zero and entropy zero, if such exists,[4] must be present in any calculated value of A, i.e., the free energy value we have calculated is $A - A_0$. If we are interested in changes that occur only when a system is taken from one thermodynamic state (T_1V_1) to another state (T_2V_2), questions about the residual E_0, A_0, etc. are avoided, e.g., $\Delta A = (A_2 - A_0) - (A_1 - A_0) = A_2 - A_1$.

From the entropy calculations we note that a particularly useful form results for a closed system change in the entropy value when the thermodynamic state of the system is changed in that, regardless of the process route followed and the statistical detail involved in changing T from T_1 to T_2 and V from V_1 to V_2, the difference in molar entropy between the initial and final (isolated) states for an ideal gas is

$$\Delta\tilde{S} = N_A k\left[\frac{3}{2}\ln\frac{T_2}{T_1} + \ln\frac{V_2}{V_1}\right] \tag{3.16a}$$

or

$$\Delta\tilde{S} = N_A k\left[\frac{3}{2}\ln\frac{T_2}{T_1} + \ln\frac{P_1}{P_2}\right]. \tag{3.16b}$$

Processes carried out at constant temperature are common concerns. In this circumstance the Eqs. (3.16a) and (3.16b) expressions for the entropy change reduce to

$$(\Delta\tilde{S})_T = N_A k \ln\frac{V_2}{V_1} \tag{3.17a}$$

$$(\Delta\tilde{S})_T = N_A k \ln\frac{P_1}{P_2} \tag{3.17b}$$

and the change in Helmholz free energy

$$\Delta A = -kT \ln \mathscr{Z}$$

[4] It is not possible to discuss the entropy zero on the basis of our present understanding.

reduces to

$$(\Delta\tilde{A})_T = N_A kT \ln \frac{V_1}{V_2}. \tag{3.18}$$

In many arguments that will develop as we extend our considerations to more complex systems changes in chemical potential will be of large concern. From our definition, $\mu = -kT \ln z/N$, we note that for the ideal gas

$$\mu = -kT \ln \left(\frac{2\pi m kT}{h^2}\right)^{3/2} \frac{V}{N}$$

which we rephrase as

$$\mu = -\frac{3}{2} kT \ln \frac{2\pi m}{h^2} - \frac{3}{2} kT \ln kT + kT \ln \frac{N}{V}$$

or, for constant temperature and with $N/V = P/kT$

$$\mu(T) = \mu^0(T) + kT \ln P(\text{atm}) \tag{3.19a}$$

where $\mu^0(T)$ is defined as the collection of constants

$$\mu^0(T) = -\left(\frac{3}{2} kT \ln \frac{2\pi m}{h^2} + \frac{5}{2} kT \ln kT + kT \ln 1.013 \times 10^6\right). \tag{3.19b}$$

We can refer to $\mu^0(T)$ as the standard chemical potential. It is the chemical potential of the ideal gas at temperature T and 1 atmosphere pressure. For a constant temperature ideal gas process going from pressure state 1 to pressure state 2 in a closed system

$$[\Delta\mu(T)]_N = kT \ln \frac{P_2}{P_1}. \tag{3.19c}$$

At any fixed temperature T and any given ideal gas pressure P the molar chemical potential can be written as

$$\tilde{\mu}(T) = N_A \mu^0(T) + N_A kT \ln \frac{P}{P_{std}} = N_A \mu^0(T) + N_A kT \ln P. \tag{3.19d}$$

When our concerns are extended to nonideal gas systems Eq. (3.19a) and the following results will require reconsideration.

In examining the numerical values of z and $\ln z$ we remarked on the importance of the ratio ϵ/kT in assessing the terms that contribute to the partition function. We could say that kT is the yardstick by which the importance of various energy levels is measured. This goes beyond the contribution of a term to the partition function since each energy level term, when multiplied by its degeneracy, determines the probability of occupancy of that level. We must be interested in the numerical values of the energy levels and in their degeneracies when we are interested in quantifying energy distributions.

In the quantum argument the translational energy levels are

$$\epsilon_n = \frac{(n_x^2 + n_y^2 + n_z^2)h^2}{8m\mathrm{V}^{2/3}} \qquad n_x, n_y, n_z = 1, 2, \ldots, \infty \qquad \mathrm{I.1}$$

and the spacing between levels is (O) $n \times 10^{-32}$ ergs where $n = (n_x^2 + n_y^2 + n_z^2)^{1/2}$. The exclusion of $n = 0$ implies that the particle cannot be at rest, which is a reflection of the uncertainty principle (at rest implies exact knowledge of both position and momentum). For translation the consideration is completely trivial, the lowest energy associated with motion in one dimension being (O) $10^{-31}/\mathrm{V}^{2/3}$ erg molecule^{-1}, i.e., (O) 10^{-33} erg molecule^{-1} for gas volumes (O) 10 liters. Any such zero point translational energy is completely trivial when compared to the 3/2 value for the average energy at even 1 K (2.07×10^{-16} erg molecule^{-1}). A reasonable range of translational energies for our present considerations might then range from $\bar{\epsilon}$ at 10 K [(O) 10^{-15}] to $\bar{\epsilon}$ at 10^4 K [(O) 10^{-11}] with spacing between levels of (O) $n \times 10^{32}$ ergs as shown earlier. The step from (O) 10^{-33} erg molecule^{-1} to (O) 10^{-16} erg molecule^{-1} is very large, suggesting that many translational levels are populated at even quite low values of T although it could be a large mistake to assume that most molecules in a large collection at specified T are in energy levels corresponding to $\bar{\epsilon}_{\mathrm{T}}$. The range of translational energies is, however, rather clear running from (O) 10^{-33} to (O) 10^{-11}. The distribution of particles within this range follows from the distribution expression

$$\frac{\mathrm{N}_i^*}{\mathrm{N}} = \frac{g_i e^{-\epsilon_i/kT}}{\sum g_i e^{-\epsilon_i/kT}}. \qquad (1.8)$$

We can obtain some feeling for the order of magnitude of the degeneracy terms g_i from the following exercise. Starting with Eq. (1.1) we write,

$$n^2 = \frac{\epsilon_i(8m)\mathrm{V}^{2/3}}{h^2}$$

and assign to ϵ_i the energy of a molecule at, say, 300 K and to V the value of, say, 10 liters. Then for helium

$$n^2 = \frac{(3)(8)(4)(300)(1.38 \times 10^{16})(10 \times 10^3)^{2/3}}{(2)(6.02 \times 10^{23})(6.63 \times 10^{-27})^2} = 3.48 \times 10^{19}.$$

Recalling that $n^2 = n_x^2 + n_y^2 + n_z^2$, we see that the number of ways in which a number of (O) 10^{20} can be constructed using the three quantum numbers relating to translational components along the three coordinate axes is enormously large. We cannot appeal to any simple arithmetic approach to obtain the g_i.

First we rewrite the distribution and related expressions in a form that focuses on the distribution numbers

$$\epsilon_n = \frac{h^2 n^2}{8m\mathrm{V}^{2/3}} \tag{3.20}$$

$$\frac{\mathrm{N}_n}{\mathrm{N}} = \frac{g_n e^{-\epsilon_n/kT}}{\sum g_n e^{-\epsilon_n/kT}} \tag{3.21}$$

$$\sum^{n} g_n e^{-\epsilon_n/kT} = z = \sum^{j} e^{-\epsilon_j/kT} = \left(\frac{2\pi m k\mathrm{T}}{h^2}\right)^{3/2} \mathrm{V}$$

where we note that our procedure for summing over energy states to produce a value for the partition function has already included the individual term degeneracies but has not resolved them. In fact, a term by term resolution is not practical or even necessary. Since the n^2 values are very large we can regard n as a continuous variable and imagine that the ϵ_n resulting from a single n value are represented by a spherical surface segment formed by sweeping a radius of length n throughout the positive octant of the sphere illustrated in Figure 3.1. This octant of the sphere includes all positive values of n_x, n_y, and n_z that can contribute to the value of n (only positive values of the quantum numbers are permitted). The degeneracies associated with an energy level we now regard as the volume of the spherical segment shell of thickness dn, i.e., in treating the numbers n and the associated energies as continuous functions we replace the energy level with an infinitesimal energy interval $d\epsilon$. The shell volume is the area of the inner surface multiplied by the thickness dn, i.e., 1/8 of the

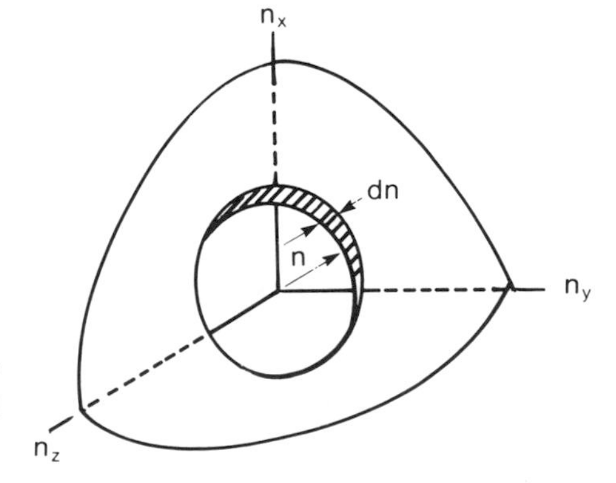

Figure 3.1. Illustration of the degeneracy to be associated with an infinitisimal energy interval $d\varepsilon$.

volume of a sphere of radius n

$$g_n = \tfrac{1}{8}(4\pi n^2 dn).$$

From Eq. (3.20)

$$n = \left(\frac{8mV^{2/3}}{h^2}\epsilon_n\right)^{1/2}$$

and

$$dn = \left(\frac{8mV^{2/3}}{h^2}\right)^{1/2} \frac{\epsilon_n^{-1/2}}{2} d\epsilon$$

then

$$g_n = \frac{\pi}{4}\left(\frac{8m}{h^2}\right)^{3/2} V\epsilon_n^{1/2} dn.$$

Insertion of this value for the degeneracy term in Eq. (1.6) gives

$$\frac{dN_n}{N} = \frac{2\pi(2m/n^2)^{3/2} V\epsilon_n^{1/2} e^{-\epsilon_n/kT}}{(2\pi mkT/h^2)^{3/2} V} d\epsilon$$

$$\frac{dN_n}{N} = \frac{2}{\pi^{1/2}} \epsilon_n^{1/2} e^{-\epsilon_n/kT} (kT)^{3/2} d\epsilon \tag{3.22}$$

for the numbers of molecules dN_n having energies in the range $\epsilon_n + d\epsilon_n$.

We obtain the most probable energy (ϵ^*) from the maximum value of

$$\frac{dN_n/N}{d\epsilon} = \frac{df(\epsilon)}{d\epsilon} = 0,$$

where

$$f(\epsilon) = \frac{2\epsilon_n^{1/2}}{\pi^{1/2}(kT)^{3/2}} e^{-\epsilon_n/kT},$$

with the result

$$\frac{\epsilon^{*1/2}}{kT} = \frac{\epsilon^{*-1/2}}{2} \quad \text{or} \quad \epsilon^* = \frac{kT}{2}. \tag{3.23}$$

The form of the distribution, given by the curves representing N_n/N vs ϵ, is obtained by calculation from

$$F(\epsilon) = \frac{2}{\pi^{1/2}} \frac{[y_n(kT)]^{1/2}}{(kT)^{3/2}} e^{-y_n} = \frac{2}{\pi^{1/2} kT} f(\epsilon)$$

where $f(\epsilon) = y_n^{1/2} e^{-y_n}$ with y_n representing ϵ as a multiple of kT. In Figure 3.2 we show $f(\epsilon)$ vs ϵ for arbitrary T, 2T, and 5T. The energy axis is ϵ/kT referenced to the selected value of T so the maximum in $f(\epsilon)$ vs ϵ is at $kT/2$, kT, and $2.5kT$ as anticipated. The $f(\epsilon)$ values are to be multiplied by $(2/\pi^{1/2})d\epsilon/kT$ in order to obtain actual distributions.

If we require the average value of ϵ, we are required to sum over all values of ϵ and divide by the number of particles, i.e.,

$$\bar{\epsilon} = \frac{\sum^{i} g_i \epsilon_i N_i^*}{N} = \sum^{i} g_i \epsilon_i \frac{N_i^*}{N}.$$

We will represent the discrete N_i^*/N values by Eq. (3.22) and integrate over all values of ϵ from zero to infinity obtaining

$$\bar{\epsilon} = \int_0^\infty \epsilon \frac{2}{\pi^{1/2}} \frac{1}{(kT)^{3/2}} \epsilon^{1/2} e^{-\epsilon/kT} d\epsilon$$

$$\bar{\epsilon} = 2\pi^{-1/2}(kT)^{-3/2} \int_0^\infty \epsilon^{3/2} e^{-\epsilon/kT} d\epsilon \tag{3.24}$$

which we recast as

$$\bar{\epsilon} = \mathrm{A} \int_0^\infty x^{3/2} e^{-ax} dx.$$

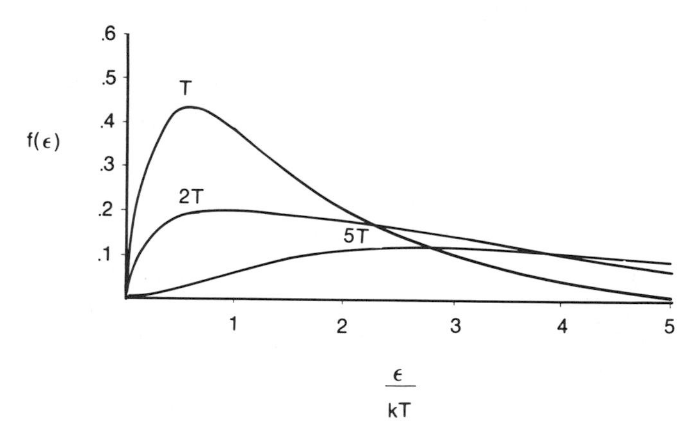

Figure 3.2. The distribution function $f(\epsilon)$ for translational energy at various temperatures.

A table of definite integrals produces

$$\int_0^\infty x^n e^{-ax} dx = \frac{\Gamma(n+1)}{a^{n+1}}.$$

The gamma function can be simplified as follows:

$$\Gamma(n+1) = n\Gamma(n) \qquad \text{if } n > 0 \text{ also } \Gamma^{1/2} = \pi^{1/2}$$

$$\Gamma(\tfrac{3}{2}+1) = \tfrac{3}{2}\Gamma(\tfrac{3}{2}) = (\tfrac{3}{2})(\tfrac{1}{2})\Gamma^{1/2} = \tfrac{3}{4}\pi^{1/2}$$

and

$$\bar{\epsilon} = \frac{2}{\pi^{1/2}(kT)^{3/2}}(\tfrac{3}{2})(\tfrac{1}{2})\pi^{1/2}kT^{5/2} = \tfrac{3}{2}kT \tag{3.25}$$

in complete accord with prior results.

It is useful to express the translational energies in terms of molecular speeds both because such reflection of the kinetic energy is more intuitively familiar and because such values can be extended to collision frequency and other kinetic considerations. Rather than repeat the arguments leading to Eq. (3.22) in the velocity framework we can convert that expression into velocity terms. For this purpose we recognize that it is the absolute value of the vector velocity, i.e. the molecular speed, which is reflected in the energy term. We write

$$\epsilon = \tfrac{1}{2}mv^2, \qquad d\epsilon = mv\,dv, \qquad \epsilon^{1/2} = \left(\frac{m}{2}\right)^{1/2} v$$

so that

$$\frac{N_v}{N} = \frac{2}{\pi^{1/2}}\left(\frac{m}{kT}\right)^{3/2} v^2 e^{-m^{3/2}kT} dv = f(v)dv.$$

For reference purposes we can rephrase $f(v)$ into a form consistent with the expression normally encountered when the derivation follows the velocity argument

$$f(v) = 4\pi v^2 \left(\frac{m}{2\pi kT}\right)^{3/2} e^{-mv^2/2kT}. \tag{3.26}$$

The most probable velocity, without arithmetic, is

$$v^* = \left(\frac{2kT}{m}\right)^{1/2} \tag{3.27}$$

following from $df(v)/dv = 0$. The general order of molecular velocities can be obtained from the most probable velocity as

$$v^* = \left[\left(\frac{2T}{M}\right)(1.38 \times 10^{-16})(6.02 \times 10^{23})\right]^{1/2} = 5.96 \times 10^4\left(\frac{T}{M}\right)^{1/2}\left(\frac{\text{erg}}{\text{g}}\right)^{1/2}$$

$$= 5.96 \times 10^4\left(\frac{T}{M}\right)^{1/2} \text{cm sec}^{-1}.$$

Common gases [M(O) 30] at room temperature [T(O) 300] will have most probable molecular speeds (O) 2×10^5 cm sec^{-1} presuming that the translational energy approximates that of the ideal gas.

The general form of the speed distribution is nonsymmetrical like that of the energy distribution and the various average speed terms that are found to be useful in kinetic argument have somewhat different values than the most probable value. We can identify two average speeds following the procedures detailed above. The mean (average) speed is

$$\bar{v} = \int_0^\infty v\text{f}(v)dv$$

and the mean square speed, i.e., that average obtained by summing over all speeds squared, then dividing by N, is

$$\overline{v^2} = \int_0^\infty v^2\text{f}(v)dv.$$

The integrals we require are

$$\int_0^\infty v^3 e^{-mv^2/2kT}dv = \frac{1}{2}\int_0^\infty xe^{-ax}dx$$

$$\int_0^\infty v^4 e^{-mv^2/2kT}dv = \frac{1}{2}\int_0^\infty x^{3/2}e^{-ax}dx$$

with, in both cases, $a = m/2kT$, $x = v^2$, $dx = 2vdv$. Then

$$\bar{v} = 4\pi\left(\frac{m}{2\pi kT}\right)^{3/2}\frac{\Gamma(2)}{2}\left(\frac{2kT}{m}\right)^2 = \left(\frac{8kT}{\pi m}\right)^{1/2} \tag{3.28}$$

since $\Gamma(2) = 1\Gamma(1) = 1$ and

$$\overline{v^2} = 4\pi\left(\frac{m}{2\pi kT}\right)^{3/2}\frac{\Gamma(5/2)}{2}\left(\frac{2kT}{m}\right)^{5/2} = \frac{3kT}{m}. \tag{3.29}$$

The root mean square speed $(\overline{v^2})^{1/2}$ is $(3kT/m)^{1/2}$.

These conclusions regarding molecular speeds can be extended by simple argument to yield information about collision frequency and mean pathlength. The exact treatment is quite involved and would lead us far away from the intuitive appreciation which we are trying to develop. We will follow the simple argument and amend the result as required. We will imagine a gaseous system of particles that can be represented as a collection of hard spheres with radius r and particle density N/V. All particles are located in the available volume by their mass center. A collision will occur when any two mass centers are separated by a distance equal to the molecule diameter, b $(=2r)$. Any single molecule in free motion would sweep out a collision volume of $\pi\bar{v}b^2\,\text{cm}^3$ each second and encounter $(\text{N/V})\pi\bar{v}b^2$ other particles each second. Each of the N particles can be considered in such argument, leading to $\frac{1}{2}\pi\bar{v}b^2(\text{N}^2/\text{V})$ separate collisions per second (the $\frac{1}{2}$ factor recognizes that each collision involves two particles). The number of collisions per second per unit volume is

$$Z_{11} = \frac{\pi}{2}\bar{v}_1 b_1^2 \frac{\text{N}^2}{\text{V}}$$

with units molecule (encounters) $\text{sec}^{-1}\,\text{cm}^{-3}$. The subscripts 11 and 1 refer to identical particle collisions and average speeds. Simple argument would insert $\bar{v}$ from Eq. (3.28) into this expression. In an exact treatment we should consider not single particle speeds, but relative velocities and we should generalize the treatment to include molecules of different diameter. Such consideration produces $\bar{v}_{12} = (8k\text{T}/\pi\mu_r)^{1/2}$ for the average velocity and $b_{12} = r_1 + r_2$ for the collision cross section rather than $(8k\text{T}/\pi m)^{1/2}$. The mass term μ_r is the reduced mass defined as $(m_1 m_2/m_1 + m_2)$, which reduces to $m/2$ for identical particles. Then,

$$Z_{11} = \frac{\pi}{\sqrt{2}} b_1^2 \left(\frac{8k\text{T}}{\pi m}\right)^{1/2} \left(\frac{\text{N}}{\text{V}}\right)^2 \tag{3.30}$$

$$Z_{12} = \frac{\pi}{2} b_{12}^2 \left(\frac{8k\text{T}}{\pi \mu_r}\right)^{1/2} \left(\frac{\text{N}_1\text{N}_2}{\text{V}^2}\right). \tag{3.31}$$

Collision cross sections $(\sigma = \pi b^2)$ for helium and argon are 2.1×10^{-6} and $2.6 \times 10^{-6}\,\text{cm}^2$, respectively. The number of collisions per second per cm^3 in a gaseous helium sample under standard conditions of temperature and pressure is

$$Z_{\text{He}} = \left(\frac{4k\text{TN}_\text{A}}{\pi M}\right)^{1/2} \left(\frac{\text{N}_\text{A}}{\text{V}^2}\right) \sigma_{\text{He}} = \left[\frac{(4)(1.38 \times 10^{-16})(6.02 \times 10^{23})(273.2)}{(3.14)(4.00)}\right]^{1/2}$$
$$\times \left(\frac{6.02 \times 10^{23}}{2.24 \times 10^4}\right)^2 (2.1 \times 10^{-6}) = 1.29 \times 10^{38} \tag{3.32}$$

and for argon under the same conditions

$$Z_{Ar} = Z_{He}\left(\frac{4.00}{13.96}\right)^{1/2}\left(\frac{3.6}{21}\right) = 0.918Z_{He}.$$

Equation (3.32) is a quite good general result, even going to mass 100 with increase in collision diameter to 10×10^{-6} cm^2 produces a similar factor (0.952) to be applied to the helium collision frequency. Under standard conditions, then, there will be (O) 10^{38} collisions per second per cm^3 for any gas behaving ideally. Temperature appears in Eq. (3.30) as $T^{1/2}$ so that even 10-fold changes in T do not really modify this conclusion if the volume remains constant. Volume appears as a squared term so that 10-fold changes in volume do influence the collision rate by factors of 10^2 (or 10^{-2}) but these are also relatively small changes in a number (O) 10^{38}.

The mean free path (λ) for a single molecule is the distance traveled between collisions, i.e., the velocity (cm sec^{-1}) divided by the collision frequency (sec^{-1}) for that particle

$$\lambda = \frac{\bar{v}}{(\pi/2^{1/2})b^2\bar{v}N/V} = \frac{2^{1/2}}{\pi b^2}\frac{V}{N}. \tag{3.33}$$

The mean free path λ_{He} for helium under standard conditions is

$$\lambda_{He} = \frac{(1.41)(2.24 \times 10^4)}{(2.1 \times 10^{-6})^2(6.02 \times 10^{23})(3.14)} = 3.78 \times 10^{-9} \text{ cm}.$$

For other conditions and other species

$$\lambda = \lambda_{He}\frac{\sigma_{He}}{\sigma}\frac{V}{V^0}.$$

One is often interested in the mean free path of a particle under reduced pressure, e.g., in a vacuum system. If we retain the number (O) 10^{-9} at 1 atm pressure then

$$\lambda = (O)\,10^{-9}\frac{\sigma_{He}}{\sigma}\frac{1}{P_{atm}}$$

producing a λ value (O) 1 cm at P (O) 10^{-9} atm [(O) 10^{-6} Torr] for small molecules in general.

It will possibly be of some utility in arriving at intuitive argument relating to molecule behavior to retain these order of magnitude generalizations on molecular energy distributions and average translational motions.

PROBLEMS

At sufficiently low pressures the vapor phase of any substance will satisfy the ideal behavior conditions and the translational features of all dilute gaseous systems can be calculated from relationships developed to this point.

3.1. Calculate the pressure in dynes cm^{-2}, $N\,m^{-2}$, and atmospheres which 2.5×10^{-5} g of magnesium would exert if it occupied a volume of 1 liter at 298 K.

3.2. Calculate the molar translational entropy and Helmholz free energy for magnesium vapor under the problem 3.1 conditions.

3.3. Calculate the change in entropy and in Helmholz free energy if the volume is expanded to 1.75 liters and the temperature increased to 398 K.

3.4. The collision cross section for oxygen and nitrogen molecules is nearly the same value (0.040 and 0.043 nm). Calculate the mean free path and the collision frequency for air molecules in a room temperature 10 liter vacuum system in which the residual pressure is 10^{-6} Torr.

SYMBOLS: CHAPTER 3

New Symbols

$\mathscr{L}$ = length
t = time
m = meter
N_A = Avogardo's number of particles
R = the gas constant ($N_A k$)
N = the Newton
J = the Joule
M = mole weight ($N_A m$)
ρ = density
g_n = degeneracy associated with a value of the translational quantum number
Å = angstrom, 10^{-8} cm
$F(\epsilon)$, $f(\epsilon)$, $f(v)$ = functions related to the distribution of particles among energy intervals $d\epsilon_n$, or velocity intervals dv
Γ = the gamma function
$\tilde{S}, \tilde{A}, \tilde{E} \cdots$ = molar value of a thermodynamic function
Z_{11}, Z_{12} = frequency of particle collision
μ_r = reduced (effective) mass
b_1, b_{12} = collision radii
σ = collision cross section

λ = mean free path

V°, S° = value of a thermodynamic function when (defined) standard conditions apply

Carryover Symbols from Chapters 1 and 2

p_j^*, ϵ_i^*, ϵ_i, E, S, A, z, $\mathscr{Z}$, P, V, T, h, k, e, μ, n_x, n_y, n_z, N_i, N, g_i, m, v, $\bar{v}$, $\overline{v^2}$

4

ENERGY EXCHANGE II: GENERALIZATIONS AND MACROSCOPIC CONSIDERATIONS

In Chapter 2 we concentrated on reversible energy exchange processess to construct computational routes between initial and final states of a thermodynamic system of interest undergoing a finite, spontaneous process. Although we are still confined to the single species ideal gas partition function in our statistical constructs, we recognize that this is only *our* present limitation; the arguments will extend to any system for which the partition function can be constructed. It is worthwhile, therefore, to summarize these statistical results to this point in a generalized structure of conclusions before the accumulating detail confuses our arguments.

The ideal gas law, which resulted from our statistical arguments, is one such general conclusion. We defined P and T in terms of $E_t[=kT^2(\partial \ln \mathscr{Z}_t/\partial T)]$ and obtained

$$P = \frac{2}{3}\frac{E_t}{V} \tag{4.1}$$

$$T = \frac{2}{3}\frac{E_t}{Nk}. \tag{4.2}$$

The equation of state concept followed with $PV = NkT$ summarizing the arguments for the ideal gas. We also obtained from statistical considerations applied to the conservation of energy principle,

$$dE = N\left[\sum^{j} \epsilon_j d\mathscr{p}_j^* + \sum^{j} \mathscr{p}_j^* dE_j\right] + \left(\sum^{j} \mathscr{p}_j \epsilon_j\right) dN \tag{4.3}$$

with the generalization following for a reversible process

$$dE = TdS - PdV + \mu dN. \tag{4.4}$$

Maximization of entropy

$$S = k \ln \mathscr{Z} + \frac{E}{T} \tag{4.5}$$

in an isolated system with E, V, and N specified, or minimization of Helmholz free energy

$$A = -kT \ln \mathscr{Z}$$

in a system with T, V, and N specified are equilibrium requirements.

Since the terms A and S, together with E, are completely defined by the partition function for the system, their change during any postulated process depends only on the initial and final equilibrium states of the system as we have demonstrated in Chapter 3. We will refer to these thermodynamic terms and others that share this property as state functions. The variables that determine these particular state functions are conveniently taken to be T, V, and N since these specifications determine the partition function. This will be generally true even when structure and attendant complication are introduced.

We have related changes in the state functions *A* and *S* to process spontaneity (i.e., directionality). In introducing changes in the free energy function A as a measure of spontaneity we have, in effect, fulfilled the isolated system constraint by holding the energy of the system constant through thermal exchange with the surroundings. We have also held the volume constant. In an isolated system E, V, and N are constant. The constraints are real and necessary. General utility is extended to finite changes in T and V and N through the reversible process concept where the finite change is accomplished as a sequence of infinitesimal steps. For each such step the system is in a state of virtual equilibrium for which the variables T, V, and N may be regarded as constant valued. The sum of such steps is a reversible process that has replicated the finite process in terms of initial and final states. If the process is one that will occur spontaneously, $dA < 0$ for every infinitesimal step of the process.

It is possible to summarize the above conclusions in a mathematical framework since state functions have the mathematical properties of perfect differentials.[1] We can write for the change in entropy assoicated with a finite

[1] See Appendix B on mathematical considerations.

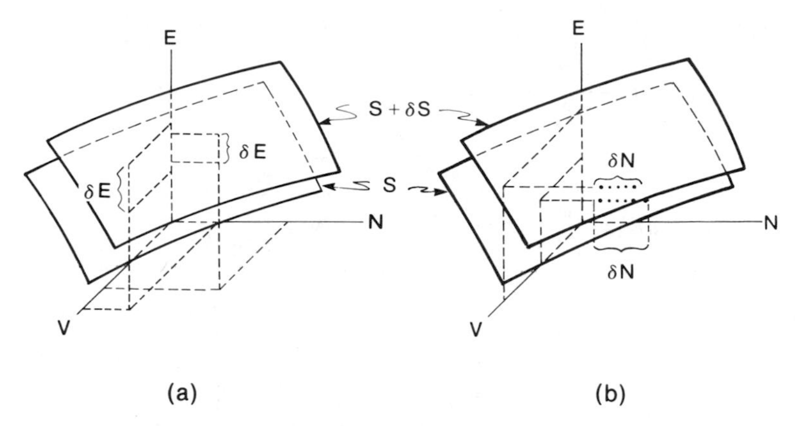

Figure 4.1. Sketches illustrating $(\delta S/\delta E)_{V,N}$ and $(\delta S/\delta N)_{E,V}$ for iso-entropic surfaces in E, V, N space.

state to state transition

$$\Delta S = S_2 - S_1 = \int_1^2 dS \tag{4.7}$$

where the dS are the infinitesimal entropy changes associated with any reversible route between states 1 and 2 regardless of how the process was actually accomplished. A further property of perfect differentials permits us to construct an expression for the total derivative of $S[=S(E, V, N)]$ as

$$dS = \left(\frac{\partial S}{\partial E}\right)_{V,N} dE + \left(\frac{\partial S}{\partial E}\right)_{E,N} dV + \left(\frac{\partial S}{\partial N}\right)_{E,V} dN. \tag{4.8}$$

The partial derivative terms, e.g., $(\partial S/\partial E)_{V,N}$ should be associated with the variation in entropy with variation in energy at particular values of V and N. From the standpoint of thermodynamics they are possibly best considered in the Figure 4.1a framework where surfaces of constant entropy, S and $S + \delta S$, are projected on the 3-space of E, V, and N. When V and N have particular constant values as indicated in the Figure 4.1a construct, a particular δE value is associated with the δS value. In general, the S and $S + \delta S$ would not be parallel surfaces throughout the variable space and the δE would vary with choice of V and N as indicated. The term $(\delta S/\delta E)_{V,N} \rightarrow (\partial S/\partial E)_{V,N}$ as the δ quantities shrink to infinitesimals. A similar construct for $(\partial S/\partial N)_{E,V}$ is shown in Figure 1b for two fixed values of E.

From Eq. (4.5) it is clear that we can regard E as a function S, V, and N or of T, V, and N. Either choice of variables leads to a particular partial derivative expression,

$$dE = \left(\frac{\partial E}{\partial S}\right)_{V,N} dS + \left(\frac{\partial E}{\partial V}\right)_{S,N} dV + \left(\frac{\partial E}{\partial N}\right)_{S,V} dN \tag{4.9}$$

or

$$dE = \left(\frac{\partial E}{\partial T}\right)_{V,N} dT + \left(\frac{\partial E}{\partial V}\right)_{T,N} dV + \left(\frac{\partial E}{\partial N}\right)_{T,V} dN \tag{4.10}$$

each of which can be productive. Comparison of Eq. (4.9) with Eq. (4.4) establishes the relationships

$$\left(\frac{\partial E}{\partial S}\right)_{V,N} = T, \quad \left(\frac{\partial E}{\partial V}\right)_{S,N} = -P, \quad \left(\frac{\partial E}{\partial N}\right)_{S,V} = \mu. \tag{4.11}$$

Consideration of Eq. (4.10) leads to several conclusions. First, we note that the conditions of constraint on the partial derivative expressions lead to quite different definitions. The term $(\partial E/\partial V)_{S,N}$ is identified as $-P$, while the term $(\partial E/\partial V)_{T,N}$ must have value zero for the ideal gas since $E = \frac{3}{2}N\mathscr{k}T$, independent of V. The terms $(\partial E/\partial N)_{S,V}$ and $(\partial E/\partial N)_{T,V}$ are equivalent and both identified as chemical potential. For the ideal gas

$$\left(\frac{\partial E}{\partial T}\right)_{V,N} = \frac{\partial}{\partial T}\left(\frac{3}{2}N\mathscr{k}T\right) = \frac{3}{2}N\mathscr{k}. \tag{4.12}$$

Both Eq. (4.12) and the zero value for $(\partial E/\partial V)_{T,N}$ would apply to any system for which the partition function is given by

$$\mathscr{Z} = \left(\frac{2\pi m\mathscr{k}T}{h^2}\right)^{3N/2}\frac{V^N}{N!}.$$

When structure, leading to particle–particle interactions is introduced, the new form for the partition function will be reflected in the partial derivative terms. If there is a significant potential energy associated with interactions between particles, $(\partial E/\partial V)_{T,N}$ will not be zero.

To continue arguments in the framework of state functions as perfect differentials, we note that with

$$A = E - TS \tag{4.13}$$

$$dA = dE - TdS - SdT \tag{4.14}$$

which, together with Eq. (4.4), provides

$$dA = -SdT - PdV + \mu dN. \tag{4.15}$$

Taken with

$$dA = \left(\frac{\partial A}{\partial T}\right)_{V,N} dT + \left(\frac{\partial A}{\partial V}\right)_{T,N} dV + \left(\frac{\partial A}{\partial N}\right)_{T,V} dN, \tag{4.16}$$

Equation (4.15) leads to the defining expressions

$$\left(\frac{\partial A}{\partial T}\right)_{V,N} = -S, \quad \left(\frac{\partial A}{\partial V}\right)_{T,N} = -P, \quad \left(\frac{\partial A}{\partial N}\right)_{T,V} = \mu \tag{4.17}$$

which are in accordance with the statistical relationships

$$S = \left[\frac{\partial}{\partial T}(kT \ln \mathscr{Z})\right]_{V,N}, \quad P = kT\left[\frac{\partial}{\partial V}(\ln \mathscr{Z})\right]_{T,N}, \quad \mu = -kT\left[\frac{\partial}{\partial N}(\ln \mathscr{Z})\right]_{T,V}. \tag{4.18}$$

We will find Eqs. (4.18) useful in our extended arguments. The equations are completely general, i.e., constrained only by our ability to construct an appropriate partition function for the system.

Returning to Eq. (4.10) we note that with $(\partial E/\partial V)_{T,N}$ equal to zero, for a closed system ideal gas,

$$dE = \left(\frac{\partial E}{\partial T}\right)_{V,N} dT. \tag{4.19}$$

We have other results for change in E associated with thermal effects. Any closed system process involving heat and work can produce changes in E and therefore in T. It will be most productive to return to the reversible closed system expression,

$$dE = dq^* + dw^*$$

and recognize that the constant volume constraint in Eq. (4.12) excludes the mechanical work process so that we can equate, without further argument, the terms dq^* and $(\partial E/\partial T)_{V,N}dT$,

$$dq^* = \left(\frac{\partial E}{\partial T}\right)_{V,N} dT. \tag{4.20}$$

We have identified thermal energy transferred reversibly to or from the system with the temperature change under the constraint of constant V and N. The reversibility constraint is, of course, intuitively acceptable since it is only for the equilibrium system that E and T as instantaneous values have thermodynamic significance. It would appear, however, that Eq. (4.12) has significance beyond the reversible, constant volume process.

For any reversible process involving only thermal energy exchange with $dE = dq^*$,

$$\int_1^2 (dE)_{V,N} = (\Delta E_{1,2})_{V,N} = \int_1^2 dq^* = Q_{V,N} \tag{4.21}$$

but any irreversible process between the same two values of E at constant V and N must be accomplished by the same quantity of heat exchanged with the surroundings. In the absence of work, i.e., at constant volume, the heat quantity associated with a given change in T in a closed system is independent of the process route. Then

$$(\Delta E_{12})_{V,N} = Q_{V,N} = \int_{T_1}^{T_2} \left(\frac{\partial E}{\partial T}\right)_{V,N} dT. \tag{4.22}$$

To the extent that $(\partial E/\partial T)_{V,N}$ is constant, and it is constant for the ideal gas and gases characterized by the ideal gas partition function,

$$Q_{V,N} = \left(\frac{\partial E}{\partial T}\right)_{V,N} \int_{T_1}^{T_2} dT = \left(\frac{\partial E}{\partial T}\right)_{V,N} \Delta T \tag{4.23}$$

so that the value of $(\partial E/\partial T)_{V,N}$ is determined by $Q_{V,N}/\Delta T$ whether or not the process is accomplished reversibly. The only difference between the two processes is whether or not the system is in continuous statistical equilibrium. The term $(\partial E/\partial T)_{V,N}$, which we will represent by C_V, is known as the constant volume heat capacity. Clearly, $Q_{V,N}/\Delta T$ is a typical process observable as defined in Chapter 2. The definition is completely general as are the statistical consequences

$$E = kT^2 \frac{\partial \ln \mathscr{Z}}{\partial T}$$

$$C_V = \left(\frac{\partial E}{\partial T}\right)_{V,N} = \frac{\partial}{\partial T}\left[kT^2\left(\frac{\partial \ln \mathscr{Z}}{\partial T}\right)_{V,N}\right]. \tag{4.24}$$

For the ideal gas $C_V = \frac{3}{2}Nk$.

The logic that leads to Eq. (4.10) can be extended to other than constant volume systems. A constant pressure process is of interest in that constant pressure can be a feature of the environment with which systems of interest will most often interact. For a system of any complexity there will be an equation of state $f(PVTN) = 0$. For the ideal gas, $PV = NkT$, but we anticipate[2] that the PV product is a reflection of the thermodynamic state for any substance in specified quantity. We then write $(dE - dw)_N = (dE + PdV)_N = q_N$ for any

[2] Our ideal gas construct would imply that all gases behave ideally in the limit of infinite volume. This suggests that the equation of state will always follow a form $PV = NkT[1 + f(V)]$.

infinitesimal closed system process in which the work is restricted to reversible mechanical work. The term $(dE + PdV)_N$ would result from $d(E + PV)_{P,N}$ so we can write[3]

$$d(E + PV)_{P,N} = dq_{P,N}. \tag{4.25}$$

The quantity (E + PV) is referred to as the enthalpy and assigned the symbol H. Since E and PV are state functions, enthalpy is a state function. We write generally

$$(dH)_{P,N} = dq_{P,N}$$

and all arguments applied to $(dE)_{V,N}$ apply to $(dH)_{P,N}$. In particular there is only one value for

$$Q_{P,N} = \int_1^2 (dq)_{P,N} = \int_1^2 (dH)_{P,N}.$$

For the ideal gas at constant T, both E and the PV product are constants so that the derivative term $(\partial H/\partial P)_{T,N}$ is zero valued for this case. Neither E nor PV will fulfil this condition in interacting particle systems and $(\partial H/\partial P)_{T,N}$ will not be zero.

In the microscopic context, if the process is reversible, the system is in a state of continuous equilibrium and has the equilibrium distribution of particles among energy states so that $(dq^*)_P = N\sum^i \epsilon_j d\not{p}_j^*$ but in this case, as opposed to $(dq^*)_V$, the ϵ_j are continuously changing throughout the process as V is continuously changing. For any reversible process element the ϵ_j are the instantaneous energy states available to the particle at that point in the process. The $d\not{p}_j^*$ are given by the Eq. (3.22) analog

$$d\not{p}_j = \frac{1}{N}\left[\frac{2}{\pi^{1/2}}\epsilon_j^{1/2} e^{-\epsilon_j/kT}(kT)^{3/2}\right] d\epsilon_j. \tag{4.26}$$

In the macroscopic context we can define a term C_P, analogous to C_V, calling C_P the constant pressure heat capacity,

$$\left(\frac{\partial H}{\partial T}\right)_{P,N} = C_P = \left(\frac{\partial(E + PV)}{\partial T}\right)_{P,N} = \left(\frac{\partial E}{\partial T}\right)_{P,N} + \left(\frac{\partial PV}{\partial T}\right)_{P,N} \tag{4.27}$$

for the ideal gas $(\partial(PV)/\partial T)_{P,N} = Nk$. Also for the closed system ideal gas with

$$\left(\frac{\partial E}{\partial T}\right)_N = C_V + \left(\frac{\partial E}{\partial V}\right)_T \left(\frac{\partial V}{\partial T}\right)_P$$

[3]In a more formal argument H = E + PV is obtained through a change in variable (Legendre transformation). Refer to Appendix B for the technique.

and $(\partial E/\partial V)_T = 0$

$$C_P = \frac{\partial}{\partial T}\left[\ell T^2\left(\frac{\partial \ln \mathscr{Z}}{\partial T}\right)_{V,N} + N\ell T\right] = \frac{5}{2}N\ell. \tag{4.28}$$

Another useful thermodynamic function follows from adding the PV term to Eq. (2.48)

$$A + PV = -\ell T\left[\ln \mathscr{Z} + V\left(\frac{\partial \ln \mathscr{Z}}{\partial V}\right)_{T,N}\right] = G \tag{4.29}$$

For the ideal gas where $PV = N\ell T$

$$A = -\ell T \ln\left(\frac{z}{N}\right)^N - N\ell T \tag{2.48}$$

and

$$G = -\ell T \ln\left(\frac{z}{N}\right)^N = N\mu. \tag{4.30}$$

Also from the definition of A $(= E - TS)$ it follows that with $H = E + PV$

$$G = H - TS \tag{4.31}$$

so that G is an energy term which is a function of P, T, and N since, from

$$\begin{aligned} dE &= TdS - PdV + \mu dN \\ dH &= TdS - PdV + \mu dN + d(PV) \\ &= TdS + VdP + \mu dN \end{aligned} \tag{4.32}$$

and

$$\begin{aligned} dG &= TdS + VdP + \mu dN - d(TS) \\ &= SdT + VdP + \mu dN. \end{aligned} \tag{4.33}$$

By analogy with the Helmholz free energy $A(V, T, N)$, $G(P, T, N)$ is known as the Gibbs free energy. We note that from Eq. (4.33) $(\partial G/\partial N)_{P,T} = \mu$ and from Eq. (4.17) $(\partial A/\partial N)_{V,T} = \mu$. These are equivalent definitions and the distinction is important for gaseous systems but for (effectively) incompressible condensed phases will prove to be trivial.

Although the ideal gas relationship has been used to obtain Eq. (4.30), the result is completely general as we will later verify. The generality derives from the consideration that the system partition function will always define A and

$(\partial A/\partial V)_{T,N}$ will always define P. Interacting particle systems will reflect the more complex PV product through a more complex partition function.

The arguments advanced in Chapter 2 through Eqs. (2.28–2.32), which lead to the introduction of the free energy function A and its minimization as a measure of equilibrium in T and V controlled circumstances, can be replicated for the free energy function G. Again, thermal energy exchange with the surroundings, in this case under conditions of constant T and P, can be regarded as a reversible process in the surroundings component of the supersystem, represented by $\Delta H_{syst}/T_{syst}$ leading to minimization of the function $H - TS$ as a measure of equilibrium and decrease of that function for all process steps as a criterion of spontaneity. For processes characterized by the variables T, P, and N, all arguments developed for the Helmholz free energy in systems characterized by T, V, and N will apply to the Gibbs free energy,

$$\Delta G = G_2 - G_1 < 0$$

for any spontaneous process occurring between states 1 and 2 and

$$dG \leqslant 0 \tag{4.34}$$

for any process element, the equality prevailing when G is at its minimum (equilibrium) value.

Since the minimization of free energy defines an equilibrium state we will often be concerned with how the point of equilibrium shifts with temperature or pressure since these are the most common process variables. Although we have observed [Eq. (4.17)] that $(\partial A/\partial T)_{V,N} = -S$, entropy is not always a convenient observable. However, from $A/T = -k \ln \mathscr{Z}$

$$\left(\frac{\partial A/T}{\partial T}\right)_{V,N} = -k\left(\frac{\partial \ln \mathscr{Z}}{\partial T}\right)_{V,N} = -\frac{E}{T^2}$$

and

$$\left(\frac{\partial G/T}{\partial T}\right)_{P,N} = -\frac{H}{T^2}, \qquad \left(\frac{\partial \Delta G/T}{\partial T}\right)_{P,N} = -\frac{\Delta H}{T^2}. \tag{4.35}$$[4]

Through its association with Q_P, ΔH is always a process observable. Also from $G = kT \ln \mathscr{Z} + PV$

$$\left(\frac{\partial G}{\partial T}\right)_{P,N} = -S, \qquad \left(\frac{\partial \Delta G}{\partial T}\right)_{P,N} = -\Delta S \tag{4.36}$$

which also follows from Eq. (4.33).

[4]The Eq. (4.35) relationship is known as the Gibbs–Helmholz equation. Its formal derivation from the statement $G = H - TS = H + T(\partial G/\partial T)_{P,N}$ and $G/T + (\partial G/\partial T)_{P,W} = H/T = -T(\partial G/T/\partial T)_{P,N}$ does not depend on the statistical argument.

For future reference we note that, since both free energy functions include the total energy E in their definitions, any work terms other than mechanical work would appear in the free energy differential expressions associated with energy exchange processes,

$$dA = dE - d(TS)$$

$$dG = dE - d(TS) + d(PV)$$

$$dE = TdS - PdV + \mu dN + dw_{\text{other}}. \tag{4.37}$$

The PdV term associated with mechanical work is retained in the derivative expression for the Helmholz free energy

$$dA = -SdT + \mu dN + (-PdV + dw_{\text{other}}) \tag{4.38}$$

but vanishes in the derivative expression for the Gibbs free energy

$$dG = -SdT + VdP + \mu dN + dw_{\text{other}}. \tag{4.39}$$

Changes in the Gibbs free energy would be the preferred framework for process considerations involving charged particles in electric fields, interfacial systems, and other such circumstances involving non-PV-related work terms.

There are features of the mechanical work process that could not be conveniently examined until the heat capacity concept was developed. We now address several such considerations. Equation (2.15) provides the general statistical condition for a reversible work element,

$$dw^* = N\sum^{1} \wp_j^* d\epsilon_j$$

where $\wp_j^*$ is the instantaneous distribution of particles among energy states and the $d\epsilon_j$ reflect the change in these states with change in volume. Any volume change process can be conceptualized in terms of a reversible work process where we only require the virtual identity of P_{ex} and P at all stages. Two work processes would appear to be of particular interest. If dE is zero at every infinitesimal stage of the process we can term the process isoenergetic or, since E and T are directly related, isothermal. If dS is zero at every infinitesimal stage of the process, we can term the process isoentropic or, since dq^* must then be zero, adiabatic.

In the closed system isothermal case ($dE = 0$) we recognize that

$$dw^* = N\sum^{j} \wp_j d\epsilon_j = -P\,dV = -dq^* = -N\sum^{j} \epsilon_j d\wp_j^* = -TdS. \tag{4.40}$$

The reversible work and heat elements are equal in magnitude and opposite in sign and the entropy change temperature product is equal in magnitude but opposite in sign to the reversible work element. In statistical terms the increase (decrease) in energy level spacing is exactly compensated for by the decrease (increase) in distribution numbers to produce a zero sum for the two terms, each of which is equal in magnitude to the statistical term $kTd\ln\Omega$. For a reversible isothermal process Eq. (4.40) would be phrased as

$$W^*_{th} = -{}_c\int_1^2 dw^*_{th} = -\int_1^2 PdV \tag{4.41}$$

with the requirement that P be the equilibrium system pressure at every point of the process or that $P = NkT/V$ apply at every such point. The integration can now be readily performed,

$$W^*_{th} = -NkT\int_{V_1}^{V_2}\frac{dV}{V} = -NkT\ln\frac{V_2}{V_1}. \tag{4.42}$$

If $V_1 > V_2$, then the reversible work associated with the process is positive and E for the system would have increased but for the fact that heat was transferred out of the system at just such a rate that E (and T) always remained constant. This heat value can only be obtained from W_{th}. Equation (4.34) could equally well be phrased in terms of P. For the ideal gas

$$W^*_{th} = -NkT\ln\frac{P_1}{P_2} = NkT\ln\frac{P_2}{P_1}. \tag{4.43}$$

In the closed system adiabatic case we recognize that

$$dW^*_{ad} = dE = N\sum^{j} p^*_j d\epsilon_j = -PdV \tag{4.44}$$

and

$$TdS = 0 = \sum^{j} \epsilon_j dp^*_j = kTd\ln\Omega = dq^*. \tag{4.45}$$

The distribution numbers are constant; if the energy levels are more widely spaced (compression), the same distribution number set leads to an increase in E (and T); if the energy levels are more closely spaced (expansion), the same distribution number set leads to a decrease in E (and T).

The work associated with the ideal gas reversible adiabatic process is, as before

$$W^*_{ad} = {}_c\int_1^2 dw^*_{ad} = -{}_c\int_1^2 \frac{NkT}{V}dV \tag{4.46}$$

but T is now constantly changing throughout the process,

$$W^*_{ad} = -N\ell \int_{c\,1}^{2} \frac{T}{V} dV. \tag{4.47}$$

The integral cannot be directly evaluated until T can be expressed in terms of V at every infinitesimal stage of the process. We can see, however, that $(N\ell T/V)dV$ is equal to dE at every stage of the process and that dE is given by $C_V dT$, where dT is the infinitesimal change in T accompanying the infinitesimal work process. It may be convenient to imagine that since dV is infinitesimal there is no problem associated with considering V as constant during the infinitesimal change in T. With this provision it is then possible to equate the terms $(N\ell T/V)dV$ and $C_V dT$ at every infinitesimal step of the process

$$C_V dT = -\frac{N\ell T}{V} dV, \qquad C_V d \ln T = -N\ell d \ln V. \tag{4.48}$$

These terms do integrate, with C_V = constant (which it certainly is for the ideal gas[5]), to provide

$$C_V \ln\frac{T_2}{T_1} = -N\ell \ln\frac{V_2}{V_1}, \qquad \left(\frac{T_2}{T_1}\right)^{C_V} = \left(\frac{V_1}{V_2}\right)^{N\ell} \tag{4.49}$$

so that the final temperature T_2, in terms of the adiabatic work process starting at T_1, V_1, and ending at T_2, V_2 is

$$T_2 = T_1\left(\frac{V_1}{V_2}\right)^{N\ell/C_V}. \tag{4.50}$$

The most direct way to obtain the adiabatic work is from the integral over dE for the process or the integral over $C_V dT$, since we now have the initial and final temperatures for the process,

$$W^*_{ad} = \int_{T_1}^{T_2} C_V dT = C_V(T_2 - T_1) = C_V T_1\left[\left(\frac{V_1}{V_2}\right)^{N\ell/C_V} - 1\right]. \tag{4.51}$$

If the ideal gas equation of state is used to represent T in Eq. (4.50), we obtain the result $(PV)^{\gamma_c}$ = constant for the process with γ_c representing the term C_P/C_V. Of course, in the isothermal work case PV = constant.

These conclusions regarding the work process do not add to our statistical understanding but relate to it as follows. In the reversible closed system isothermal work process $d\sum^j \not{p}^*_j \epsilon_j = 0$ at every infinitesimal step of the process. In the reversible closed system adiabatic process with $dS = 0 = d(\ell \ln \Omega)$ at every

[5] We anticipate that C_V will be some function of T for structured particles and that the integral will be more complex.

infinitesimal step of the process $d\sum^j p_j^* \ln p_j^* = 0$ so that we require that $d\sum^j p_j^* \epsilon_j = d(kT \ln z)$ at every step. The macroscopic statement

$$C_V d \ln T = Nk d \ln V$$

reflects this requirement.

The work and heat relationships involving the ideal gas are relatively simple to project graphically providing experience that may prove useful in more complex circumstances. The ideal gas relationship would provide a curved surface in PVT space which can be represented as in Figure 4.2a. All equilibrium PVT points for the ideal gas lie on the PVT surface, those P and V values which lie on the line intersection of a constant T plane with the surface are said to represent an ideal gas isotherm. This is also illustrated in Figure 4.2a. A reversible isothermal work path follows such an isotherm exactly and the integral over the work path is given by the area under the curve between initial V_1 and final V_2 volumes (Figure 4.2b). In mathematical expression we would say that the line integral for the work process has been reduced to a Reimann integral for which the area under the curve argument applies. If a work process between the same initial and final volumes had been carried out at a constant external pressure P_2, the final state would be the same but the work would be only that included in the double hatched area in Figure 4.2b. Only the initial and final points lie on the equation of state isotherm. Both work processes are expansions.

If the ideal gas system is returned to the original state by the route of reversible compression, we would retrace, in the opposite direction, the isotherm to produce exactly the same area with the work term sign reversed. A single step constant pressure compression to the original state must be carried out at pressure P_1, and involves the very much larger total area under the dashed line rectangle in Figure 4.2b. Any sequence of spontaneous constant pressure expansions will produce a work term smaller than the reversible expansions

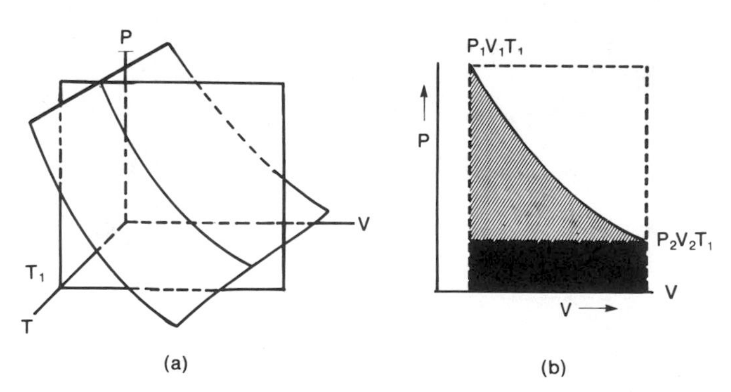

Figure 4.2. Sketch illustrating (a) a reversible isothermal path on an ideal gas PVT surface and (b) the PV work associated with reversible and irreversible expansion processes between two points on the isotherm.

between the same initial and final states. Any sequence of spontaneous constant pressure compressions will produce a larger work term. The equilibrium path is a "reversible" path.

Since the original and final PV values in all above cases lie on the same isotherm, the energy change for all processes is zero. The thermal energy ΔE transferred into the system during either expansion process was exactly equal to the mechanical energy expended by the system. The thermal energy transferred out of the system in either compression process was exactly equal to the mechanical energy absorbed by the system. Either constant pressure process occurred spontaneously and under nonequilibrium conditions. We do not attempt to describe the state of the system during spontaneous processes since none of our equilibrium arguments applies.

To complete arguments related to the kinds of processes discussed above we note that expansion of the ideal gas into a vacuum would be described as expansion against zero external pressure and the work term would be zero. One of the features of our ideal gas is that there is no potential energy that relates to the density of particles within the system. This density can be changed by expansion, e.g., into a vacuum, with no accompanying change in the energy of the system. In terms of our statistical arguments, energy levels must change since the volume has changed. If the system energy is constant it is because populations have changed in such a manner that $\sum^j \not{p}_j^* \epsilon_j$ is constant. We have already observed in connection with Eq. (4.10) that we can express this microscopic feature by the macroscopic observation $(\partial E/\partial V)_{T,N} = 0$.

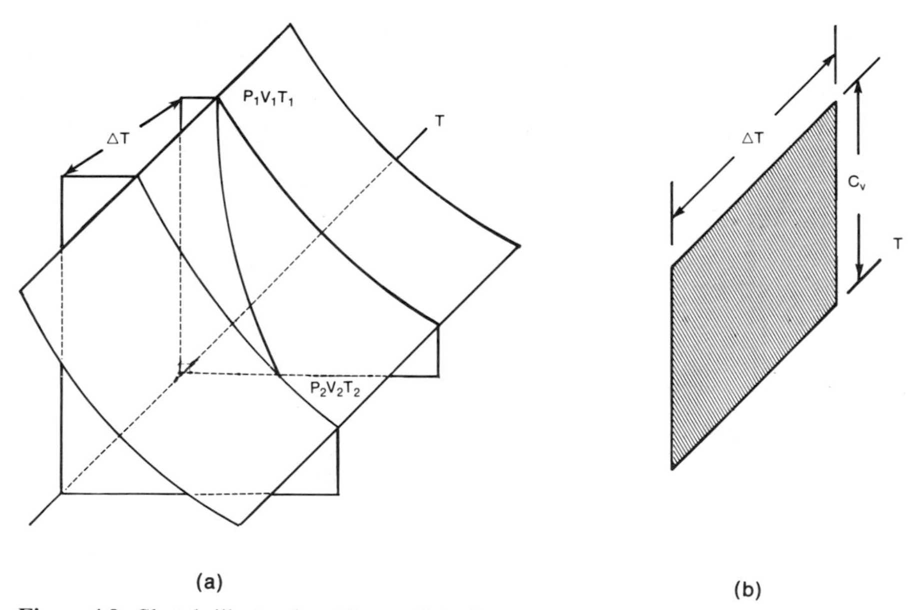

Figure 4.3. Sketch illustrating (a) an adiabatic expansion process and (b) the associated PV work.

If we consider the adiabatic work process in the same format that we have utilized above, we would recognize that during reversible adiabatic work processes the PVT description of the ideal gas must also lie on the PVT surface at every point in the process, the route being that which steps infinitesimally between isotherms with the value of T always increasing (expansion) or always decreasing (compression). The path form is shown in Figure 4.3a. For reference, isothermal paths which include the $P_1V_1T_1$ and the $P_2V_2T_2$ points are shown.

The line integral for adiabatic work does not reduce to a Reimann integral and cannot be represented by a simple area on the PVT plot. It has the value, from Eq. (4.52) of $C_V\Delta T$, which is represented in Figure 4.3b; C_V has no geometrical relationship to the PVT surface. There is one and only one reversible adiabatic path between any two specified volume values given an initial temperature. The final temperature is determined by these specifications. It is easy to see that any excursion away from the most direct line between the two points still lying within the PVT surface must represent an excursion, however short, along an intermediate isotherm, i.e., an isothermal deviation from the adiabatic process. Any excursion which does not lie in the PVT plane is an irreversible deviation.

The true utility of the idealized work processes is that ΔE for any process will follow from Q and W values for the reversible route and a reversible route can be constructed for any gas phase process between $(T_1V_1)_N$ and $(T_2V_2)_N$ even if such values are not directly accessible by one step isothermal or adiabatic processes. From Eq. (4.42), when an initial volume together with initial and final values of T are specified, the final volume is fixed for the single reversible adiabatic path. Any final volume at T_2 can, however, be reached by a reversible isothermal process. Two such routes to volumes, in one case larger and in the other case smaller than the adiabatically reversible accessible volume, are shown in Figure 4.4. The ΔE value for the reversible path (1) + (2) is $\frac{3}{2}Nk(T_2 - T_1)$ obtained from

$$\Delta E_1 = W^*_{ad} = C_V(T_2 - T_1) = \tfrac{3}{2}Nk(T_2 - T_1)$$
$$\Delta E_2 = 0.$$

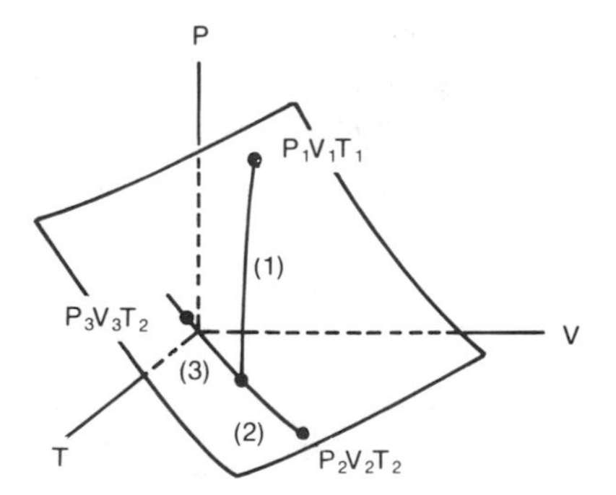

Figure 4.4. Sketch illustrating a reversible process route between arbitrary points on an ideal gas PV surface.

Also ΔE for the reversible path (1) + (3) is $C_V(T_2 - T_1)$, i.e., ΔE for any path from T_1 to T_2 is $C_V \Delta T$ regardless of the final volume and this must be true whether or not the actual route follows the reversible path. The result is general but C_V will not be generally $\frac{3}{2}N\mathscr{k}$.

We have noted that, since dq is zero at every stage of a reversible adiabatic process, with $dS = dq^*/T$, dS is zero at every stage of the process. For the reversible isothermal process with $dq^* = -dw^* = -N\mathscr{k}d\ln V$ for every stage of the (ideal gas) process,

$$dS = \frac{dq^*}{T} = N\mathscr{k}d\ln V \tag{4.52}$$

and ΔS for the process is $N\mathscr{k}\ln(V_2/V_1)$. This is entirely consistent with our Chapter 3 statistical calculation starting with $S = \mathscr{k}T\ln\mathscr{Z} + (E/T)$. For any process ΔS is $S_2 - S_1$, independent of the process route. Therefore

$$\Delta S = \mathscr{k}\ln\mathscr{Z}_2 - \mathscr{k}\ln\mathscr{Z}_1 + \left(\frac{E_2}{T} - \frac{E_1}{T}\right) \tag{4.53}$$

with the final bracketed term being zero for any isothermal ideal gas process. Then

$$\Delta S = \mathscr{k}\ln\frac{z_2^N}{N!} - \mathscr{k}\ln\frac{z_1^N}{N!} = N\mathscr{k}\ln\frac{V_2}{V_1} \tag{4.54}$$

since the common term $(2\pi m\mathscr{k}T/h^2)^{3/2}$ in both numerator and denominator cancels at constant temperature. The Eq. (4.54) result is a formal confirmation of the arithmetic exercise that leads to Eq. (3.17).

Also we note that for any change in T in a constant volume closed system the final term is again zero for the ideal gas with each E being $(\frac{3}{2})N\mathscr{k}T$ and

$$\Delta S = N\mathscr{k}\ln\left(\frac{T_2}{T_1}\right)^{3/2} = \frac{3}{2}N\mathscr{k}\ln\frac{T_2}{T_1}$$

since the common term $(2\pi m\mathscr{k}/h^2)^{3/2}V$ in both numerator and denominator cancels at constant V. The same result, in a more general form, is obtained from the consideration that, regardless of the route actually followed, dS is dq^*/T for the reversible route

$$\Delta S = \int_1^2 dS = \int_1^2 \frac{dq^*}{T} = \int_{T_1}^{T_2} \frac{C_V}{T}dT \tag{4.55}$$

and C_V for the (structureless) ideal gas is a constant ($= (\frac{3}{2})N\mathscr{k}$). We also recognize that any energy exchange process involving changes in both T and V could be

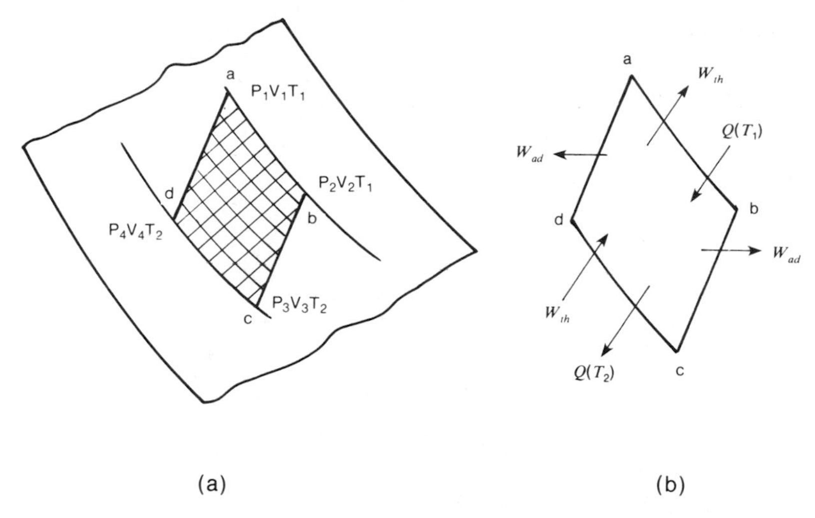

Figure 4.5. Sketch illustrating (a) a reversible cyclic work process and (b) the associated work and heat quantities for each step.

visualized as reversibly accomplished in a two-step sequence, first changing temperature reversibly at constant volume then isothermally changing volume. For this process ΔS must be the sum of the entropy changes for each process.

$$\Delta S = C_V \ln \frac{T_2}{T_1} + Nk \ln \frac{V_2}{V_1} \tag{4.56}$$

There are many features of energy exchange processes that we could explore in terms of Q and W. Cyclic processes are of particular interest in that we would immediately understand that the change in any state function must be zero when the process begins and ends at the same point, e.g., around any cycle

$$\oint dE = 0 = \Delta E_{cycle}.$$

For reasons relating to the classical argument introducing entropy, a reversible ideal gas work process known as the Carnot cycle is important. Figure 4.5a represents such a process in terms of isothermal (*ab*, *cd*) and adiabatic (*bc*, *da*) work paths in the PVT surface. Assume that the processes begin and end at point a with the arrows indicating the (arbitrary) clockwise path direction with $T_1 > T_2$.[6] we summarize the system energy together with the thermal and

[6]The clockwise direction does not control the therodynamic result but has historical relevence to heat engines where thermal energy is absorbed at a high temperature and partially rejected at a lower temperature.

mechanical energy exchanges as

$$\Delta E_{ab} = 0 \qquad W_{ab} = Nk T_1 \ln \frac{V_2}{V_1} \qquad Q_{ab} = -W_{ab}$$

$$\Delta E_{bc} = C_V(T_2 - T_1) \qquad W_{bc} = C_V(T_2 - T_1) \qquad Q_{bc} = 0$$

$$\Delta E_{cd} = 0 \qquad W_{cd} = -Nk T_2 \ln \frac{V_4}{V_3} \qquad Q_{cd} = -W_{cd}$$

$$\Delta E_{da} = C_V(T_1 - T_2) \qquad W_{da} = C_V(T_1 - T_2) \qquad Q_{da} = 0.$$

Then $\Delta E_{\text{cycle}} = \sum \Delta E_{\text{process step}} = 0$ as required for any state function, but W_{cycle} (cross-hatched area) $\neq 0$ and therefore $Q_{\text{cycle}} \neq 0$,

$$Q_{\text{cycle}} = 0 + Nk T_1 \ln \frac{V_2}{V_1} + 0 + Nk T_2 \ln \frac{V_4}{V_3}.$$

The adiabatic steps serve to define and relate the volume ratios when T_1 and T_2 are specified since

$$\left(\frac{T_2}{T_1}\right)^{C_V/Nk} = \frac{V_2}{V_3}, \qquad \left(\frac{T_1}{T_2}\right)^{C_V/Nk} = \frac{V_4}{V_3}$$

so that

$$\frac{V_2}{V_3} = \frac{V_1}{V_4}, \qquad \frac{V_2}{V_1} = \frac{V_3}{V_4}.$$

Then $Q_{\text{cycle}} = (T_1 - T_2)Nk \ln(V_3/V_4)$. However if each of the reversible heat components is divided by the temperature at which that heat was exchanged,

$$0 + \frac{Q_{ab}}{T_1} + 0 + \frac{Q_{cd}}{T_2} = Nk\left(\ln \frac{V_2}{V_1} + \ln \frac{V_4}{V_3}\right) = 0$$

i.e., there is a function, defined by Q^*/T, that does sum to zero for the cyclic process. This somewhat unsatisfactory introduction of a state function related to reversible thermal energy exchange is an often practiced route to the introduction of the entropy concept. Other consequences of this cyclic process provide the basis for other classical statements of the Second Law of thermodynamics.

If we summarize the cyclic process as in Figure 4.5b, a quantity of heat is absorbed by the system at T_1, net work is performed , and a quantity of heat is rejected at T_2. No cyclic process that produces net work can avoid this constraint, which leads to Second Law statements such as "it is impossible to construct a cyclic process that will produce no other effect than the extraction of heat from a reservoir and the performance of an equivalent amount of work"

(Kelvin) or the converse, "it is impossible by a cyclic process to transfer heat from a cold to a hot reservoir with no other effect" (Clausius). These statements and their correlary phrased as denial of the possibility of perpetual motion of the second kind (complete conversion of heat into work) are not terribly useful to us. The relationship

$$\frac{dq^*}{T} = dS$$

which can be phrased[7] as "$1/T$ is an integrating factor for the imperfect differential dq^*," does have utility although that utility is obscured by the route to its construction via the Carnot cycle. In particular for any change in temperature taking place in a constant volume closed system

$$(\Delta S)_V = \int_{T_1}^{T_2} \frac{(dq^*)_V}{T} = \int_{T_1}^{T_2} \frac{C_V dT}{T}$$

or in a constant pressure closed system

$$(\Delta S)_P = \int_{T_1}^{T_2} \frac{(dq^*)_P}{T} = \int_{T_1}^{T_2} \frac{C_P}{T} dT.$$

Also in a reversible isothermal closed system process $dq^* = -dw^*$ and

$$(\Delta S)_{th} = -\int \frac{(dw^*)}{T} = \int_{V_1}^{V_2} Nk d\ln V \qquad \text{(the ideal gas)}.$$

These are, of course identical to results which we obtained by calculating ΔS as the difference in entropy between two different (isolated) equilibrium states.

In this chapter we have touched on most of the classical thermodynamic argument based on heat and work cycles. Such emphasis is required when the classical route is followed making it sometimes seem that such processes are the subject. The route which we follow and the applications of chemical interest are independent of such process. Understanding of them is largely an historical exercise.

PROBLEMS

4.1. In terms of the translational partition function show that

$$(\partial A/\partial T)_{V,N} = (\partial G/\partial T)_{P,N} = -S.$$

[7] This route to the concept of entropy is known as the Caratheodory Principle.

4.2. Beginning with recognition that the energy of a closed system is a function of V and T, show that, independent of any particular equation of state,

$$C_P - C_V = \left(\frac{\partial V}{\partial T}\right)_P \left[\left(\frac{\partial E}{\partial V}\right)_T + P\right]$$

4.3. Compare the work of reversible expansion of an ideal gas at 350 K from 1 liter to 10 liters to the work of adiabatic expansion from 1 to 10 liters, again starting the process at 350 K.

4.4. Compare the entropy changes associated with the problem 4.3 processes using macroscopic arguments relating to the process observables Q^* and W^*.

4.5. One method of measuring the heat capacity of a gas would involve measuring the temperature drop associated with an adiabatic expansion. In such an experiment with an unknown gas the gas was expanded 2-fold with a resulting temperature change 298.6 K to 237.4 K. Calculate the heat capacity. What can you say about the structure and behavior of this gas?

4.6. Calculate the change in enthalpy when 1 mole of argon is expanded from a pressure of 2 atm to a pressure of 1 atm.

4.7. Using the argument that any thermodynamic state function can be constructed from the difference between statistical expressions for the function in the two thermodynamic states, calculate the change in enthalpy when 1 mole of an ideal gas is heated at constant volume from 300 K to 350 K.

4.8. If the statistical expressions for C_P and C_V are constructed for an ideal gas, show that they reduce to $C_P - C_V = R$.

4.9. From the defining equation for Helmholz free energy $A = -kT \ln \mathscr{Z}$, for entropy $S = k \ln \mathscr{Z} + E/T$, for the partition function and for Gibbs free energy $G = A + PV$, show that

$$\Delta G = -NkT \ln \frac{V_2}{V_1} = -T\Delta S$$

and

$$\Delta H = \Delta(E + PV) = 0$$

for the isothermal expansion of an ideal gas from $P_1V_1T_1$ to $P_2V_2T_2$.

4.10. How would these expressions differ if the expansion were adiabatic from $P_1V_1T_1$ to $P_2V_2T_2$?

CHAPTER CONTENT SOURCES AND FURTHER READING

1. G. N. Lewis and M. Randall, *Thermodynamics*, 2nd ed., revised by K. S. Pitzer and L. Brewer. McGraw-Hill, New York, 1961.
2. N. A. Glokcen, *Thermodynamics*. Techscience, Hawthorne, CA, 1975.
3. I. M. Klotz, *Chemical Thermodynamics*. W. A. Benjamin, New York, 1964.

SYMBOLS: CHAPTER 4

E_t = translational energy
$E_{V,N}$ = value of a thermodynamic function in a constant volume closed system
W_{th} = work along an isothermal path
W_{ad} = work along an adiabatic path
$E_1, E_2 \cdots$ = value of a thermodynamic function associated with a thermodynamic state
$\Delta E_{12}, \ldots, \Delta E_{ab}$ = value of the change in a thermodynamic function associated with a state to state transition or a process along a path $(ab\cdots)$
C_V = constant volume heat capacity
H = enthalpy
C_P = constant pressure heat capacity
G = Gibbs free energy
γ_c = ratio of constant pressure to constant volume heat capacity

Carryover Symbols

μ_j^*, ϵ_j^*, Q^*, dq^*, W^*, dw^*, k, P, V, T, N, S, $\mathscr{Z}$, z, A

5

STRUCTURED PARTICLE SYSTEMS: INTERNAL MOTIONS

In the ideal gas framework our collection of unstructured particles has neither potential energy associated with high paticle density nor modes of motion other than translation. Even the simplest structured particle systems do have energies associated with their electronic structure so they could exist in excited states and there will be a potential energy in the particle collection at sufficiently high density. We ignored these considerations in Chapter 3 in treating argon and helium as ideal gas analogs in dilute systems with all particles in the ground electronic state. Many dilute atomic particle systems can be similarly treated but our expressions should provide for electronic excitation to include those few cases for which such states are accessible. The extension to diatomic and polyatomic structures must also provide for electronic excitation and, in addition, for rotation and vibration as normal modes of motion.

Since we cannot know the ground state electronic energy of atoms in any useful way, we will take this to be an energy baseline. This is not a problem in that we have developed our arguments to this point in terms of energy states that are thermally accessible. Only the ground state can be occupied at 0 K and we do not need to know its value. The precise atomic electronic energy levels above the ground state cannot be expressed in simple, compact, analytical form but they are available from spectroscopic measurements that, in the simplest case, can be expressed as

$$h\nu_1 = \epsilon_1 - \epsilon_0$$
$$h\nu_2 = \epsilon_2 - \epsilon_0$$
$$\vdots$$
$$h\nu_i = \epsilon_i - \epsilon_0$$

where ν_1, ν_2, etc., are photon energies associated with ground to excited state transitions. Each electronic energy level $\epsilon_{i,e}$ will be ω_i fold degenerate with ω determined by quantum considerations. We will reference the ground molecular state to the ground states of the constituent atoms through the dissociation energy to obtain a universal energy zero. This is illustrated in Figure 5.1 for an $\mathscr{AB}$ molecule with a single dissociation energy term D_0. If there are L atoms involving K bonds in the molecule, the ground state electronic energy will reflect all terms L and K,

$$\eta_0 = \sum^{\mathrm{L}} \epsilon_{0,\mathrm{L}} - \sum^{\mathrm{K}} D_{0,\mathrm{K}}. \tag{5.1}$$

The enumeration of energy levels associated with molecular motion is readily accomplished by the quantum expressions set down in the Introduction and developed in Appendix A. For ready reference they are

$$\epsilon_{n,t} = \frac{h^2(n_x^2 + n_y^2 + n_z^2)}{8m\mathrm{V}^{2/3}} \qquad n_x, n_y, n_z = 1, 2, 3, \ldots, \infty \tag{5.2}$$

$$\epsilon_{j,r} = \frac{j(j+1)h^2}{8\pi^2 I} \qquad j = 0, 1, 2, \ldots \infty \tag{5.3}$$

$$\epsilon_{v,v}(n + \tfrac{1}{2})h\nu_e \qquad v = 0, 1, 2, \ldots, \infty \tag{5.4}$$

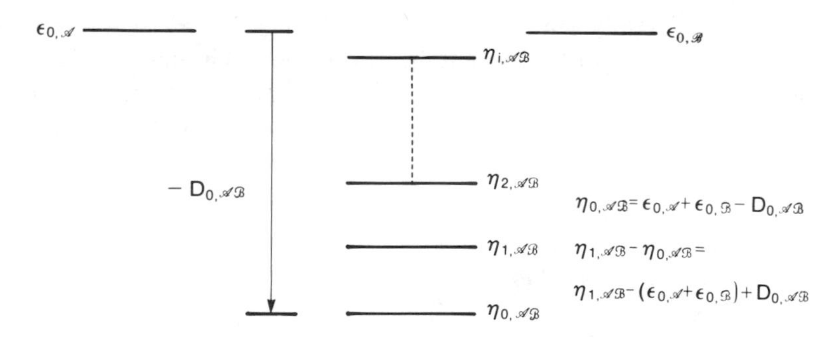

Figure 5.1. Illustration of ground level and excited molecular energy states with reference to the ground level energy states of the constituent atoms.

where mass (m), moment of inertia (I), and vibrational frequency (ν_e, each independent vibration) are unique for a molecule species. The degeneracy for translational levels is not explicit in terms of quantum numbers but we obtained a functional form through the rather laborious Chapter 3 exercise. Electronic energy states may be degenerate following quantum argument with the explicit degeneracy (ω_i) identified from spectroscopic data. The rotational degeneracy is explicit ($= 2j + 1$ from quantum argument) in terms of the energy level. The enumeration of rotational energy levels for nonlinear molecules tends to require rather involved expressions. We will develop those forms as extensions to Eq. (5.2). There is no vibrational degeneracy in terms of energy levels although ν_e values could be identical for two or more independent vibrations.

The statistical arguments developed for the distribution of particles among translational energy states for unstructured particles would apply to any set of energy states; however, we must recognize that a single molecule will simultaneously be in any available electronic state, any available translational state, and in any available rotational–vibrational states since we take each such set of states to be independent. It is required that we couple each distribution of particles among any single set of states with all distributions among other sets of energetic states,

$$\Omega = \Omega_t \Omega_e \Omega_r \Omega_v. \tag{5.5}$$

Since, however, the individual state occupancies are independent, each value can be maximized separately and the product of those maximum valued terms will produce Ω. The statistical argument is indifferent to the origin of any particular energy level (characterized as ϵ_i with general degeneracy g_i) so, for each (κ) mode of motion,

$$N_{i\kappa} = \frac{N g_{i,\kappa} e^{-\epsilon_{i\kappa}/kT}}{\sum_i g_i e^{-\epsilon_i/kT}} = \frac{N g_{i,\kappa} e^{-\epsilon_{i\kappa}/kT}}{z_\kappa} \tag{5.6}$$

where the general equivalence of β and $1/kT$ has been recognized and incorporated into the distribution number expression. The independent sum over each (κ) set of states to provide z_κ is an important feature since the probability $p^*_{i(t,r,v,e)}$, that a given particle is in total energy state $\epsilon_{i(t,r,v,e)}$ is the product of the probability set,

$$p^*_{i(t,r,v,e)} = p^*_{i,t} p^*_{i,r} p^*_{i,v} p^*_{i,e}$$

$$p^*_{i(t,r,v,e)} = \frac{g_{i,t} e^{-\epsilon_{i,t}/kT}}{z_t} \frac{(2j+1) e^{-\epsilon_{i,r}/kT}}{z_r} \frac{e^{-\epsilon_{i,v}/kT}}{z_v} \frac{\omega_{i,e} e^{-\epsilon_{i,e}/kT}}{z_e} \tag{5.7}$$

and the N particle partition function for indistinguishable (nonlocalized) particles is

$$\mathscr{Z}_{\mathrm{I}} = \frac{1}{\mathrm{N}!}(z_t z_r z_v z_e)^{\mathrm{N}}. \tag{5.8}$$

The distribution functions ($p^*_{i,\mathrm{K}}$) and the partition functions ($z_{i,\mathrm{K}}$) for each energy set (K) will be of quite different form since the energy levels in each set are very different with respect to the measure kT. For reference we construct Table 5.1. In the absence of very large degeneracies, Table 5.1A indicates that ϵ_i values greater than about 10 kT will make effectively no contribution to the partition function sum and that the probability of occupancy of their associated energy states is vanishingly small. In the absence of degeneracy and where we take $\epsilon_i = kT$ as a rough median measure of the term importance with (Table 5.18) kT ranging from 10^{-13} to 10^{-14} erg for the range of common thermodynamic interest, energy level values greater than (0) 10^{-13} erg are not important and those (0) 10^{-15} erg or smaller will make an effectively constant term contribution as $e^{-\epsilon_i/kT} \to 1$.

As we have discovered, translational degeneracies can be enormous and they have a correspondingly large influence on the individual term contributions. The individual states are so closely spaced ($n \times 10^{-32}$ ergs) as to represent a continuum and this is the only appropriate way to envision that mode of motion. The partition function in the absence of interactions is

$$z_t = \left(\frac{2\pi m k\mathrm{T}}{h^2}\right)^{3/2} \mathrm{V} \tag{3.6}$$

and the distribution of particles is of the Figure 3.2 form.

The spacing between rotational levels for linear molecules is obtained from Eq. (5.3) as

$$\Delta \mathrm{E}_{j,r} = [(j+1)(j+2) - j(j+1)]\frac{h^2}{8\pi^2_v I}$$

$$\Delta \epsilon_{j,r} = 2(j+1)\frac{h^2}{8\pi^2 I}.$$

TABLE 5.1. A Scale for Relative Values of Energetic Terms Appearing in the Partition Function Sum (A) and Their Relationship to the Temperature Scale (B)

			(A)		
ϵ_i/kT	10	1	10^{-1}	10^{-2}	10^{-3}
$e^{-\epsilon_i/kT}$	4×10^{-5}	0.368	0.904	0.990	0.999

		(B)		
T(K)	10^3	10^2	10	1
kT(erg)	10^{-13}	10^{-14}	10^{-15}	10^{-16}

Moments of inertia[1] are (O) 10^{-39} g cm^2 so the $\Delta\epsilon_{j,r}$ are (O) $2(j+1)\times 10^{-16}$, i.e., small compared to 10^{-13}, so that j can run to very large values before the spacing becomes significant. In almost all cases the rotational energy levels can be regarded as an energy continuum. When j is large the $2(j+1)$ degeneracy will have a significant impact on the distribution numbers producing as p_j vs ϵ_j curve of the same form as that for the translational energy states.

The spacing between vibrational levels from Eq. (5.4) is

$$\Delta\epsilon_{iv} = [(v+1+\tfrac{1}{2})-(v+\tfrac{1}{2})]h\nu_e = h\nu_e.$$

With ν_e(O) 10^{13} sec^{-1}, the energy level spacing is (O) 10^{-13} erg meaning that, under normal circumstances, few vibrational levels would be occupied except at the upper end of our range of interesting temperatures.

The energy spacing of excited electronic states for atoms is (O) 10^{-12}. For molecules there are cases in which it is somewhat smaller but exponent -12 ± 1 is a reasonable generalization. Normal degeneracies are $1,2,3,\ldots$ and will not influence the conclusion that, at most, two or three excited states are thermally accessible.

With these considerations in mind we can formulate the partition function components for the various sets of energetic states as follows. For atoms the partition function will be the product of translational and electronic partition functions. The translational component of the partition function for dilute systems is given by Eq. (3.6). We construct the electronic partition function for atoms as the sum over some few levels,

$$z_e = \omega_0 e^{-\epsilon_0/kT} + \omega_1 e^{-\epsilon_1/kT} + \omega_2 e^{-\epsilon_2/kT} + \cdots .$$

We then restructure this expression as

$$z_e = \omega_0 e^{-\epsilon_0/kT}\left[1 + \frac{\omega_1}{\omega_0} e^{-(\epsilon_1-\epsilon_0)/kT} + \frac{\omega_2}{\omega_0} e^{-(\epsilon_2-\epsilon_0)/kT} + \cdots\right] \tag{5.9}$$

or in logarithmic form as

$$\ln z_e = \ln\omega_0 + \ln\left[1 + \frac{\omega_1}{\omega_0} e^{-\Delta\epsilon_1/kT} + \cdots\right] - \frac{\epsilon_0}{kT} \tag{5.10a}$$

with $\Delta\epsilon_1 = \epsilon_1 - \epsilon_0$, etc.

For molecules, referring to our Figure 5.1 notation, the above arithmetic with $\Delta\eta = \eta_1 - \eta_0$, etc., and η_0 given by Eq. (5.1) produces

$$z_e = \omega_0 e^{-\eta_0/kT}\left[1 + \frac{\omega_1}{\omega_0} e^{-(\eta_1-\eta_0)kT} + \frac{\omega_2}{\omega_0} e^{-(\eta_2-\eta_0)kT} + \cdots\right] \tag{5.11}$$

[1] See Appendix A on Energy.

$$\ln z_e = \ln \omega_0 + \ln\left[1 + \frac{\omega_1}{\omega_0} e^{-\Delta\eta_i/kT} + \cdots\right] + \sum^{K} \frac{D_0}{kT} - \sum^{L} \frac{\epsilon_{0,L}}{kT}. \tag{5.12a}$$

In most cases Eq. (5.10) and (5.12) reduce to

$$\ln z_e = \ln \omega_0 - \epsilon_0/kT \tag{5.10b}$$

and

$$\ln z_e = \ln \omega_0 + \sum^{K} \frac{D_{0,K}}{kT} - \sum^{L} \frac{\epsilon_{0,L}}{kT}. \tag{5.12b}$$

These terms will always be present and can be conveniently incorporated, together with the translational partition function and the $(N!)^{-1}$ nonlocalization feature, into a single term. On the basis of Eq. (5.10b) the entire partition function for dilute atomic gaseous systems simplifies to

$$z_{t,e} = \left(\frac{2\pi m kT}{h^2}\right)^{3/2} V\omega e^{-\epsilon_0/kT} \tag{5.13a}$$

$$\ln \mathscr{Z} = \ln \frac{z_{t,e}^{N}}{N!} = \ln\left(\frac{2\pi m kT}{h^2}\right)^{3N/2} \frac{(V_e\omega_0)^N}{N!} - E_0/kT. \tag{5.13b}$$

The thermodynamic functions for atomic systems reduce to those for non-structured particle systems when ω_0 is one except that we have formally recognized the existence of an $E_0(= N\epsilon_0)$ associated with the electronic ground state. Since

$$E = NkT^2\left(\frac{\partial \ln \mathscr{Z}}{\partial T}\right)_{V,N} = \tfrac{3}{2}NkT + E_0$$

we see that it is $E - E_0$ that we can formally calculate. The E_0 terms cancel in the expression for entropy,

$$S = k \ln \mathscr{Z} + kT\left(\frac{\partial \ln \mathscr{Z}}{\partial T}\right)_{V,N},$$

but E_0 is present in other thermodynamic functions. For enthalpy

$$H = NkT^2\left(\frac{\partial \ln \mathscr{Z}}{\partial T}\right)_{V,N} + PV = \tfrac{3}{2}NkT + E_0 + PV.$$

It is $H - [E_0 + (PV)_0]$, i.e. $H - H_0$, that we can formally calculate. The same argument shows that we can calculate $A - E_0$ and $G - H_0$, not A or G.

Equivalent arguments will apply to the ground state electronic contribution for molecule systems except that

$$\ln \mathscr{Z} = \ln \frac{(z_{t,e})^{N}}{N!} + \ln(z_r z_v)^{N}$$

and the translational/electronic term produces terms $N(\sum^{L} \epsilon_{0,L} + \sum^{K} D_{0,K})$. Also reference to Eq. (5.4), which represents the vibrational energy levels, reveals that there is a nonzero $\epsilon_{0,v}$ value of $\frac{1}{2}h\nu_e$ associated with vibrational quantum number $v = 0$. We represent this contribution as $\sum^{K}(h\nu_{e,K}/2)$ deferring for the moment a discussion of the number (K) of independent vibrations. The ground state energy for molecule systems is

$$E_0 + \sum^{K} D_{0,K} = N\left(\sum^{L} \epsilon_{0,L} + \sum^{L} \frac{h\nu_{e,K}}{2} + \sum^{K} D_{0,K}\right) \tag{5.14}$$

and this term will appear in representations of E, H, A and G just as it does for monatomic particle systems. Although, for many applications, $\sum^{K} D_{0,K}$ would be incorporated into the E_0 term it is often preferable to keep it separate, e.g., for application to chemical reaction systems (Chapter 9).

The rotational component for linear molecules is given by

$$z_r = \sum^{j} 2(j+1)e^{-\epsilon_j/kT}$$

and can be regarded as classical, i.e., a continuous function of ϵ_j, for $T > 10\,K$ excluding H_2, which has a very small moment of inertia. Then

$$z_r = \int_0^{\infty} (2j+1)e^{-j(j+1)h^2/8\pi^2 IkT}\, dj. \tag{5.15a}$$

For notational convenience we define $\Theta_r = h^2/8\pi^2 Ik$ and write

$$z_r = \int_0^{\infty} (2j+1)e^{-j(j+1)\Theta_r/T}\, dj \tag{5.15b}$$

which can be expressed as

$$z_r = \int_0^{\infty} e^{-j(j+1)\Theta_r/T} d[j(j+1)]. \tag{5.15c}$$

For $x = j(j+1)$, $d[j(j+1)] = (2j+1)$, and $a = \Theta_r/T$

$$z_r = \int_0^{\infty} e^{-ax} dx = -\frac{1}{a}[e^{-ax}]_0^{\infty} = \frac{1}{a} = \frac{T}{\Theta_r}. \tag{5.16a}$$

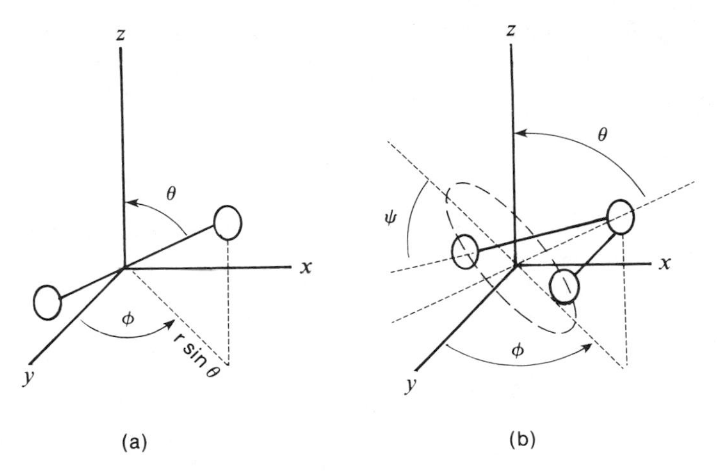

Figure 5.2. Illustration of the rotational orientation specifications for (a) linear and (b) nonlinear molecules.

The term Θ_r has dimension T and is referred to as the characteristic temperature for rotation.

Equation (5.3), which provided the quantized energy levels, is appropriate for linear molecules that are characterized by a single moment of inertia value. This derives from the absence of an energetic contribution from rotation about the molecule axis leaving two identical modes of motion. Nonlinear molecules, even if symmetry considerations dictate that all moments are identical, still have three, not two, moments of inertia and the partition function expression will be of somewhat different form. Since nonlinear molecules will be polyatomic by definition, the moments of inertia will be large and classical considerations will unambiguously apply over the range of normal thermodynamic concern. The framework for the construction of the appropriate integral[2] for rotation follows from Figure 5.2.

For linear molecules (Figure 5.2a) all orientations in space are described by rotation of the vector defined by axial orientation θ (as θ assumes all values $0 < \theta < \pi$) through all values ϕ over $0 < \phi < 2\pi$. There is no recognizable orientation feature associated with spinning on the molecule axis. When the molecule is nonlinear there is a recognizable orientation associated with that motion (Figure 5.1b) that is described by the angle ψ as it assumes values $0 < \psi < 2\pi$. Rotations with respect to the remaining two axes are unchanged from Figure 5.2a.

The classical energy of a rigid rotator is

$$\mathrm{E} = \frac{J_x^2}{2I_x} + \frac{J_y^2}{2I_y} + \frac{J_z^2}{2I_z} \tag{5.17a}$$

[2] See Appendix F on Classical Considerations.

where J_x, J_y, J_z are angular momentum components and I is the moment of inertia. The integral taken over all orientations in space, when normalized to recognize that elements of space smaller than h^3 have no physical significance,[3] is equivalent to the sum over all rotational states and is therefore the partition function. The integral taken over all values of J^2 is

$$z_r = \frac{1}{h^3}\int_0^{\infty}\int_0^{\infty}\int_0^{\infty}\int_0^{2\pi}\int_0^{2\pi}\int_0^{\pi} e^{-(J_x^2/2I_x + J_y^2/2I_y + J_z^2/2I_z)/kT}\sin\theta\, d\theta\, d\phi\, d\psi\, dJ_x\, dJ_y\, dJ_z \tag{5.17b}$$

which produces

$$z_r = \pi^{1/2}\left(\frac{8\pi^2 I_x kT}{h^2}\right)^{1/2}\left(\frac{8\pi^2 I_y kT}{h^2}\right)^{1/2}\left(\frac{8\pi^2 I_z kT}{h^2}\right)^{1/2} \tag{5.17c}$$

$$z_r = \pi^{1/2}T^{3/2}/(\Theta_{r,x}\Theta_{r,y}\Theta_{r,z})^{1/2} \tag{5.18a}$$

with each Θ_r defined as $h^2/8\pi^2 Ik$. Except for the additional $\pi^{1/2}$ term, the form closely replicates that of Eq. (5.16a) for a molecule with two rather than three moments of inertia. The Eq. (5.18a) form should be used with all nonlinear polyatomic molecules. Again the partition function depends only on temperature and those intrinsic molecule structural features reflected by the moments of inertia.

There is one additional feature of rotation that we must recognize. If a molecule is symmetrical there are duplicate orientations over much of the space included in the phase integral. In particular, with reference to Figure 5.1a, it is immediately clear that a rotation through $0 < \theta < \pi$ produces a set of configurations that is indistinguishable from some of those produced by the $\pi < \phi < 2\pi$ rotation if the molecule is linearly symmetrical (Cl_2, CO_2, etc.). We have overcounted states by a factor of 2. For nonlinear molecules symmetry considerations are more complex but in general an appropriate symmetry factor is just the number of like orientations that can be produced by rotation. The symmetry number of methane is 12 since the tetrahedron can be placed with three hydrogens in a plane and rotated into three identical configurations. This can be repeated in three additional ways utilizing all choices of the coplanar hydrogens. In ammonia we have only one choice for the three coplanar hydrogens and three identical configurations for the rotation through 2π. The symmetry factor is 3. Equations (5.16a) and (5.18a) must be modified to include these symmetry considerations. The modification is just multiplication by $1/\sigma_r$ where σ_r is the symmetry factor

$$z_r = \frac{1}{\sigma_r}\left(\frac{T}{\Theta_r}\right) \qquad \text{linear} \tag{5.16b}$$

[3] See Appendix F.

$$z_r = \frac{\pi^{1/2}}{\sigma_r} \frac{T^{3/2}}{(\Theta_{r,x}\Theta_{r,y}\Theta_{r,z})^{1/2}} \qquad \text{nonlinear.} \tag{5.18b}$$

The rotational component of all thermodynamic functions, being related to the partition function logarithm, follows from Eqs. (5.16b) or (5.18b)

$$\ln z_r(\text{linear}) = \ln \frac{8\pi^2 k}{h^2} + \ln \frac{I}{\sigma_r} + \ln T \tag{5.19}$$

$$\ln z_r(\text{nonlinear}) = \ln \frac{8\pi^2 k}{h^2} + \ln \frac{(\pi I_x I_y I_z)^{1/2}}{\sigma_r} + \tfrac{3}{2} \ln T. \tag{5.20}$$

With pressure defined as $(\partial A/\partial V)_{TN}$ and $A = NkT \ln \mathscr{Z}$, there being no volume dependence in $\ln z_r$, it follows that there is no contribution to gas phase pressure from rotational terms in the partition function. For linear molecules the rotational component of the various thermodynamic functions of interest follows by straightforward arithmetic:

$$E_r = H_r = NkT \frac{\partial \ln z}{\partial T} = NkT^2 \frac{\partial \ln T}{\partial T} = NkT \tag{5.21}$$

$$C_{v,r} = C_{p,r} = \left(\frac{\partial E_r}{\partial T}\right)_{v,N} = Nk \tag{5.22}$$

$$A_r = G_r = -NkT \ln z_r = -NkT \ln\left(\frac{8\pi^2 I}{h^2}\right) + NkT \ln \sigma_r. \tag{5.23}$$

$$\mu_r = kT \ln \frac{8\pi^2 IkT}{h^2} + kT \ln \sigma_r \tag{5.24}$$

$$S_r = \frac{E_r - A_r}{T} = Nk + Nk \ln \frac{8\pi^2 IkT}{h^2} + Nk \ln \sigma_r \tag{5.25a}$$

or most compactly,

$$S_r = Nk \ln \frac{Te}{\sigma_r \Theta_r}. \tag{5.25b}$$

For nonlinear molecules (polyatomics)

$$E_r = \tfrac{3}{2} NkT, \qquad C_v = \tfrac{3}{2} Nk \tag{5.26}$$

$$A_r = G_r = NkT\left[\ln\left(\frac{8\pi^3 k}{h^2}\right)^{3/2} + \ln \frac{\pi^{1/2}}{\sigma_r} + \tfrac{1}{2} \ln I_x I_y I_z\right] \tag{5.27}$$

$$\mu_r = G_r/N \tag{5.29a}$$

$$S_r = \frac{E_r - A_r}{T} \tag{5.29a}$$

or most compactly,

$$S_r = Nk \ln \frac{\pi^{1/2}}{\sigma_r}\left(\frac{T^3 e^3}{\Theta_x \Theta_y \Theta_z}\right)^{1/2}. \tag{5.29b}$$

When the vibrational energy level expression from Eq. (5.4) is used to construct a vibrational partition function,

$$z_v = \sum^{v} e^{-(v+1/2)h\nu_e/kT} = e^{-h\nu_e/2kT} \sum^{v} e^{-vh\nu_e/kT},$$

the individual terms are not closely spaced at normal temperatures and do not form a continuum of energy. It is, therefore, fortunate that the summation can be carried out explicitly to provide $(1 - e^{-h\nu_e})^{-1}$ so that[4]

$$z_v = \frac{e^{-h\nu_e/2kT}}{1 - e^{-h\nu_e/kT}}. \tag{5.30}$$

It is again convenient to construct a term $\Theta_v = h\nu_e/k$, which has dimension T and to which we will refer as the characteristic temperature for vibration.

There will be several modes of vibration in a polyatomic molecule and

$$z_v = \prod^{K} \frac{e^{-\Theta_{v,K}/2kT}}{1 - e^{-\Theta_{v,K}/kT}}. \tag{5.31}$$

The term $\exp(-\Theta_{v,K}/2kT)$ derives from the nonzero ground vibrational state. Normally it would not be included in the partition function expression. The contribution is empirically carried in the E_0 term [Eq. (5.14)]. The usual expression for the partition function is

$$z_v = (1 - e^{-h\nu_e/kT})^{-1} \tag{5.32}$$

for each vibrational mode summing only over energy levels above the ground state.

[4] Other algebraic forms are often encountered:

$$z_r = (e^{\Theta_v/2kT} - e^{-\Theta_v/2kT})^{-1}$$

$$z_v = \left(2 \sinh \frac{\Theta_v}{T}\right)^{-1}.$$

With no volume dependence in the partition function for vibration $E_v = H_v, A_v = G_v, C_{P,v} = C_{V,v}$. Since all thermodynamic functions relate to $\ln z$, we construct (for a single vibrational mode)

$$\ln z = \frac{\Theta_v}{2T} - \ln(1 - e^{-\Theta_v/T}) \tag{5.33}$$

(retaining the ground state term for the moment in order to observe its complete role) to produce,

$$\begin{aligned} E_v = H_v &= NkT^2\left[-\frac{\Theta_v}{2}\left(-\frac{1}{T^2}\right) - \left(\frac{1}{1 - e^{-\Theta_v/T}}\right)\left(\frac{\partial e^{-\Theta_v/T}}{\partial T}\right)\right. \\ &= Nk\left[\frac{\Theta_v}{2} + \frac{\Theta_v e^{-\Theta_v/T}}{1 - e^{-\Theta_v/T}}\right] = Nk\left[\frac{\Theta_v}{2} + \frac{\Theta_v}{e^{\Theta_v/T} - 1}\right] \end{aligned} \tag{5.34}$$

$$\begin{aligned} C_{V_v} = C_{P,v} &= Nk\frac{\partial}{\partial T}\left[\frac{\Theta_v}{2} + \frac{\Theta_v}{e^{\Theta_v/T}}\right] = Nk\Theta_v\frac{\partial}{\partial T}\left(\frac{\Theta_v}{e^{\Theta_v/T} - 1}\right) \\ &= Nk\Theta_v\left[\frac{\Theta_v}{T^2}e^{\Theta_v/T}(e^{\Theta_v/T} - 1)^{-2}\right] = Nk\left(\frac{\Theta_v}{T}\right)^2\frac{e^{\Theta_v/T}}{(e^{\Theta_v/T} - 1)^2} \end{aligned} \tag{5.35}$$

$$A_v = G_v = -NkT\ln z_v = -Nk\left[\frac{\Theta_v}{2} - T\ln(1 - e^{-\Theta_v/T})\right] \tag{5.36}$$

$$\mu_v = k\left[-\frac{\Theta_v}{2} + T\ln(1 - e^{-\Theta_v/T})\right] \tag{5.37}$$

$$S_v = \frac{E - A}{T} = Nk\left[\frac{\Theta_v/T}{e^{\Theta_v/T} - 1} + \ln(1 - e^{-\Theta_v/T})\right]. \tag{5.38}$$

Each unique Θ_v value will contribute independently to each term. We note that the ground state vibrational energy appears as $\Theta_v/2$ in all expressions except that for S_v and would not have been present had we used the Eq. (5.32) expression for z_v rather than the Eq. (5.31) expression.[5] Having identified the contribution of the zero point vibrational energy we can omit it from our operational expressions with awareness that we are including this zero point energy in our E_0 term. As with the rotational partition function, the vibrational partition

[5] Since entropy is the difference between two state functions each of which include E_0, it will always have absolute value.

function depends only on T and intrinsic molecule features as reflected by the various ν_e.

In obtaining the partition function $\mathscr{Z}$ for molecule systems we have simply extended the statistical arguments applied to the nonstructured particle system with recognition that several distribution functions must be simultaneously satisfied. Mathematically there is no problem; in a physical sense there is a difficulty. When a single distribution function (atoms in their ground electronic state) must be satisfied, we can depend on the single specification of total energy in a system of stated V/N to govern the distribution through what we called very weak interactions. These interactions are molecule pair collisions and it is difficult to see that there is sufficient communication in such encounters to ensure that the several distributions are met within the constraint of a specified total energy. However, all of the distribution functions are determined by a fixed set of energy levels and the value of T. Clearly T is as sufficient a specification of the distributions as is E and T for a molecule system can be both the specification that controls all distributions and the feature that controls averaging when the system is in diathermal contact with an infinite thermal reservoir.

Reference to Eq. (3.32) and (3.27) demonstrates that, while (STP) there are many more particle collisions than wall collisions [Z_{AA}/Z_{wall} is (0) 10^7] there are many [(0) $10^{23}\,\text{sec}^{-1}\,\text{cm}^{-2}$] wall collisions so that while fluctuations[6] in E may occur they should be averaged out through system–environment interactions. We should regard $\mathscr{Z}(T)$ as a constant temperature system partition function.[7] This feature is also convenient in arguments relating to free energies since these are defined for constant temperature systems. Naturally we are not confined to fixed values of T since any macroscopic change can be visualized in terms of a reversible process for which all statistical features are defined at every step of the process. If possible perturbations in the energy level set in very dense systems are neglected in a first approximation, all z_K depend only on species character and on T.

We have noted [Eq. (4.2)] that the value of E_t is $\frac{3}{2}N\ell T$ for the nonstructured particle system. For our dilute structured particle system it is still $\frac{3}{2}N\ell T$, deriving from $[(T/\Theta_t)^{1/2}]^3$ as before. A similar temperature dependence follows for the rotational motions described by Eqs. (5.21) and (5.26). For a linear molecule when $T \gg \Theta_r$, $E_r = N\ell T$ deriving from $[(T/\Theta_r)^{1/2}]^2$ and for a nonlinear molecule $E_r = \frac{3}{2}N\ell T$ deriving from $[(T/\Theta_r)^{1/2}]^3$. We would refer to these motions as classical and to the $\ell T/2$ contribution from each independent mode of motion as the equipartition of energy. The distribution of energy levels is effectively continuous and very many levels are populated. As we have noted in the preliminary discussion the distribution among vibrational levels will be classical only at very high values of T and among electronic energy levels never.

[6] Such fluctuations are (0) $1/N^{1/2}$. See Appendix C.

[7] In a more formal development (see Appendix E on Ensembles) $\mathscr{Z}(T)$ is referred to as the canonical ensemble partition function. Our particle partition function is microcanonical.

[8] In analogy to Θ_r and Θ_v, here we define a Θ_t as $(h^2/2\pi \ln \ell)$.

Table 5.2 lists characteristic temperatures, bond lengths, and disssociation energies for common diatomic molecules. With such information it is possible to obtain values for thermodynamic functions of interest and to obtain distribution numbers N_i^*/N. The distribution numbers are particularly informative when various levels of Θ/T are referenced to Table 5.2 for particular molecule species. Such data are represented in Figure 5.3 for the populations of rotational states.

From Figure 5.3 and Table 5.2 we note that since Θ_r is (0) 1 to 10 for most molecules, the Θ_r/T range of common concern is (0) 10^{-2} to 10^{-3}. In the former case about 0.99 of the system molecules occupy the first 25 quantum levels while in the latter case only 0.81 of the system molecules occupy the first 40 quantum levels. Significant occupancy in the higher quantum levels increases very rapidly with increasing temperature ($\sum^n$ represents the sum of individual state population fractions over n terms). There is a well-defined maximum for each Θ_r/T curve deriving from energy level degeneracy.

Equations (5.4) for vibrational energy levels is the Schroedinger equation solution for the harmonic oscillator model of a vibrator. The potential energy, $\mathscr{V}(R)$, for a harmonic oscillator is proportional to the square of the displacement,

$$\mathscr{V}(R) = \tfrac{1}{2}\kappa(R - R_e)^2 \tag{5.39}$$

where R_e is the equilibrium bond length. The potential energy associated with displacement in real molecules is not symmetric for other than small values of the displacement as is illustrated by the solid line in Figure 5.4. The dashed

TABLE 5.2. Characteristic Temperatures, Bond Lengths, and Dissociation Energies for Common Diatomic Molecules

	Θ_v, °K	Θ_r, °K	r_e, A	D_0, ev
H_2	6210	85.4	0.740	4.454
N_2	3340	2.86	1.095	9.76
O_2	2230	2.07	1.204	5.08
CO	3070	2.77	1.128	9.14
NO	2690	2.42	1.150	5.29
HCl	4140	15.2	1.275	4.43
HBr	3700	12.1	1.414	3.60
HI	3200	9.0	1.604	2.75
Cl_2	810	0.346	1.989	2.48
Br_2	470	0.116	2.284	1.97
I_2	310	0.054	2.667	1.54

From T. L. Hill, *An Introduction to Statistical Thermodynamics*. Addison-Wesley, Reading, MA, 1960.

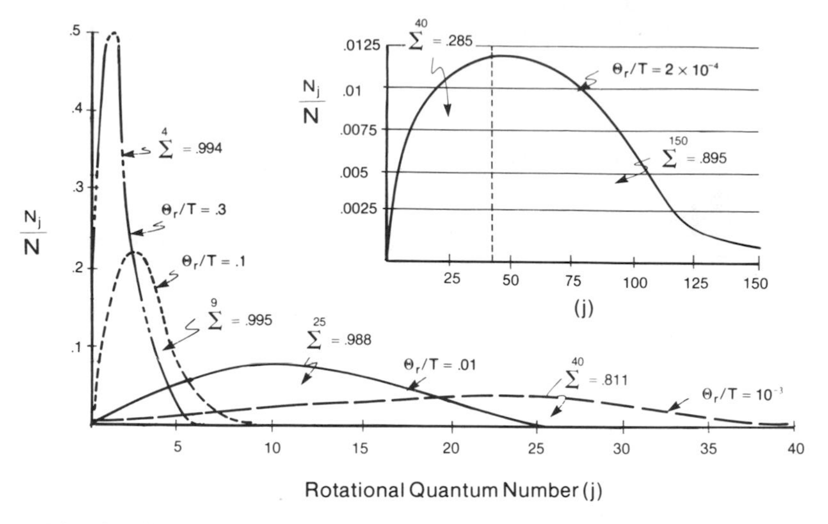

Figure 5.3. The distribution of particles among rotational states and the fraction of particles included in the sum over n terms for various Θ_v/T values.

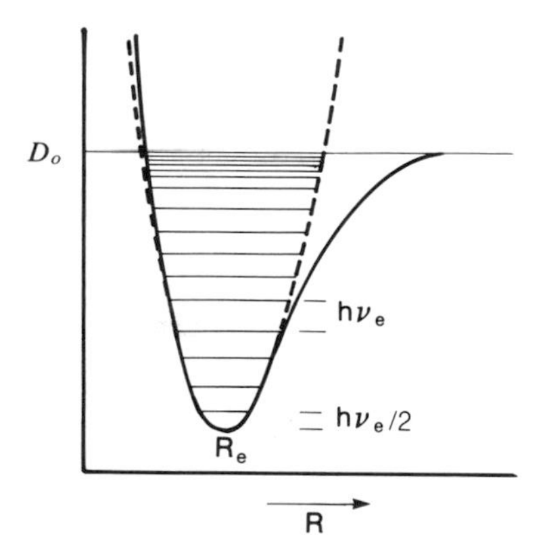

Figure 5.4. Sketch illustrating the harmonic oscillator (dashed line) model for a vibrational state and (solid line) anharmonic deviation from the model.

line in Figure 5.4 is the harmonic oscillator approximation. When a realistic potential energy expression is used, e.g.,

$$\mathscr{V}(\mathrm{R}) = \mathrm{D_0}(1 - e^{-a(\mathrm{R}-\mathrm{R_e})})^2, \qquad a = \left(\frac{\mu_r}{2D_0}\right)^{1/2} \nu_e \tag{5.40}$$

where $\mathrm{D_0}$ is the dissociation energy and μ_r is reduced mass $[m_1 m_2/(m_1 + m_2)]$

the Schroedinger equation solution is

$$\epsilon_v = (v + \tfrac{1}{2})h\nu_e - (v + \tfrac{1}{2})^2 hx_e\nu_e \tag{5.41}$$

where $x_e = ha^2/m_r\nu_e$. Energy levels are not uniformly spaced at intervals of $h\nu_e$ for large displacements. The ever closer spacing for larger and larger displacements approaches an energy continuum only as the dissociation energy D_0 is approached. Anharmonicity, as reflected by the constant X_e, is not really an important consideration for us in that few, if any, energy levels above the ground level are occupied at temperatures of thermodynamic concern. This is reflected in Figure 5.5 where the populations of vibrational states are represented. $\sum^n$ represents the sum of individual state population fractions over n terms.

Reference to Table 5.2 demonstrates that Θ_v/T values (0) 1 to 10 are those of common interest where, in the worst case, $\Theta_v/T = 1$, 0.998 of the system molecules occupy the first seven quantum states with 0.892 of them being in the first two quantum states. For $\Theta/T > 5$ effectively all molecules are in the ground vibrational state.

Reference to Eqs. (5.21)–(5.29), which provide the rotational contribution to thermodynamic functions, reveals that although rotational energy and rotational heat capacity will be species independent, other thermodynamic functions require species specific values including the moment or moments of inertia. Such values will often be available for polar molecules from unique absorptions in the microwave spectral region. In the absence of a dipole moment the rotational spectrum is obtained from vibration–rotation transitions or from

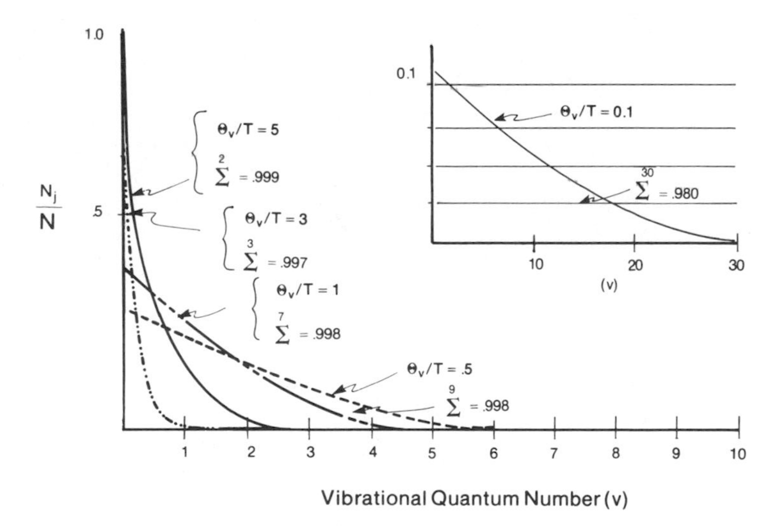

Figure 5.5. The distribution of particles among vibrational states and the fraction of particles included in the sum over terms for various Θ_v/T values.

Raman spectra. In the simplest case for a linear molecule with energy level spacing of $(h^2/8\pi^2 I)2(j+1)$ and in terms of the commonly encountered spectroscopic constant B in wave numbers, cB is frequency

$$\mathrm{B} = \frac{h}{8\pi^2 Ic}\left(= \frac{k}{hc}\Theta_{\mathrm{r}}\right) \tag{5.42}$$

$$\epsilon_{j,\mathrm{r}} = hc\mathrm{B}[j(j+1)].$$

Then, as in Figure 5.6, the absorption lines are spaced at intervals of $2\mathrm{B}(\mathrm{cm}^{-1})$. As an example, the following rotational absorption lines have been observed for $^{1}H^{35}Cl$: 83.32, 104.13, 124.73, 145.37, 165.89. The spacing is at intervals of 20.81, 20.60, 20.64,..., i.e., the (average) value of B is 10.34 and the moment of inertia is

$$I = \frac{h}{8\pi^2 c\mathrm{B}} = \frac{6.63 \times 10^{-27}}{8(3.14)^2(3 \times 10^{10})(10.34)}$$

$$I = 2.71 \times 10^{-40}\,\mathrm{g\,cm}^2.$$

If spectroscopic data are not available it is relatively easy to calculate the moment of inertia for simple molecules, i.e., those that are rigidly linear or polyatomics that have particular symmetry elements. For this purpose we would recognize that the moment of inertia of a body that can be considered as composed of mass points $m_1, \ldots, m_i$ is $I = \sum^{i} m_i r_i^2$ where r_i is the perpendicular distance from the mass point to a molecule axis. If the axis is one which passes through the center of mass, the internal moments sum to zero ($\sum m_i r_i = 0$). To illustrate the calculation we consider a heteronuclear molecule represented by masses m_1 and m_2 separated by distance R.

If X is the distance from atomic mass 1 to the mass center the sum of internal moments is

$$\sum m_i r_i = 0 = m_1 x - m_2(\mathrm{R} - x)$$

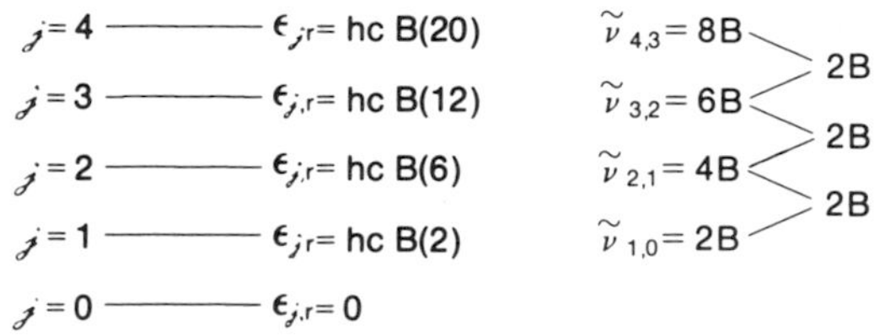

$j = 4$ ——— $\epsilon_{j,r} = hc\,\mathrm{B}(20)$ $\tilde{\nu}_{4,3} = 8\mathrm{B}$
2B
$j = 3$ ——— $\epsilon_{j,r} = hc\,\mathrm{B}(12)$ $\tilde{\nu}_{3,2} = 6\mathrm{B}$
2B
$j = 2$ ——— $\epsilon_{j,r} = hc\,\mathrm{B}(6)$ $\tilde{\nu}_{2,1} = 4\mathrm{B}$
2B
$j = 1$ ——— $\epsilon_{j,r} = hc\,\mathrm{B}(2)$ $\tilde{\nu}_{1,0} = 2\mathrm{B}$
$j = 0$ ——— $\epsilon_{j,r} = 0$

Figure 5.6. Illustration of the calculation of moments of inertia and rotational energy levels from spectroscopic data.

the r_i being of different sign. Then

$$m_1 x = m_2(\mathrm{R} - x)$$

$$x = \frac{m_2}{m_1 + m_2}\mathrm{R}$$

and

$$I = m_1 x^2 + m_2(x - \mathrm{R})^2.$$

Inserting the required value of x and simplifying produces

$$I = \frac{m_1 m_2}{m_1 + m_2}\mathrm{R}^2 = \mu_r \mathrm{R}^2 \qquad (5.43)$$

where $m_1 m_2/(m_1 + m_2) = \mu_r$ is the same reduced mass term encountered in Eq. 5.40. Expressions for the calculation of moments of inertia for several simple molecule forms are collected in Tables 5.3.

Applying Eq. (5.44) together with the appropriate Table 5.2 value to the calculation of the moment of inertia for $^1H^{35}Cl$ produces

$$I = \frac{(35)(1)(6.02 \times 10^{23})^2}{(35 + 1)(6.02 \times 10^{23})}(1.275)^2(10^{-8})^2 = 2.63 \times 10^{-40}\,\mathrm{g\,cm}^2.$$

This is close to but not identical to the $2.71 \times 10^{-40}\,\mathrm{g\,cm}^2$ value obtained from the spectroscopic constant. Clearly there is some centrifugal stretching in the molecule that is reflected by the spectroscopic data. The rigid rotator model is a good but not exact representation.

There are several operational features that we need to address as regards vibration. We will require v_e values from spectral data and the data are straightforward; there is a single absorption frequency associated with each mode of vibration. The number of such vibrational modes is readily determined.

We can completely specify all motions of a molecule by identifying the coordinates of each atom and noting how this coordinate set changes with time. This requires $3n$ coordinate specifications if there are n atoms in the molecule. If the molecule is a rigid rotator we have another choice. We can specify the center of mass motion (translation, three coordinates) and either two (linear) or three (nonlinear) orientations with respect to axes passing through the center of mass. Vibrations can be described by the single coordinate specification of distance (R) between two atomic masses. Since we cannot avoid the total of $3n$ coordinate specifications, there are $3n - (3 + 2)$ such independent specifications required for a linear molecule and $3n - (3 + 3)$ for a nonlinear molecule. The $3n - 5$ or $3n - 6$ coordinate specifications are not required to be vibrations; they can include free or hindered internal rotations that would require other treatment.

TABLE 5.3. Moments of Inertia for Common Molecule Forms[a]

The expressions below give the moments of inertia for molecules of the type shown.

1. *Diatomics*

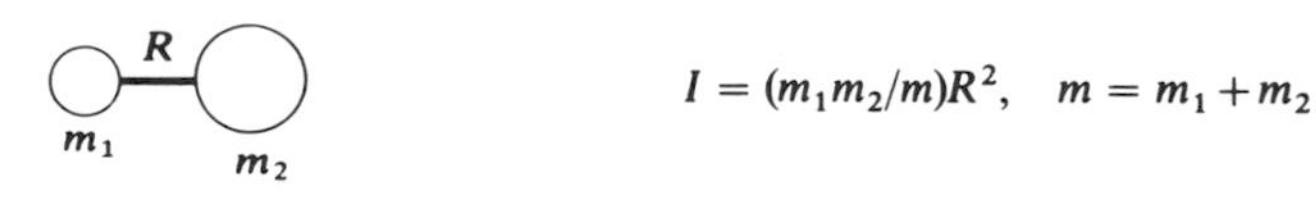

$$I = (m_1 m_2/m)R^2, \quad m = m_1 + m_2$$

2. *Linear triatomics*

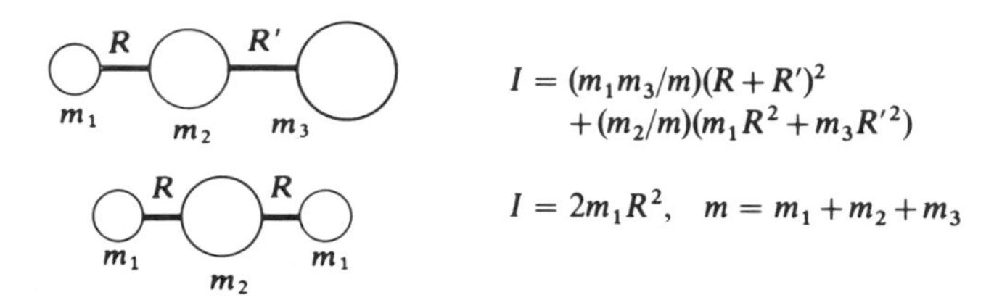

$$I = (m_1 m_3/m)(R+R')^2 + (m_2/m)(m_1 R^2 + m_3 R'^2)$$

$$I = 2m_1 R^2, \quad m = m_1 + m_2 + m_3$$

3. *Symmetric tops*

$$\begin{aligned} I_{\parallel} &= 2m_1 R^2(1-\cos\theta), \\ I_{\perp} &= m_1 R^2(1-\cos\theta) \\ &\quad + (m_1/m)(m_2+m_3)R^2(1+2\cos\theta) \\ &\quad + (m_3/m)R'\{(3m_1+m_2)R' \\ &\qquad + 6m_1 R\sqrt{[\tfrac{1}{3}(1+2\cos\theta)]}\} \\ m &= 3m_1 + m_2 + m_3 \end{aligned}$$

$$\begin{aligned} I_{\parallel} &= 2m_1 R^2(1-\cos\theta), \quad m = 3m_1 + m_2 \\ I_{\perp} &= m_1 R^2(1-\cos\theta) \\ &\quad + (m_1 m_2/m)R^2(1+2\cos\theta) \end{aligned}$$

$$\begin{aligned} I_{\parallel} &= 4m_1 R^2 \\ I_{\perp} &= 2m_1 R^2 + 2m_3 R'^2 \end{aligned}$$

4. *Spherical tops*

$$I = (8/3)m_1 R^2$$

$$I = 4m_1 R^2$$

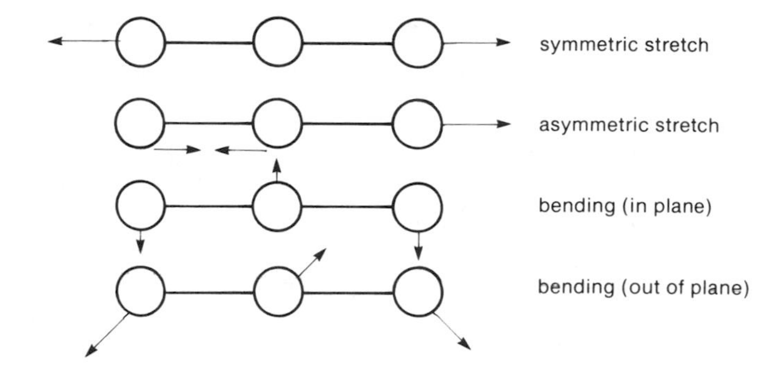

Figure 5.7. Normal vibrational modes for a linear triatomic molecule.

Since we require that all such motions be independent (called normal modes) they must not interfere with each other, i.e., no motion should excite another motion. This requirement can be illustrated readily with a linear triatomic structure. With $n = 3$ there are four independent internal motions and these are easy to visualize. They are shown in Figure 5.7. The bending modes have the same frequencies and this mode is stated to be doubly degenerate. Operationally such degenerate internal modes will be reflected as two terms in the Eq. (5.31) product.

Finally, it will sometimes be convenient when applying Eqs. (5.34)–(5.38) to use a table of functions which simplify the arithmetic operations. Writing

$$x = \frac{\Theta_v}{T} = \frac{h\nu_e}{kT} = \frac{hc\tilde{\nu}_e}{kT}$$

we construct, with reference to Eqs. (5.34), (5.35), and (5.36),

$$\frac{\Theta_v/T}{e^{\Theta_v/T} - 1} = \frac{x}{e^x - 1} = f_1(x) \tag{5.44}$$

$$\frac{(\Theta_v/T)^2 e^{\Theta_v/T}}{(e^{\Theta_v/T} - 1)^2} = \frac{x^2 e^x}{(e^x - 1)^2} = f_2(x) \tag{5.45}$$

$$-\ln(1 - e^{-\Theta_v/T}) = -\ln(1 - e^{-x}) = f_3(x). \tag{5.46}$$

Taken from P. W. Atkins, *Physical Chemistry*, 2nd ed. W. H. Freeman, San Francisco CA, 1982.
[a]Molecules represented here are characterized by either a single value for the moments of inertia or by $I_x = I_y \neq I_z$ the two values being represented by $I_\perp$ and $I_\parallel$. See Appendix A for further discussion.

Then

$$E_v = N k T f_1(x) = H_v \tag{5.47}$$

$$C_{V,v} = Nk f_2(x) = C_{P,v} \tag{5.48}$$

$$A_v = Nk f_3(x) = G_v \tag{5.49}$$

$$S_v = Nk[f_3(x) - f_1(x)]. \tag{5.50}$$

Table 5.4 provides a sufficiently closely spaced tabulation of these (Einstein) functions for most purposes.

Since the N particle total partition function ($\mathscr{Z}_I$) via Eq. (5.8) is the N-fold product of the partition function for electronic excited states and the various modes of motion that are available to the particle species and the thermodynamic functions all involve $\ln \mathscr{Z}_I$, the various contributions to any such function are additive as we would anticipate. In particular:

$$E_\Sigma - E_0 = \sum^{K} NkT^2\left(\frac{\partial \ln z_K}{\partial T}\right)_{V,N} \tag{5.51}$$

$$S_\Sigma = \sum^{K}\left[Nk \ln z_K + NkT\left(\frac{\partial \ln z_K}{\partial T}\right)_{V,N}\right] \tag{5.52}$$

from which the other functions follow. The experimentally accessible quantity which will measure molar changes in E or in S is the appropriate integral over $C_V dT$

$$\Delta E_V = \int_{T_1}^{T_2} C_V dT \tag{4.24}$$

$$\Delta S_V = \int_{T_1}^{T_2} \frac{C_V}{T} dT. \tag{4.55}$$

The heat capacity C_V is also the sum of terms relating to electronic excitation and to the various modes of motion $C_V = \sum^K C_{V,K}$ with each $C_{V,K} = (\partial E_K/\partial T)_V$. The heat capacity C_P for a dilute gas system behaving ideally is $C_V + Nk$ [Eq. (4.28)] so, for present needs, it is sufficient to address C_V.

We have obtained values for $C_{V,K}$ for the dilute gas system for all modes of motion

$$C_{V,t} = \tfrac{3}{2}Nk$$

$$C_{V,r} = Nk(\text{linear}) \quad \text{or} \quad \tfrac{3}{2}Nk(\text{nonlinear})$$

$$C_{V,v} = Nk\left(\frac{\Theta_v}{T}\right)^2 \frac{e^{\Theta_v/T}}{(e^{\Theta_v/T} - 1)^2}(\text{each vibration}).$$

TABLE 5.4. Einstein Functions

x	$f_1(x)$	$f_2(x)$	$f_3(x)$	x	$f_1(x)$	$f_2(x)$	$f_3(x)$
0.000	1.0000	1.0000	∞	1.90	0.3342	0.7465	0.1620
0.001	0.9995	1.0000	6.9083	1.95	0.3235	0.7354	0.1535
0.005	0.9975	1.0000	5.3008	2.00	0.3130	0.7241	0.1454
0.010	0.9950	1.0000	4.6102	2.10	0.2930	0.7013	0.1306
0.02	0.9900	1.0000	3.9220	2.20	0.2741	0.6783	0.1174
0.03	0.9851	0.9999	3.5215	2.30	0.2563	0.6552	0.1055
0.04	0.9801	0.9999	3.2388	2.40	0.2394	0.6320	0.0951
0.05	0.9752	0.9998	3.0206	2.50	0.2236	0.6089	0.0854
0.10	0.9508	0.9992	2.3522	2.60	0.2086	0.5859	0.0772
0.15	0.9269	0.9981	1.9712	2.70	0.1945	0.5631	0.0695
0.20	0.9033	0.9967	1.7078	2.80	0.1813	0.5405	0.0627
0.25	0.8802	0.9948	1.5087	2.90	0.1689	0.5182	0.0565
0.30	0.8575	0.9925	1.3502	3.00	0.1572	0.4963	0.0511
0.35	0.8352	0.9898	1.2197	3.10	0.1462	0.4747	0.0451
0.40	0.8133	0.9868	1.1096	3.20	0.1360	0.4536	0.0415
0.45	0.7918	0.9833	1.0151	3.30	0.1264	0.4330	0.0376
0.50	0.7707	0.9794	0.9328	3.40	0.1174	0.4129	0.0339
0.55	0.7501	0.9752	0.8603	3.50	0.1090	0.3933	0.0307
0.60	0.7298	0.9705	0.7958	3.60	0.1011	0.3743	0.0277
0.65	0.7100	0.9655	0.7382	3.70	0.0938	0.3558	0.0250
0.70	0.6905	0.9601	0.6863	3.80	0.0870	0.3380	0.0226
0.75	0.6714	0.9544	0.6394	3.90	0.0806	0.3207	0.0205
0.80	0.6528	0.9483	0.5965	4.00	0.0746	0.3041	0.0185
0.85	0.6345	0.9419	0.5576	4.20	0.0639	0.2276	0.0151
0.90	0.6166	0.9351	0.5218	4.40	0.0547	0.2435	0.0124
0.95	0.5991	0.9281	0.4890	4.60	0.0467	0.2170	0.0101
1.00	0.5820	0.9207	0.4587	4.80	0.0398	0.1928	0.0088
1.05	0.5652	0.9130	0.4307	5.00	0.0339	0.1707	0.0068
1.10	0.5489	0.9050	0.4048	5.5	0.0226	0.1246	0.0041
1.15	0.5329	0.8967	0.3807	6.0	0.0149	0.0897	0.0025
1.20	0.5172	0.8882	0.3584	6.5	0.0098	0.0637	0.0015
1.25	0.5019	0.8794	0.3376	7.0	0.0064	0.0448	0.0009
1.30	0.4870	0.8703	0.3182	7.5	0.0042	0.0311	0.0006
1.35	0.4725	0.8610	0.3001	8.0	0.0027	0.0215	0.0003
1.40	0.4582	0.8515	0.2832	8.5	0.0017	0.0147	0.0002
1.45	0.4444	0.8418	0.2673	9.0	0.0011	0.0100	0.0001
1.50	0.4308	0.8318	0.2525	9.5	0.0007	0.0068	0.0001
1.55	0.4176	0.8217	0.2386	10.0	0.0005	0.0045	0.0000
1.60	0.4048	0.8114	0.2255	11.0	0.0002	0.0020	0.0000
1.65	0.3922	0.8010	0.2133	12.0	0.0001	0.0009	0.0000
1.70	0.3800	0.7903	0.2017	13.0	0.0000	0.0004	0.0000
1.75	0.3681	0.7796	0.1909	14.0	0.0000	0.0002	0.0000
1.80	0.3565	0.7689	0.1807	15.0	0.0000	0.0001	0.0000
1.85	0.3452	0.7577	0.1711				

$$f_1(x) = \frac{x}{e^x - 1}. \quad f_2(x) = \frac{x^2 e^x}{(e^x - 1)^2}. \quad f_3(x) = -\ln(1 - e^{-x}).$$

Taken from S. M. Blinder, *Advanced Physical Chemistry*. MacMillan, Toronto, 1969.

We did not obtain a heat capacity expression relating to electronic excitation since, via Eqs. (5.10b) and (5.12b), the ln z_e term normally involves only constants and $C_{V,e} = 0$. The term $C_{V,\Sigma}$, normally written as C_V, understood to refer to all contributions other than electronic, is then

$$C_V = Nk\left[\frac{5}{2} + \left(\frac{\Theta_v}{T}\right)^2 \frac{e^{\Theta_v/T}}{(e^{\Theta_v/T} - 1)^2}\right] \qquad \text{(linear, diatomic)} \tag{5.53}$$

$$C_V = Nk\left[3 + \sum^{K}\left(\frac{\Theta_{v,K}}{T}\right)^2 \frac{e^{\Theta_{v,K}/T}}{(e^{\Theta_{v,K}/T} - 1)^2}\right] \qquad \text{(nonlinear, polyatomic).}$$

A few molecules will have thermally accessible electronic states and in these cases $C_{V,e} \neq 0$. The heat capacity contribution will be temperature dependent, just as the vibrational contribution is temperature dependent. Double differentiation of Eq. (5.12a) will produce

$$\tilde{C}_{V,e} = N_A k\left[\frac{\omega_1}{\omega_0} \frac{(\Delta\eta_1/kT)^2 e^{\Delta\eta_1/kT}}{(1 - (\omega_1/\omega_0)e^{-\Delta\eta_1/kT})^2} + \cdots\right] \tag{5.54}$$

for the electronic contribution to molar heat capacity.

At any value of T, the importance of any term involving Θ_v depends on the fundamental vibrational frequency, i.e., on the stiffness of the chemical bond. At temperatures at which many excited states are populated, classical behavior (as defined for translation and rotation) would produce a contribution of Nk for each vibration ($Nk/2$ for the kinetic energy term and $Nk/2$ for the potential energy term). This will effectively never be observed since the molecules would thermally dissociate. At high T with weak bonds, some nonzero contribution would be expected and that contribution would be temperature dependent. If vibrational contributions become large there would also be a substantial anharmonicity component that would place our calculated values in some doubt. We should rely on experiment to the extent that it is available.

The quantity normally measured is C_P and the result is commonly fitted to the empirical form

$$\tilde{C}_P = a + bT + \frac{c}{T^2}.$$

Table 5.5 lists the constants for several common gases at one atmosphere pressure.

We can now calculate all thermodynamic parameters that we have identified as relevant to process considerations for substances in the dilute gas phase. In Chapters 2 through 4 we recognized the controlled process circumstances from which observable features emerge (ΔP, ΔV, ΔT, Q, W) that we identify with changes in the thermodynamic functions (ΔE, ΔH, ΔA, ΔG, ΔS). To fully

TABLE 5.5. Temperature Dependence of Heat Capacities, $\tilde{C}_P = a + bT + cT^{-2}$

	a ($J\,K^{-1}\,mol^{-1}$)	$b \times 10^3$ ($J\,K^{-2}\,mol^{-1}$)	$c \times 10^5$ ($J\,K\,mol^{-1}$)
Gases (298–2000 K)			
He, Ne, Ar, Kr, Xe	20.78	0	0
H_2	27.28	3.26	0.50
O_2	29.96	4.18	−1.67
N_2	28.58	3.77	−0.50
F_2	34.56	2.51	−3.51
Cl_2	37.03	0.67	−2.85
Br_2	37.32	0.50	−1.26
CO_2	44.22	8.79	−8.62
H_2O	30.54	10.29	0
NH_3	29.75	25.10	−1.55
CH_4	23.64	47.86	−1.92

From *Thermodynamics*, G. N. Lewis and M. Randall, revised by K. S. Pitzer and L. Brewer, McGraw-Hill, New York (1961). Reproduced with permission from McGraw Hill.

implement the thermodynamic structure that has developed we must extend our arguments to nonideal gases and condensed phases. We anticipate that the introduction of potential energies of interaction will certainly influence translation in the gas phase and largely eliminate that mode of motion in condensed phases. We would expect that such interactions, even when they lead to condensation, will have a lesser effect on rotational motion and little or no effect on vibrational motions. Much of the understanding we have acquired in this chapter will have general applicability. We begin in the next chapter with dense gas systems.

PROBLEMS

5.1. The tellurium atom has a number of low-lying excited states. Calculate the electronic partition function for atoms at 298 K and 500 K from the following information.

State	Degeneracy	Wave Number (cm^{-1})
Ground	5-fold	—
First	3-fold	4751
Second	1	4707
Third	5-fold	10559

5.2. The NO molecule has a low-lying doubly degenerate excited state at 121.1 cm^{-1} above the doubly degenerate ground state. The dissociation energy for the molecule is 5.30 eV (1 eV = 1.602×10^{-19} J). Calculate the electronic partition function for the molecule at 298 K and at 800 K.

5.3. Calculate the rotational partition function for CO at 298 K and at 500 K, using data from Table 5.2.

5.4. Calculate the rotational partition function for H_2O at 500 K given the rotational constants $A = 27.33\ \text{cm}^{-1}$, $B = 14.58\ \text{cm}^{-1}$, and $C = 9.50\ \text{cm}^{-1}$.

5.5. Calculate the vibrational partition function for CO at 298 K and at 500 K using data from Table 5.2.

5.6. Using the Table of Einstein functions and the fundamental vibrations for H_2O (cm^{-1}) $\tilde{v}_1 = 3651.7$, $\tilde{v}_2 = 1290$, $\tilde{v}_3 = 1408$, calculate the vibrational contribution to the heat capacity for H_2O at 500 K.

5.7. Obtain the standard molar entropy for CO gas at 298 K.

5.8. Obtain the Gibbs free energy $(G - E_0)$ for 0.25 mol of H_2 gas when it occupies a volume of 10 liters at 298 K.

5.9. Calculate the constant volume heat capacity for N_2 gas at 298 K using data from Table 5.2 for a vibrational frequency and a rotational constant. What is the constant pressure heat capacity? How does this value compare with experiment as it is reflected in the Table 5.5 data?

CHAPTER CONTENT SOURCES AND FURTHER READING

1. T. L. Hill, *Introduction to Statistical Thermodynamics*, Chapters 8 and 9. Addison-Wesley, Reading MA, 1960.
2. E. A. Moelwyn-Hughes, *Physical Chemistry*, 2nd ed., Chapters VIII through XIII. Pergamon Press, London, 1961.
3. N. Davidson, *Statistical Mechanics*, Chapter 11. McGraw-Hill, New York, 1962.
4. J. G. Aston and J. J. Fritz, *Thermodynamics and Statistical Thermodynamics*, Chapters 19 and 20. Wiley, New York, 1959.

DATA SOURCES

Sources of Spectroscopic Data

1. C. E. Moore, *Atomic Energy Levels*, Vols. I, II, III. NBS Circ. 467, U.S. Government Printing Office, Washington, D.C., 1949, 1952, 1958.
2. G. Herzberg, *Molecular Spectra and Molecular Structure*, Vols. I, III. Van Nostrand, New York, 1950, 1956.
3. B. Rosen, Ed., *Spectroscopic Data Relative to Diatomic Molecules*, Int. Tables of Selected Constants, #17. Pergamon Press, Oxford, 1970.

4. T. Shimanouchi, *Tables of Molecular Vibrational Frequencies*, Consolidated Volume. National Bureau of Standards Office of Standard Reference Data, NSRDS-NBS, Circ. 39, U.S. Government Printing Office, Washington D.C. [See also T. Shimanouchi, *Tables of Molecular Vibrational Frequencies, J. Phys. Chem. Ref. Data*, **1**, 189 (1972).]

SYMBOLS: CHAPTER 5

$\epsilon_{0,\mathscr{A}}, \epsilon_{0,\mathscr{B}}$ = ground state atomic energies, atoms $\mathscr{A}$, $\mathscr{B}$

$\eta_{0,\mathscr{A}\mathscr{B}}, \eta_{1,\mathscr{A}\mathscr{B}} \cdots$ = ground state and excited state energies for the $\mathscr{A}\mathscr{B}$ molecule

$D_{0,\mathscr{A}\mathscr{B}}$ = the $\mathscr{A}\mathscr{B}$ molecule dissociation energy from the ground state

K, L = number of bonds and number of atoms in a molecule

$\epsilon_{n,t}, \epsilon_{j,r}, \epsilon_{v,v}$ = designation for an energy level associated with translational, rotational, vibrational motion

$\Omega_t, \Omega_e, \Omega_r, \Omega_v$ = combinatorial terms for the distribution of particles among available energy states

$p_{i,t}, p_{i,v} \cdots p_i(t, r, v, e)$ = probabilities related to energy level occupancy

z_t, z_r, z_v, z_e = particle partition function relating to the several sets of energy levels

$\omega_0, \omega_1, \ldots$ = degeneracy of electronic energy levels

x, y, z = axes in a coordinate system

ψ, ϕ, θ = orientation angles in a coordinate system

I_x, I_y, I_z = moments of inertia for rotation about a particular axis

J_x, J_y, J_z = angular momentum components

σ_r = symmetry number

Θ_r = characteristic temperature for rotation

$E_r, E_v, \ldots, C_{P,v}, \ldots, G_r \cdots$ = components of the total energy or other thermodynamic function deriving from rotation, vibration...

Θ_v = characteristic temperature for vibration

$\mathscr{V}_R$ = potential energy for an oscillator associated with displacement from a rest position

a, R, R_e, x_e = parameters associated with oscillation

A, B, C = spectroscopic rotational constants

n = number of atoms in a molecule

$f_1(x), f_2(x), f_3(x)$ = Einstein functions

$E_\Sigma, S_\Sigma \cdots$ = total energy, entropy...

a, b, c = constants in an empirical expression for heat capacity

Carryover Symbols

k, h, v, ϵ_i, I, v_e, m, Ω, g_i, z, $\mathscr{Z}$, N, E, E_0, V, T, P, S, H, G, A, C_P, C_V.

6

INTERACTING PARTICLE SYSTEMS I: THE NONIDEAL GAS

Our preliminary argument deriving from the independent, structureless, indistinguishable particle set that we called "the ideal gas" has now been expanded to include some of the effects of introducing structure. In terms of our important process parameters Q and W, introducing structure adds energetic options and, therefore, influences thermal response but the additional modes of motion do not contribute to the pressure of the system or to PV related work terms. However, structured particles do have individual electric fields and interactions involving the collection of these internal fields can influence translation in the gas phase and therefore the gas phase pressure.

The interactions are electrostatic in nature and we would anticipate both attractive and repulsive forces. Attractive forces produce a potential energy of interaction that we take to be dependent on the inverse sixth power of the distance of separation.[1] Repulsive forces produce a potential energy of interaction that must be infinite at some distance of closest approach and that diminishes rapidly with increasing distance of separation. Any power that completely dominates the r^{-6} dependence will produce approximately the same potential energy function. An r^{-12} dependence is often taken for mathematical convenience. The 6–12 potential, referred to as the Leonard–Jones potential, produces an interaction energy expression

$$\phi = Ar^{-12} - Br^{-6} \tag{6.1}$$

[1] See Appendix D on Intermolecular Forces.

where A and B are species-specific constants. Attractive forces should inhibit the free motion of molecules and the effect should increase as the average particle separation decreases (i.e., as V decreases) and diminish as the average translational energy increases (i.e., as T increases). The most obvious effects of repulsive interactions should be evident in high-density systems but there should always be a reduction in free volume by the summed molecule volumes.

The general form of the Leonard–Jones potential, as indicated in Figure 6.1, leads to a minimum in the potential energy of interaction between two particles. The depth of the potential energy well ϕ_e, the particle separation r_e at that potential energy minimum, and the distance of closest approach σ are species specific.

Qualitatively, in terms of the above potential function, we anticipate the following behavior. In a dilute gas phase the average separation distance is large compared to the distance of separation r_e when the potential energy is minimized so that attractive interactions are effective only during collisions between molecules. When the kinetic energy of a molecule pair in collision is large compared to the depth of the potential well, ϕ_e, collision produces only a small perturbation in the translational energy. As the gas density increases, increasing the frequency of collision and even leading to a substantial probability of three (or more) body collisions, and/or as the average kinetic energy of colliding molecules decreases, the perturbations on translational energy become larger and, when extended over many simultaneous molecule interactions, can lead to entrapment in a potential well, the form of which is indicated by the Figure 6.1 minimum. Each molecule moves in a potential well the boundaries of which are determined by its interaction with all other molecules in the collection. Translation throughout the entire volume would be replaced by motion within the confines of the potential well, and the system would occupy a near minimum volume configuration. Such (condensed phase) particles are clearly subject to different constraints on their motion than are even strongly interacting gas phase particles. We concentrate in this chapter

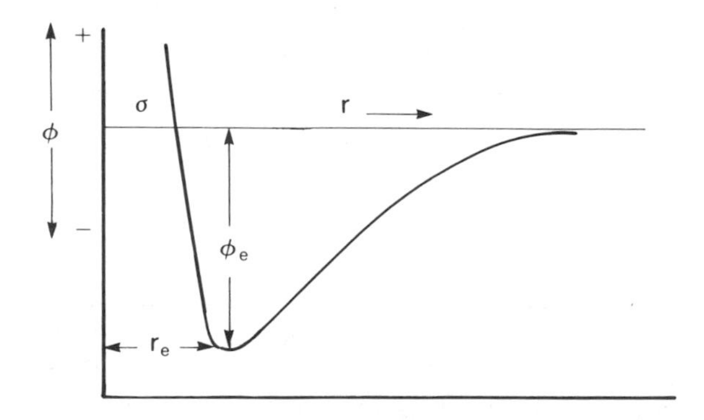

Figure 6.1. Leonard–Jones potential for molecule–molecule interactions.

on interactions in the gas phase recognizing that a different framework will be required for condensed phases.

In dilute (low-pressure) systems at moderate to high temperature any gas should behave ideally. At moderate temperature and pressure, strong attractive forces will reduce the pressure below the ideal value ($PV/N\not{k}T < 1$). At high temperature, with small attractive forces, the gas might behave ideally even at high pressure except for replusive forces that are reflected as a decrease in free volume, $[P(V - Nb]/N\not{k}T = 1]$ where b is an effective molecule volume and $PV/N\not{k}T > 1$. Repulsive forces will eventually dominate the system in all cases at sufficiently high pressure. We summarize the condensation free gas phase behavior in Figure 6.2 in which $PV/N\not{k}T$ is plotted against pressure for $T_1 < T_2 < T_3$. We would imagine that Figure 6.2 would equally well reflect the behavior of gases *A*, *B*, and *C*, all at the same temperature, with attractive forces increasing in the order $A > B > C$. With these general remarks as background we now consider the framework for quantifying the effect of particle–particle interactions using our statistical methods.

The most general expression for gas phase behavior will always be the equation of state and we have obtained a route to its construction via the translational partition function,

$$A_t = -\not{k}T \ln \mathscr{Z}_t \tag{2.48}$$

$$P = -\left(\frac{\partial A}{\partial V}\right)_{T,N}. \tag{2.52}$$

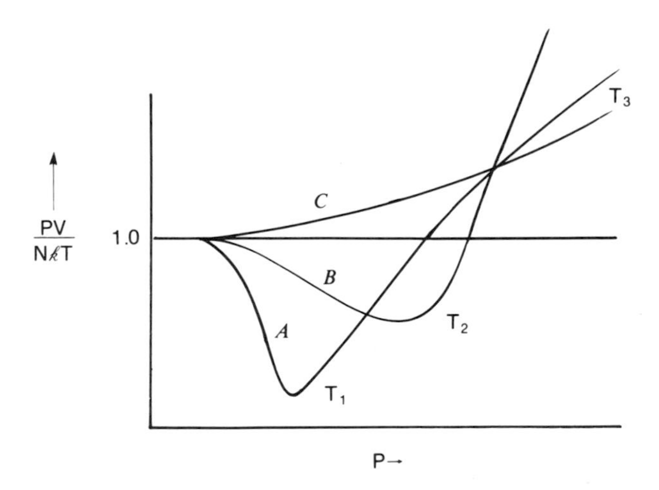

Figure 6.2. Illustration of nonideal gas phase behavior in terms of compressibility for various values of T or for different gases at a common value of T.

Applied to the nonstructured particle system

$$A_t = -kT \ln\left(\frac{2\pi mkT}{h^2}\right)^{3N/2}\left(\frac{Ve}{N}\right)^N \tag{6.2}$$

$$P = NkT\left[\frac{\partial}{\partial V}(\ln V)\right]_{T,N} = NkT/V. \tag{6.3}$$

Since attractive forces are quite short range all gases conform to this behavior in very dilute systems. Generally,

$$\lim_{N/V \to 0} \frac{PV}{NkT} = 1. \tag{6.4}$$

The approach to ideal behavior as the density is descreased will be species specific. The partition function that we construct must reduce to the ideal gas form as density decreases.

In principle the partition function for an N-fold collection of interacting particles cannot be obtained as the N-fold product of a particle partition function. The total energy of the system must include the potential energy arising from the simultaneous interaction of every particle with every other particle. The formal approach to the construction of an appropriate partition function is through the ensemble concept[2] for which we write directly

$$\mathscr{Z} = \sum^{j} e^{-E_j/kT} \tag{6.5}$$

where E_j is any one of the total energy states available to the N-particle system when T and V are specified. This is effectively the constant temperature partition function that we introduced in the previous chapter. We can still think in single system terms since all E_j are effectively the same, i.e., the ensemble is a large collection of energetically identical systems. The partition function of present concern can be regarded as representing the sum over all translational states and all particle–particle interactions in an N particle system at constant temperature and specified volume. All motions other than translation are defined by T and unperturbed by interactions.

Since only translation is involved, we do not have to utilize the quantum structure unless it is most convenient and in this case it is not. We construct the classical N-particle Hamiltonian for the total energy of the system

$$H(p, q) = \frac{1}{2m}\sum_{K=1}^{3N} p_K + Q(q_1, \ldots, q_{3N}) \tag{6.6}$$

[2] See Appendix E on Ensembles.

where the p_K are individual particle momentum components and where Q, a function of the positional coordinates (q) of all particles, represents the potential energy.[3] We will assume that the potential energy can be represented as the sum of pairwise interactions, i.e., each particle interacting independently with every other particle in the system. Then

$$Q(q_1, \ldots, q_{3N}) = \sum_{J>K} U(r_{JK}) \tag{6.7}$$

where r_{JK} is the interparticle distance and the notation $J > K$ in the summation avoids double counting. The dependence of the potential energy on only r_{JK} ignores orientation effects and, in principle, limits the treatment to structurally simple systems. The partition function is a 6-N-fold integral over momentum and coordinate spaces, corrected for quantum considerations (factor of h^{-3N}) and indistinguishability [factor of $(N!)^{-1}$],

$$\mathscr{Z}_t = \frac{1}{N!}\frac{1}{h^{3N}} \int_{6N}\cdots\int e^{-\sum^{3N} p_K/2mkT} e^{-\sum^{J>K} u(r_{JK})/kT} dp_1, \ldots, dp_{3N} dq_1, \ldots, dq_{3N}. \tag{6.8}$$

Integration over the momentum variables should be from $-\infty$ to $+\infty$. When these integrations are carried out over momentum space, the familiar translational expression emerges as a factor and we can write

$$\mathscr{Z}_t = \frac{1}{N!}\Lambda^{-3N} \int_{N}\cdots\int e^{-U(r_{JK})/kT} d\tau_1, \ldots, d\tau_N \tag{6.9}$$

where two notational simplifications have been introduced. First we write

$$\Lambda = \left(\frac{h^2}{2\pi mkT}\right)^{1/2} \tag{6.10}$$

and second we introduce the volume element $d\tau$ as

$$d\tau_1 = dq_1 dq_2 dq_3, \qquad d\tau_2 = dq_3 dq_4 dq_5, \ldots.$$

This notation is commonly encountered.

We now define the configurational integral Q_N in terms of the integral over coordinate space,

$$Q_N = \int_{N}\cdots\int e^{-U(r_{JK})} d\tau_1, \ldots, d\tau_N \tag{6.11}$$

[3] See Appendix F on Classical Considerations.

and consider methods for its evaluation. One such route is classical and follows from the application of the Virial Theorem of Clausius.[4] Another route is based on the evaluation of cluster integrals following a statistical approach.[5] The cluster integrals have a physical meaning in that they represent two body, three body, etc. collisions but the arithmetic is quite tedious and the final form is not intuitively meaningful. Our immediate interests are best served by a much more simple approach that, although not providing the highest level of exactness, does replicate the behavior of real gas systems to a surprising extent and produces useful physical insight.

We assume that the pair-interaction energy between two molecules can be represented by a form such as Eq. (6.1) having a repulsive and an attractive term

$$\phi = \phi_r - \phi_a \tag{6.12}$$

and that the configurational integral [Eq. (6.11)] can be represented as

$$Q_N = \int_N \cdots \int e^{-\phi_r/kT} d\tau_1, \ldots, d\tau_N \int_N \cdots \int e^{\phi_a/kT} d\tau_1, \ldots, d\tau_N. \tag{6.13}$$

The effect of the repulsive energy term in gas phase considerations is to reduce the free volume in which molecules can move.

If a molecule is represented as a hard sphere of radius $r_e/2$ with positional coordinates identified with the center of the sphere, as indicated in Figure 6.3, no other molecule center can enter the spherical volume of radius r_e. The volume excluded to another molecule by any given molecule is $\frac{4}{3}\pi r_e^3$ and the total excluded volume for the N molecule system is $\frac{1}{2}(\frac{4}{3}\pi N r_e^3)$ since each molecule of any pair excludes the other from a like volume. We replace the first N-fold integral by the term $(V - (2\pi/3)N r_e^3) = V_f$.

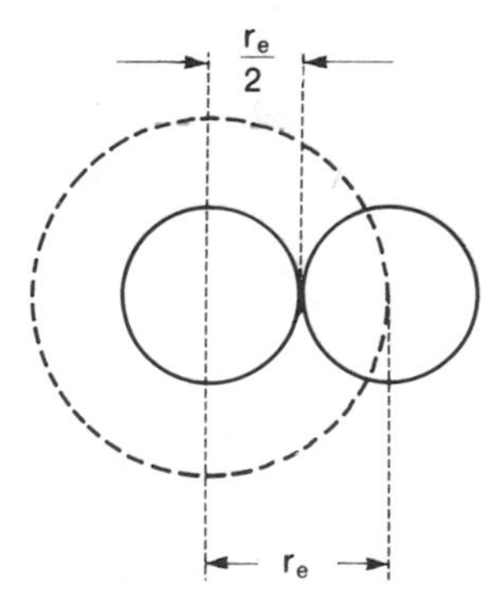

Figure 6.3. Illustration of the volume excluded by one molecule to occupancy by another molecule.

[4] See Appendix D on Intermolecular Forces.
[5] See Appendix C on Statistical Considerations.

The second N-fold integral can be addressed as follows. The potential energy of interaction between two particles separated by distance r is[6]

$$\phi = \frac{\phi_e}{6}\left[12\left(\frac{r_e}{r}\right)^{12} - 6\left(\frac{r_e}{r}\right)^{6}\right] \tag{6.14}$$

if the Leonard–Jones potential is assumed to apply. The parameters r_e and ϕ_e are defined by Figure 6.1. We require only the attractive component of the expression taken over all interactions.

There are no molecules between $r = 0$ and $r = r_e$ away from any single molecule. If the distribution is random between r_e and ∞, there are $(N/V)dV$ molecules, where N/V is the average molecule density, in any volume element in this region. In terms of r, there are $(N/V)d(\frac{4}{3}\pi r^3) = (N/V)4\pi r^2 dr$ other molecules between r and $r + dr$ for $r > r_e$. The attractive potential energy for a single molecule interacting with all others is

$$\phi_{a_{JK}} = \int_{r_e}^{\infty} \phi_e\left(\frac{r_e}{r}\right)^6\left(\frac{N}{V}\right)4\pi r^2 dr. \tag{6.15}$$

Integration produces

$$\phi_{a_{JK}} = \frac{4\pi N}{3V}\phi_e r_e^3 \tag{6.16}$$

and there are N/2 such terms.

The partition function is, then,

$$\mathscr{Z}_t = \frac{1}{N!}\left(\frac{1}{\Lambda}\right)^{3N} V_f^N e^{-N\phi_{a_{JK}}/2kT} \tag{6.17}$$

in which the term $V_f^N e^{-N\phi_{a_{JK}}/2kT}$ replaces V^N in the ideal gas expression. The Helmholz free energy is $(A_t = -kT \ln \mathscr{Z}_i)$,

$$A_t = -kT\left[-N\ln N - N - 3N\ln\Lambda + N\ln\left(V - \frac{2\pi N r_e^3}{3}\right)\right] + \frac{2\pi N^2 \phi_e r_e^3}{3V} \tag{6.18}$$

from which

$$\left(\frac{\partial A_t}{\partial V}\right) = -NkT\left\{\frac{\partial}{\partial V}\left[\ln\left(V - \frac{2\pi N r_e^3}{3}\right)\right]\right\} + \frac{2\pi N^2 \phi_e r_e^3}{3}\frac{\partial(1/V)}{\partial V} \tag{6.19}$$

[6] See Appendix D on Intermolecular Forces.

and

$$P = NkT\left(V - \frac{2\pi N r_e^3}{3}\right) - \frac{2\pi N^2 \phi_e r_e^3}{3V^2}. \tag{6.20}$$

We can simplify this by writing

$$\frac{2\pi\phi_e r_e^3}{3} = a^2, \qquad \frac{2\pi r_e^3}{3} = b \tag{6.21}$$

to obtain

$$\left(P + \frac{N^2 a^2}{V^2}\right)(V - Nb) = NkT \tag{6.22}$$

for an equation of state. Since we would like our final form to reduce to the ideal gas form in the limit of low density we expand Eq. (6.22) in powers of 1/V to obtain

$$P = \frac{NkT}{V} + \frac{N^2 kTb}{V^2} + \frac{N^3 bkT}{V^3} + \cdots - \frac{N^2 a^2}{V^2}$$

$$P = \frac{NkT}{V}\left[1 + \frac{N}{V}\left(b - \frac{a^2}{kT}\right) + \frac{N^2}{V^2}\frac{b^2}{kT} + \cdots\right]. \tag{6.23}$$

The Eq. (6.23) form is now that of the so-called virial equation with

$$P = \frac{NkT}{V}\left[1 + \frac{B(T)}{V} \times \frac{C(T)}{V^2} + \cdots\right] \tag{6.24}$$

with the second virial coefficient $B(T) = N[b - (a^2/kT)]$ and the third virial coefficient $C(T) = N^2 b^2/kT$. These virial coefficient expressions are to be regarded as applying to a gas obeying the pair interaction potential

$$U(r_{JK}) = \infty \qquad r < r_r$$

$$U(r_{JK}) = \phi_e\left(\frac{r_e}{r}\right)^6 \qquad r \geqslant r_e.$$

This particular form (hard sphere, exponential 6) is often referred to as the Southerland potential. The simplified approach that we have followed[7] to produce Eq. (6.22), known as the van der Waals equation of state, is adequate

[7] From end of chapter reference 4, p 286 et. seq.

for portraying the general behavior of nonideal gases and has the advantage of being intuitively satisfying. There is a temperature dependence for the second virial coefficient, and also for the higher virial coefficients. The third virial coefficient is defined for this model but the limitations of our potential function do not lend confidence to its usefulness. The sign of the second virial coefficient depends on both T and the relative magnitudes of the constants a and b so that the equation can represent all of the behavioral features summarized in Figure 6.1. In the form

$$P = \frac{N\ell T}{V - Nb} - \frac{N^2 a^2}{V^2} \tag{6.25}$$

$(V - Nb)$ is the total volume less a term closely associated with the total molecule volume ($b = 4v$) and, from kinetic argument the term $Na/V = a\rho_{qas}$ reflects a correction to the ideal gas pressure, which depends on decreased average molecule velocity due to interaction forces. Since both the frequency of collision with bounding walls and the momentum exchange with them are decreased, the term $a\rho_{qas}$ is squared in the correction to P.

When Eq. (6.22) is expanded, a cubic in volume is obtained,

$$V^3 - \left(Nb + \frac{N\ell T}{P}\right)V^2 + \frac{N^2 a^2 V}{P} + \frac{N^2 a^2 b}{P} = 0. \tag{6.26}$$

We expect solutions to produce three volume values for each pressure value although some of the values may be imaginary or otherwise lack physical meaning. Isothermal pressure–volume plots as shown in Figure 6.4 illustrate a region (e.g., T_1) below T_2 where the cubic equation has three real roots. At $T_2(=T_c)$ the three roots are identical. Above T_2 there is only one real root.

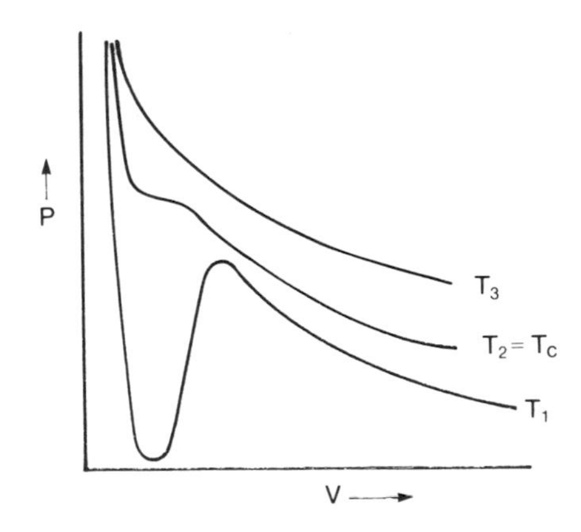

Figure 6.4. Isothermal pressure–volume plots for a van der Waals gas.

The T_2 isotherm is unique ($=T_c$) while the T_1 and T_3 isotherms are representative of general forms at temperatures below and above T_c.

At all values of T there is a region of small volumes in which $(\partial V/\partial P)_T \rightarrow 0$. Below $T = T_c$ we would regard this virtually incompressible region as a condensed phase. Above T_c, which we will refer to as the critical temperature, there is no clear breaking point between gas phase behavior (isotherm forms at large V) and the condensed phase.[8] We should regard this (postcritical) region as gaseous even where the volume requirements match those for condensed phases. Below the critical temperature there are three values of volume for any fixed pressure. One of these values lies on the condensed phase isotherm and the other lies on the gas phase isotherm. The third volume value is associated with the region of the isotherm in which $(\partial V/\partial P)_T > 0$. This is physically unreal, reflecting the inability of any general model to describe the condensation process. However, Eq. (6.26) also fails to provide the most crucial feature of the process, i.e., the single pressure value that would characterize the two-phase system at a given temperature. Fortunately there are thermodynamic generalizations relating to the equilibrium state that will resolve this question. Also there are mathematical techniques which will produce useful extensions for the critical state. We will first address the two-phase circumstance.

Any equation of state applies only to an equilibrium system. Although we will consider the general problem of phase equilibrium in extended detail in a later chapter, we already understand that an equilibrium state must be characterized by uniform and constant values of P, T, and V and the free energy for the equilibrium system is minimum valued, i.e.,

$$(dA)_{T,N} = -P dV = 0.$$

For any arbitrary infinitesimal variation in volume at any point within the Figure 6.5a dashed line (two-phase) region $dA = 0$ and ΔA for the gas phase to condensed phase transition must be zero, i.e.,

$$\Delta A = \int_g^c dA = -\int_g^c P dV = 0.$$

This requirement is met when P is valued such that positive and negative contributions to the integral cancel, i.e., when area (1) Figure 6.5a and area (2) are equal and P_1 is the unique pressure value that accomplishes this result.

A series of such constructs at increasing values of T will produce the general form shown in Figure 6.5b. The two-phase region with be represented by an envelope with form characterized by the requirement that condensation occur at increasingly higher gas phase density (smaller volume) as the temperature (kinetic energy) increases given a fixed value of the attractive force constant

[8] Experimentally one can observe the liquid–vapor interface vanish and reappear as a near liquid filled sealed glass tube is cycled through the critical temperature.

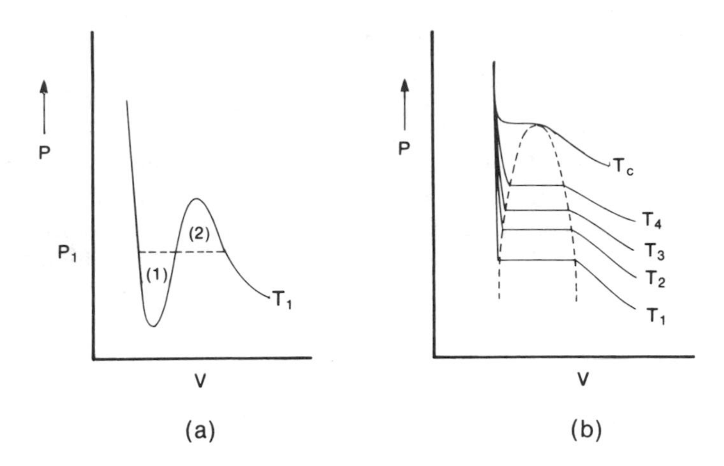

Figure 6.5. Illustration of (a) the two-phase equilibrium requirements and (b) the envelope of two-phase equilibria for various values of T within the region of two-phase stability.

"a." The two-phase region vanishes in a point at $T = T_c$ reflecting the circumstance that attractive interactions cannot produce condensation when the kinetic energy of the molecule system is sufficiently large even though the system volume is effectively at the minimum value ($V \rightarrow Nb$). The critical isotherm is a curve with point of inflection and exhibits all anticipated mathematical properties. We examine these as follows.

When a point of inflection is encountered, both the first and second derivatives are zero. In this circumstance

$$(\partial P/\partial V)_{T_c} = 0 = (\partial^2 P/\partial V^2)_{T_c}$$

and from Eq. (6.25)

$$\left(\frac{\partial P}{\partial V}\right)_{T_c} = -\frac{N k T_c}{(V_c - Nb)^2} + \frac{2N^2 a^2}{V_c^2} = 0 \tag{6.27}$$

$$\left(\frac{\partial^2 P}{\partial V^2}\right)_{T_c} = \frac{2N k T_c}{(V_c - Nb)^3} - \frac{6N^2 a^2}{V_c^4} = 0. \tag{6.28}$$

Solving each expression for T_c and dividing one equation by the other produces

$$\frac{T_c}{2T_c} = \left[\frac{2N^2 a^2}{V_c^3}\frac{(V_c - Nb)^2}{N k}\right]\frac{V_c^4 N k}{(V_c - Nb)^3\, 6N^2 a^2}. \tag{6.29}$$

Solving, $V_c = 3Nb$ and, from Eq. (6.27),

$$T_c = \frac{8}{27}\frac{N^2 a^2}{(Nb)(Nk)} = \frac{8}{27}\frac{a^2}{kb}. \tag{6.30}$$

Also from $V_c = 3Nb$ together with Eqs. (6.25) and (6.30),

$$P_c = \frac{N^2a^2}{27N^2b^2} = \frac{1}{27}\frac{a^2}{b^2}. \tag{6.31}$$

The critical parameters are fixed by the two interaction constants.

This result has interesting implications in that, if reduced parameters are defined such that $P_r = P/P_c$, $T_r = T/T_c$, and $V_r = V/V_c$, then for any gas $P = P_cP_r$, $T = T_cT_r$, and $V = V_cV_r$ and for the van der Waals gas

$$P = \frac{NkT}{V - Nb} + \frac{N^2a^2}{V^2} = \frac{NkT_cT_r}{V_cV_r - Nb} + \frac{Na^2}{(V_cV_r)^2}$$

$$P_cP_r = \frac{Nk\frac{8}{27}\frac{a^2}{kb}T_r}{3NbV_r - Nb} + \frac{N^2a^2}{(3Nb)^2V_r^2}$$

$$\frac{a^2}{27b^2}P_r = \frac{8T_ra^2}{27b^2(3V_r - 1)} - \frac{a^2}{9b^2V_r^2}$$

$$P_r = \frac{8T_r}{3V_r - 1} - \frac{3}{V_r^2}. \tag{6.32}$$

As anticipated the interaction constants do not appear in this dimensionless expression. All van der Waals gases at the same reduced pressure and temperature occupy the same reduced volume. We refer to this circumstance as the principle of corresponding states.

For many purposes it will be useful to define a compressibility factor that, in its deviation from unity, reflects the nonideal behavior of the gas (as in Figure 6.2)

$$Z = \frac{PV}{NkT} \tag{6.33}$$

where P and V are measured parameters. For a van der Waals gas $Z_r = \frac{3}{8}(P_rV_r/NkT_r)$. Figure 6.6 indicates the extent to which real gases, even in those cases for which the simplifying assumptions that are part of the above treatment do not hold, conform to the principle of corresponding states.

Many more elaborate equation of state forms can be constructed both theoretically and empirically and vastly more elaborate statistical treatments can be constructed. Such treatments would be regarded as the preferred origin for general theoretical extension. They produce a solution for the second virial coefficient of the form

$$B(T) = 2\pi \int_0^\infty r^2(1 - e^{-U(r_{JK})})dr \tag{6.34}$$

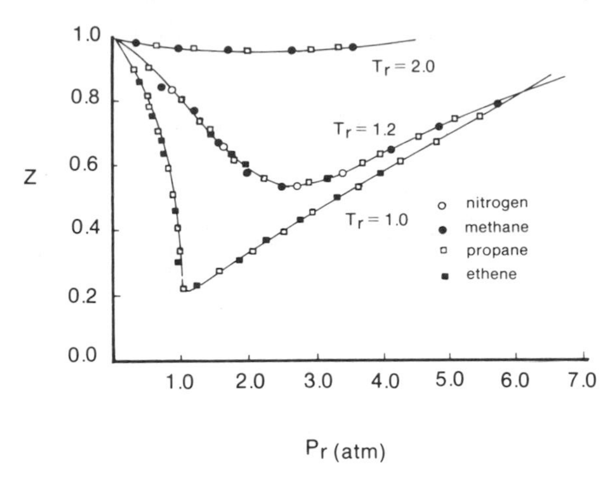

Figure 6.6. Experimental compressibility factors in reduced parameter form for various gases. From *Physical Chemistry* 2/E. By P. W. Atkins. Copyright(c) 1978, 1982 by P. W. Atkins. Reprinted with permission by W. H. Freeman and Company.

where $U(r_{JK})$ is again the pair interaction energy. Unfortunately, close examination using any reasonable potential function to obtain the virial coefficients shows that, for gases at moderate densities, successive terms in the expansion diminish quickly and the series converges. For gases at high density and in the liquid condensed phase region the series does not converge. Other treatment is required.

Other simple and not so simple equations of state forms can be constructed on the basis of other potential functions or from empirical arguments. The common equations of state are given in Table 6.1 together with their critical dependencies and their reduced forms. Table 6.2 presents van der Waals constants and critical parameters for a few common gases to illustrate orders of magnitude.[9] From such data we could obtain general expressions for thermodynamic functions for, say, a van der Waals gas and numerical values for any specific substance for which the constants or the critical parameters are available. In the material that immediately follows we concentrate on the role of the equation of state in extending thermodynamic calculations. But, as we will see, in the empirical framework, there are other reflections of nonideal gas behavior that are sometimes more effective.

First we note that if Eq. (6.18) is rewritten in terms of the van der Waals constants we have

$$\mathrm{A_t} = 3\mathrm{N}k\mathrm{T}\ln\Lambda - \mathrm{N}k\mathrm{T}\ln\left(\frac{\mathrm{V}}{\mathrm{N}} - b\right) + \frac{\mathrm{N}^2 a^2}{\mathrm{V}} - \mathrm{N}k\mathrm{T} \tag{6.35}$$

[9] The Table 6.2 van der Waals constants are molar values. Our derivations are in terms of molecular values.

TABLE 6.1. Some Equations of State and Reduced Forms Where Applicable[a]

Berthelot Equation

$$P = RT(\tilde{V} - b) - a/T\tilde{V}$$

$$P_c = \tfrac{1}{112}(2aR/3b^3), \qquad \tilde{V}_c = 3b, \qquad T_c = \tfrac{2}{3}(2a/3bR)^{1/2}$$

$$P_r = 8T_r(3V_r - 1) - 3T_r/V_r$$

Dieterici Equation

$$P = [RT/\tilde{V} - b]\exp(-aRT/\tilde{V})$$

$$P_c = a/4e^2b^2, \qquad \tilde{V}_c = 2b, \qquad T_c = a/4bR$$

$$P_r = e^2T_r(2V_r - 1)\exp(-2T_rV_r)$$

Beattie–Bridgeman Equation

$$P = (1 - \gamma)RT(\tilde{V} - \beta)\tilde{V}^2 - \alpha\tilde{V}^2$$

$$\alpha = a_0(1 - a/\tilde{V}), \qquad \beta = b_0(1 - b\tilde{V}), \qquad \gamma = c_0/\tilde{V}T^3$$

[a]The empirical constants a_0, a, b, b_0, and c_0 are independent of T and V.

TABLE 6.2. Van der Waals Constants and Critical Parameters for Selected Gases

Substance	a (liter^2 atm mol^{-1})	$b \times 10^2$ (liter mol^{-1})	T_c(°C)	P_c(atm)
He	0.03412	0.02370	−269.82	1.18
Ar	1.345	0.03219	−122.4	48.8
H_2	0.2444	0.02661	−240.0	13.0
N_2	1.390	0.03912	−146.9	34.0
O_2	1.360	0.03183	−118.5	50.8
NH_3	4.170	0.03707	132.4	113.1
H_2O	5.464	0.03049	374.2	218.3
CH_4	2.253	0.04278	−97.6	45.6

From a collection of data in *The Handbook of Chemistry and Physics*, 67th ed., Chemical Rubber Press.

or with

$$N\not{k}T\ln\left(\frac{V}{N} - b\right) = N\not{k}T\ln\left(1 - \frac{Nb}{V}\right) + N\not{k}T\ln\frac{V}{N}$$

and expanding $\ln[1 - (Nb/V)]$ as $\ln(1 - x) = x - x^2 + x^3 + \cdots$

$$A_t = 3N\not{k}T\ln\Lambda - N\not{k}T\ln\frac{V}{N} - N\not{k}T\left[1 + \frac{N}{V}\left(b - \frac{a^2}{\not{k}T}\right) + \frac{N^2b^2}{V^2\not{k}T} + \cdots\right]$$

$$= \left[-N\not{k}T\ln\left(\frac{2\pi m\not{k}T}{n^3}\right)^{3/2}\frac{V}{N}\right] - PV \tag{6.36a}$$

with PV given by Eq. (6.23). The conclusion is general; expressions involving the PV product will deviate from the ideal gas expressions by the terms

$$\mathrm{N}k\mathrm{T}\left[\frac{\mathrm{N}}{\mathrm{V}}\left(b-\frac{a^2}{k\mathrm{T}}\right)+\frac{\mathrm{N}^2b^2}{\mathrm{V}^2k\mathrm{T}}+\cdots\right] \tag{6.36b}$$

for a van der Waals gas or generally by

$$\mathrm{N}k\mathrm{T}\left[\frac{B(\mathrm{T})}{\mathrm{V}}+\frac{C(\mathrm{T})}{\mathrm{V}^2}+\cdots\right] \tag{6.36c}$$

where $B(\mathrm{T})$ and $C(\mathrm{T})$ are virial coefficients obtained by any route.

There will be other deviations from ideal gas expressions when energy is involved,

$$\mathrm{E}=k\mathrm{T}^2\left(\frac{\partial \ln \mathscr{Z}}{\partial \mathrm{T}}\right)_{\mathrm{V,N}}$$

and, where the van der Waals constants do not have temperature dependence,

$$\begin{aligned}\mathrm{E_t}&=k\mathrm{T}^2\left(\frac{\partial \ln \mathrm{T}}{\partial \mathrm{T}}\right)^{3\mathrm{N}/2}_{\mathrm{V,N}}-k\mathrm{T}^2\left(\frac{\partial \mathrm{N}^2a^2/k\mathrm{TV}^2}{\partial \mathrm{T}}\right)_{\mathrm{V,N}}\\&=\frac{3}{2}\mathrm{N}k\mathrm{T}+\frac{\mathrm{N}^2a^2}{\mathrm{V}^2}.\end{aligned} \tag{6.37}$$

From the definition $\mathrm{H}=\mathrm{E}+\mathrm{PV}$ and the PV product from Eq. (6.23) together with Eq. (6.37) produces

$$\mathrm{H_t}=\frac{3}{2}\mathrm{N}k\mathrm{T}+\frac{\mathrm{N}^2a^2}{\mathrm{V}}+\mathrm{N}k\mathrm{T}\left[1-\frac{\mathrm{N}}{\mathrm{V}}\left(b-\frac{a^2}{k\mathrm{T}}\right)+\cdots\right] \tag{6.38}$$

for the van der Waals gas. The translational entropy follows from $k\ln \mathscr{Z}_{\mathrm{t}}+(\mathrm{E_t}/\mathrm{T})$ as

$$\mathrm{S_t}=3\mathrm{N}nk\ln\Lambda^{-1}+\mathrm{N}k\ln\left(\frac{\mathrm{V}}{\mathrm{N}}-b\right)+\frac{5}{2}\mathrm{N}k. \tag{6.39}$$

From definition,

$$\mathrm{C_V}=\left(\frac{\partial \mathrm{E}}{\partial \mathrm{T}}\right)_{\mathrm{V,N}}$$

and

$$\mathrm{C_{V,t}}=\frac{\partial}{\partial \mathrm{T}}\left[\frac{3}{2}\mathrm{N}k\mathrm{T}+\frac{\mathrm{N}^2a^2}{\mathrm{V}^2}\right]=\frac{3}{2}\mathrm{N}k$$

as for the ideal gas. This will be true for any model in which the interaction constants are not temperature dependent. Also

$$C_P = \left(\frac{\partial H}{\partial T}\right)_{P,N}$$

and

$$C_{P,t} = \frac{\partial}{\partial T}\left[\frac{3}{2}N\ell + \frac{N^2 a^2}{V} + N\ell T\left(1 - \frac{N}{V}\left(b - \frac{a^2}{\ell T}\right) + \cdots\right)\right]$$

$$= \frac{5}{2}N\ell - \frac{N^2 \ell b}{V}. \tag{6.40}$$

These statistical considerations are reflected in the adiabatic work related term γ_t as

$$\frac{C_{P,t}}{C_{V,t}} = \gamma_c = \frac{\frac{5}{2}N\ell + (N^2 \ell b/V)}{\frac{3}{2}N\ell} = \frac{5}{3} + \frac{2}{3}\frac{Nb}{V} \tag{6.41}$$

and in $C_{P,t} - C_{V,t}$ as $N\ell[1 - (Nb/V)]$ rather than just $N\ell$.

From definition, $G = A + PV (= H - TS)$

$$A_t + PV = -N\ell T \ln\left(\frac{2\pi m \ell T}{n^2}\right)^{3/2}\frac{V}{N} = G_t \tag{6.42}$$

$$\mu_t = \frac{G_t}{N} = -\ell T \ln\left(\frac{2\pi m \ell T}{n^2}\right)^{3/2} - \ell T \ln\frac{V}{N} \tag{6.43}$$

as with the ideal gas but V/N can no longer be represented by a simple pressure relationship. To provide a general expression for chemical potential in a structured particle system in which nonideal behavior would be anticipated we can still collect the obvious temperature dependent structure related terms and a pressure volume conversion factor into a standard chemical potential form

$$N_A\mu = \tilde{\mu}(T) = N_A\mu^0(T) + N_A \ell T \ln\frac{\ell T N_A}{V}$$

$$N_A\mu^0(T) = -N_A \ell T \ln\left[(\ell T)^{-1}\Lambda^{-3/2}(1.013\times 10^6)\omega_{0,e}\sigma_r^{-1} f\left(\frac{\Theta_r}{T}\right)\prod^{K}(1 - e^{-\Theta_{V,K}/\ell T})\right]. \tag{6.44}$$

It will be recalled that the molar value construction was adopted to introduce pressure in a meaningful way into the simplified expression for chemical potential. Since real gas behavior is a feature of the particle collection, it will be even more important that we phrase our chemical potential arguments in molar terms. For this purpose Eq. (6.44) contains all temperature-dependent terms for ideal gases and these will also be present in the same form for real gases. The expression can be retained as a convenience which reflects temperature dependence other than that reflected in nonideal behavior.

To represent V/N for the real gas system we require a virial type expansion in powers of P rather than V^{-1}. Then

$$\frac{V}{N} = \frac{kT}{P}(1 + B'(T)P + C'(T)P^2 + \cdots) \tag{6.45}$$

where, again, the virial coefficients can be related to interaction constants for any potential function that we have the patience to pursue. For the van der Waals potential, as we have defined it, simple arithmetic will demonstrate that $B' = B/N_A kT$ and, to a first approximation, $C' = (C - B^2)(N_A kT)^2$. For most purposes of argument the first virial coefficient is sufficient. In all cases, the second virial coefficient suffers from model deficiencies; an approximation is as good as the circumstances warrant.

With these arguments in mind, we can construct an expression for the real gas molar chemical potential,

$$\tilde{\mu}(T) = N_A\mu^0(T) - N_A kT \ln\left[\frac{kT}{P}(1 + B'(T)P + C'(T)P^2 + \cdots)\right] \tag{6.46}$$

with the real gas differing from the ideal gas in pressure dependence of chemical potential by the term

$$-kT \ln[B'(T)P + C'(T)P^2 + \cdots]$$

and with $\mu^0(T)$ defined by Eq. (6.44) but no longer reflecting the entire temperature dependence.

If we wish to retain a general formulation that will be adaptable to all systems of concern we can always start with the defining expression

$$G = N\mu - kT \ln \mathscr{Z} + PV.$$

For the general nonideal gas, from Eq.(6.17)

$$\tilde{\mu}(T) = N_A\mu^0(T) + N_A kT\left[\ln\frac{V_f}{N_A} - \frac{\phi_{JK}}{2kT}\right] \tag{6.47}$$

we retain the choice of attractive potential although we are still limited to the hard sphere constraint. As a general formulation this approach will introduce some problems in dealing with phase equilibria. Operationally we should construct a route that relies on experimental data but it would be a large convenience to construct a formulation that retains the simple ideal gas form. The activity concept accomplishes this objective.

Consider the implications of retaining the expression

$$\mu(T) = \mu^0(T) + kT \ln P$$

where P is the real system pressure. If the pressure is near zero the gas behaves ideally and P is the ideal gas pressure. At higher pressures $P \neq NkT/V$ but $(NkT/V)[1 + f(V/N)]$. However we can write

$$\tilde{\mu}(T) = \tilde{\mu}^0(T) + N_A kT \ln \gamma P \tag{6.48}$$

for all circumstances where we refer to γ as an activity coefficient. It is then the requirement that $\gamma \to 1$ as $P \to 0$ and that throughout the range of pressure and for all temperatures Eq. (6.48) describes the chemical potential. The activity coefficient, which is species specific and temperature dependent, is defined by real gas chemical potential and by the standard state definition that is unchanged from Eq. (6.44) and only a function of T. In common practice γ is further defined by the expression

$$\gamma = \frac{\mathfrak{f}}{P} \tag{6.49}$$

so

$$\tilde{\mu} = \tilde{\mu}^0(T) + N_A kT \ln \mathfrak{f}. \tag{6.50}$$

The term $\mathfrak{f}$, which is called fugacity, is that representation of the Eq. (6.47) bracketed expression which, coupled with $\tilde{\mu}^0(T)$, produces the real gas chemical potential at T and P. It cannot be a simple function of P.

For the N_A particle ideal gas from Eq. (6.44)

$$d\tilde{\mu}(T) = N_A d\mu(T) = N_A kT d \ln P = \tilde{V}_{\text{Ideal}} dP. \tag{6.51}$$

For the nonideal gas we require that

$$d\tilde{\mu}(T) = N_A kT d \ln \mathfrak{f} = \tilde{V}_{\text{real}} dP = \frac{N_A kT}{P} Z dP \tag{6.52}$$

where Z is the compressibility factor. Then

$$N_A kT d \ln \frac{\mathfrak{f}}{P} = \left[\frac{N_A Z kT}{P} - \frac{N_A kT}{P} \right] dP$$

$$\int_0^P d \ln \frac{\mathfrak{f}}{P} = \ln \frac{\mathfrak{f}}{P} \bigg]_0^P = \ln \frac{\mathfrak{f}}{P} = \int_0^P \frac{1}{P} (Z-1) dP = \int_0^P (Z-1) d \ln P \tag{6.53}$$

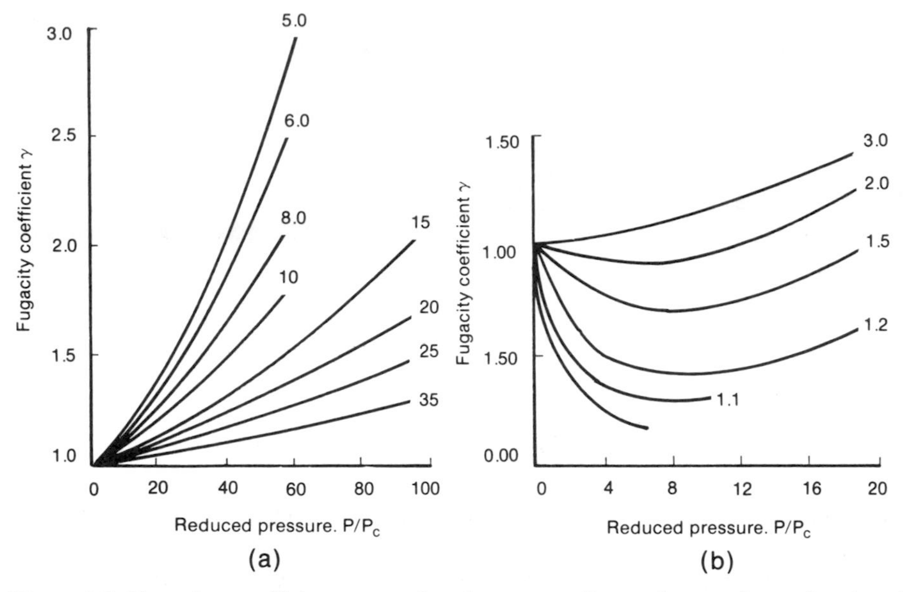

Figure 6.7. Fugacity coefficients vs reduced pressure for various values of reduced temperature. From *Physical Chemistry* 2/E. By P. W. Atkins. Copyright(c) 1978, 1982 by P. W. Atkins. Reprinted with permission by W. H. Freeman and Company.

since $\mathfrak{f}/P \to 1$ as $P \to 0$. The fugacity is, therefore, defined in terms of P as

$$\mathfrak{f} = P \exp\left[\int_0^P (Z - 1) d \ln P\right]. \tag{6.54}$$

Compressibility data in terms of reduced parameters are generally available. Appropriate graphic integration of a curve such as Figure 6.6 can produce fugacity values that are adequate for most purposes. The term that we have called the activity coefficient is often referred to as the fugacity coefficient. Typical γ vs P plots in reduced parameter form are shown in Figure 6.7 for illustrative purposes.

One of the real advantages of having a reasonably reliable equation of state is that derivative functions involving P, V, T, and N can be readily obtained in analytic form. Such expressions have particular utility when they are related to the system entropy through a set of equalities known as Maxwell relationships. Since our variables E, H, A, V, and G are state functions and therefore conform to the mathematical criteria applied to perfect differentials, the Euler criterion for differentiability[10] must apply. Then, from (closed system)

$$dE = TdS - PdV$$

[10] See Appendix B on Mathematical Considerations.

it is required that

$$(\partial T/\partial V)_S = -(\partial P/\partial S)_V.$$

When also applied to (closed system)

$$dH = TdS + VdP$$

$$dA = -SdT - PdV$$

$$dG = -SdT + VdP$$

the Table 6.3 collection of differential equalities follows. The relationships in Table 6.3 are not confined to gas phases and they have the further advantage of being experimentally accessible. Actually, not exactly the above set of equation of state terms, but various compressibilities are the normally measured quantities.

TABLE 6.3. Maxwell Relationships

$(\partial T/\partial V)_S = -(\partial P/\partial S)_V$	$(\partial P/\partial T)_V = (\partial S/\partial V)_T$
$(\partial T/\partial P)_S = (\partial V/\partial S)_P$	$(\partial V/\partial T)_P = -(\partial S/\partial P)_T$

We define, for isothermal, isobaric and adiabatic systems

$$\alpha_P = \frac{1}{V}(\partial V/\partial T)_P \tag{6.55}$$

calling α_P the isobaric compressibility,

$$\kappa_T = -\frac{1}{V}(\partial V/\partial P)_T \tag{6.56}$$

calling κ_T the isothermal compressibility, and

$$\kappa_S = -\frac{1}{V}(\partial V/\partial P)_S \tag{6.57}$$

calling κ_S the adiabatic compressibility. The Maxwell term $(\partial V/\partial T)_P$ is $V\alpha_P$. The remaining three terms are related to compressibilities through heat capacities and the general properties of partial derivatives.[11] Applied to the variables P,

[11] See Appendix B on Mathematical Considerations.

V and T in a closed system,

$$\left(\frac{\partial P}{\partial V}\right)_T \left(\frac{\partial V}{\partial T}\right)_P \left(\frac{\partial T}{\partial P}\right)_V = -1$$

and, $(\partial P/\partial V)_T = 1/(\partial V/\partial P)_T$ so that

$$\left(\frac{\partial T}{\partial P}\right)_V = -\frac{(\partial V/\partial P)_T}{(\partial V/\partial T)_P} = \frac{\kappa_T}{\alpha_P}. \tag{6.58}$$

Following the same arithmetic procedures, with

$$\left(\frac{\partial S}{\partial T}\right)_V \left(\frac{\partial T}{\partial V}\right)_S \left(\frac{\partial V}{\partial S}\right)_T = -1$$

the Maxwell identity $(\partial S/\partial V)_T = (\partial P/\partial T)_V$, the Eq. (6.58) result and the equality $(\partial S/\partial T)_V = C_V/T$

$$\left(\frac{\partial T}{\partial V}\right)_S = -\frac{\alpha_P T}{\kappa_T C_V}. \tag{6.59}$$

Also with

$$\left(\frac{\partial S}{\partial T}\right)_P \left(\frac{\partial T}{\partial P}\right)_S \left(\frac{\partial P}{\partial S}\right)_T = -1$$

the Maxwell identity $(\partial S/\partial P)_T = -(\partial V/\partial T)_P$, Eq. (6.55) defining α_P and the equality $(\partial S/\partial T)_P = C_P/T$

$$\left(\frac{\partial T}{\partial P}\right)_S = \frac{V\alpha_P T}{C_P}. \tag{6.60}$$

The adiabatic compressibility follows from the defining expression, Maxwell relationships, Eqs. (6.59) and (6.60) together with

$(\partial S/\partial P)_V(\partial P/\partial V)_S(\partial V/\partial S)_P = -1$

$$\kappa_S = \frac{1}{V}\left[\frac{(\partial S/\partial P)_V}{(\partial S/\partial V)_P}\right] = -\frac{1}{V}\frac{(\partial V/\partial T)_S}{(\partial P/\partial T)_S} = \kappa_T\frac{C_V}{C_P}. \tag{6.61}$$

The Maxwell relationships and derived expressions can be directly expressed in terms of volume and the equation of state constants. Where we use the van der Waals form,

$$P = \frac{NkT}{V - Nb} - \frac{N^2a^2}{V^2}$$

κ_T follows from differentiation

$$\kappa_T = -\frac{1}{V}[1/(\partial P/\partial V)_T]$$
$$= \frac{1}{V}\left[\frac{NkT}{(V - Nb)^2} - \frac{2N^2a^2}{V^3}\right]^{-1} \tag{6.62}$$

and α_P follows from differentiation of

$$T = \frac{1}{Nk}\left[P(V - Nb) + \frac{N^2a^2}{V^2}(V - Nb)\right]$$

to give

$$\alpha_P = \frac{1}{V}[1/(\partial T/\partial V)_P]$$
$$= \frac{1}{NkV}\left[\frac{NkT}{(V - Nb)} - \frac{2N^2a^2}{V^2} - \frac{2N^3a^2b}{V^3}\right]^{-1}. \tag{6.63}$$

The Maxwell relationship term $(\partial P/\partial T)_V$ follows from either Eq. (6.22) or Eq. (6.35)

$$(\partial P/\partial T)_V = \frac{Nk}{V - Nb}. \tag{6.64}$$

From the closed system, first and second law expression

$$dE = TdS - PdV$$

we have

$$(\partial E/\partial V)_T = T(\partial S/\partial V)_T - P = T(\partial P/\partial T)_V - P \tag{6.65}$$

which reduces to N^2a^2/V^2 when the van der Waals form is inserted for P and to zero when NkT/V is inserted for P. The Eq. (6.65) form may be regarded as a true, thermodynamic equation of state, valid for any substance. On the basis of Eq. (6.58) it reduces to

$$(\partial E/\partial V)_T = T\left(\frac{\alpha_P}{\kappa_T}\right) - P. \tag{6.66}$$

Also, from

$$dE = (\partial E/\partial T)_V\, dT + (\partial E/\partial V)_T dV$$

$$(\partial E/\partial T)_P = (\partial E/\partial T)_V + (\partial E/\partial V)_T(\partial V/\partial T)_P = C_V + (\partial E/\partial V)_T V\alpha_P \tag{6.67}$$

with $(\partial E/\partial V)_T$ given by Eq. (6.66).

A general expression for the difference in constant pressure and constant volume heat capacities follows from

$$C_P = \left(\frac{\partial H}{\partial T}\right)_P = \left(\frac{\partial E}{\partial T}\right)_P + \left(\frac{\partial (PV)}{\partial T}\right)_P = \left(\frac{\partial E}{\partial T}\right)_P + P\left(\frac{\partial V}{\partial T}\right)_P$$

with

$$\left(\frac{\partial E}{\partial T}\right)_P = \left(\frac{\partial E}{\partial T}\right)_V + \left(\frac{\partial E}{\partial V}\right)_T\left(\frac{\partial V}{\partial T}\right)_P = C_V + \left(\frac{\partial E}{\partial V}\right)_T\left(\frac{\partial V}{\partial T}\right)_P$$

so that

$$C_P - C_V = \left(\frac{\partial V}{\partial T}\right)_P\left[P + \left(\frac{\partial E}{\partial V}\right)_T\right]. \tag{6.68}$$

The square bracketed terms are equivalent to $T(\partial P/\partial T)_V$ via Eq. (6.65), so that

$$C_P - C_V = T\left(\frac{\partial V}{\partial T}\right)_P\left(\frac{\partial P}{\partial T}\right)_V$$

which, with Eq. (6.58), provides

$$C_P - C_V = \frac{\alpha_P^2 VT}{\kappa_T}. \tag{6.69}$$

The ratio $C_P/C_V = \gamma$ we have obtained [Eq. (6.61)] as the ratio of isothermal and adiabatic compressibilities.

Actually, expressions involving enthalpy are of more interest than those involving energy since many thermodynamic concerns will center on systems in which temperature and pressure are process variables. From

$$dH = C_P dT + (\partial H/\partial P)_T dP$$

we obtain

$$\left(\frac{\partial H}{\partial T}\right)_V = C_P + \left(\frac{\partial H}{\partial P}\right)_T\left(\frac{\partial P}{\partial T}\right)_V. \tag{6.70}$$

Using Eqs. (6.58) and (6.66) and with $H = H(T, P)$

$$\left(\frac{\partial H}{\partial P}\right)_T = \frac{(\partial H/\partial T)_P}{(\partial P/\partial T)_H} = -C_P\mu_{JT} \quad (6.71)$$

$$\left(\frac{\partial H}{\partial T}\right)_V = C_P\left[1 + \frac{\alpha_P}{\kappa_T}\mu_{JT}\right] \quad (6.72)$$

where we have introduced the notation $(\partial T/\partial P)_H = \mu_{JT}$, which is referred to as the Joule–Thompson coefficient. From the defining expression for enthalpy

$$\left(\frac{\partial H}{\partial T}\right)_V = C_V + V\left(\frac{\partial P}{\partial T}\right)_V \quad (6.73)$$

and $(\partial P/\partial T)_V$ is, from Eq. (6.58), α_P/κ_T so that Eq. (6.72) can be rephrased as

$$(C_V - C_P)\frac{\kappa_T}{\alpha_P} = C_P\mu_{JT}.$$

The term $(C_V - C_P)$ is, from Eq. (6.69), $\alpha_P^2 VT/\kappa_P$ so that

$$\mu_{JT} = (\alpha_P T + V)/C_P.$$

This term is a very direct reflection of nonideality in the gas phase in the same way that $(\partial E/\partial V)_T$ is a direct reflection of nonideality but μ_{JT} can be measured directly with reasonable accuracy. Since the physical meaning of $(\partial T/\partial P)_H$ is not immediately obvious, the experiment requires some discussion.

Consider the following process. A real gas is confined adiabatically at $P_1V_1T_1$ then expanded slowly through a porous boundary to $P_2V_2T_2$ as indicated in Figure 6.8 (P_2 is very slightly lower than P_1). Since, throughout the expansion, P_1 and T_1 are constants in system 1, the work done on system 2, is $-P\Delta V = P_1V_1$. Similarly the work done by system 1 is $-P_2V_2$ and, in the overall adiabatic system, $\Delta E_{process} = \Delta W_{process} = P_1V_1 - P_2V_2$; $E_2 + P_2V_2 = E_1 + P_1V_1$; $H_2 = H_1$. The ΔT associated with the change in internal energy can be associated

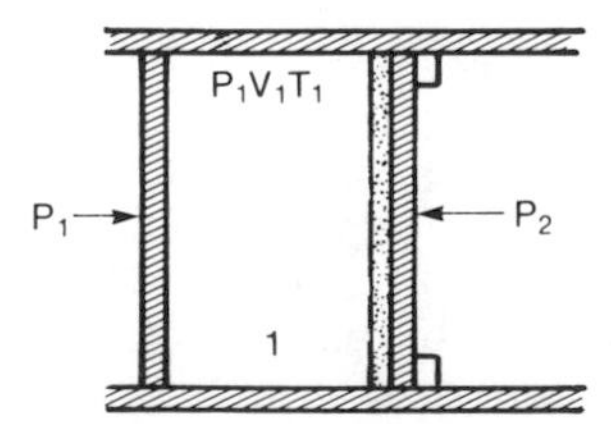

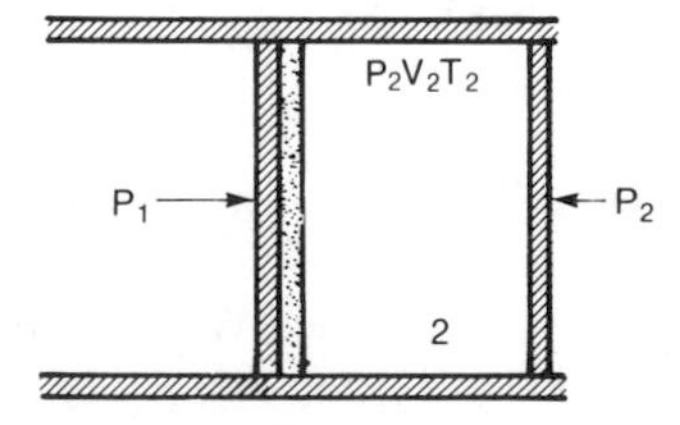

Figure 6.8. Illustration of the Joule–Thompson expansion process.

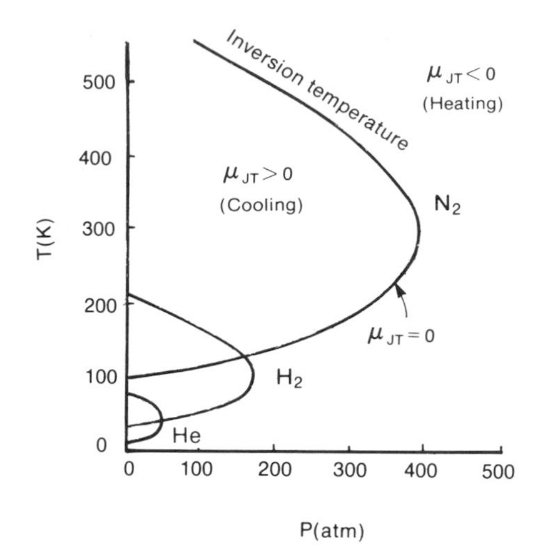

Figure 6.9. Domains of positive and negative Joule–Thompson coefficient in PT space. From *Physical Chemistry* 2/E. By P. W. Atkins. Copyright(c) 1978, 1982 by P. W. Atkins. Reprinted with permission by W. H. Freeman and Company.

with the given ΔP and this process is carried out at constant enthalpy. Then

$$\mu_{JT} = \left(\frac{\Delta T}{\Delta P}\right)_H = \left(\frac{\partial T}{\partial P}\right)_H. \tag{6.74}$$

We expect that real gases may show both positive and negative values of μ_{JT} depending on whether we are in a region where repulsive forces dominate or in a region where attractive forces dominate. Reference to Figure 6.2 shows that, for any real gas, there is a P, T domain in which repulsive forces dominate and P, T domain in which attractive forces dominate. At the crossover point $PV = N\not{k}T$, $\mu_{JT} = 0$. The temperature at which this occurs for any real pressure value P is referred to as the inversion temperature. It is species specific. Figure 6.9 illustrates this behavior for typical gases.

In closing this chapter, it should be noted that, where possible, experimental values for the various derivative terms would be used in calculation. Model-based equations of state are useful in many ways but are somewhat weak with respect to the accuracy of derivative expressions. The exercises, however, are informative in providing insight for the general effects of deviation from ideal behavior and would, in most cases, provide order of magnitude deviation quantities.

PROBLEMS

6.1. Using the van der Waals a and b values from Table 6.2, calculate P_c and T_c for He, N_2, and NH_3. Compare the calculated values with those listed.

6.2. Estimate the diameter of He, Ar, and hydrogen molecules from the van der Waals b.

6.3. Compare the calculated pressure for CH_4 at 298 K when 1 mole of the gas is confined to a 10 liter volume when (a) it is assumed to behave perfectly and (b) when it is assumed to behave as a van der Waals gas.

6.4. Compare calculation of pressure for 1 mole N_2 occupying a 5.0 liter volume at 500 K using (a) the van der Waals equation and (b) the Dieterici equation. Use the Table 6.2 constants.

6.5. Obtain a value for $(C_P - C_V)$ for 1 mole of oxygen at 298 K occupying (a) 5 liters and (b) 15 liters, assuming in each case that oxygen can be treated as a van der Waals gas.

6.6. Obtain an expression for the compressibility factor for a general van der Waals gas and use the expression to calculate the compressibility factor for 1 mole of H_2 occupying a 2 liter volume at 200 K and at 500 K.

6.7. Estimate the fugacities of nitrogen, hydrogen, and oxygen at 500 K and 100 atmospheres pressure using the Table 6.2 data and Figure 6.7.

6.8. Obtain a value for the Joule–Thompson inversion temperature of nitrogen behaving as a van der Waals gas when 1 mole of the gas occupies a volume of 10 liters.

CHAPTER CONTENT SOURCES AND FURTHER READING

1. J. Kestin and J. R. Dorfman, *A Course in Statistical Thermodynamics*, Chapters 7 and 8. Academic Press, New York, 1971.
2. T. M. Reed and K. E. Gubbins, *Applied Statistical Mechanics*, Chapters 4, 5, 6, and 7. McGraw-Hill, New York, 1973.
3. J. M. Klotz, *Chemical Thermodynamics*, Chapter 16. W. A. Benjamin, New York, 1964.
4. T. L. Hill, *Introduction to Statistical Thermodynamics*, Chapters 14 and 15. Addison-Wesley, Reading MA, 1960.
5. J. D. Hirschfelder, C. F. Curtiss, and R. B. Bird, *Molecular Theory of Gases and Liquids*, Chapters 3, 4, and 12. Wiley, New York, 1954.
6. J. M. Prausnitz, R. N. Lichtenthaler, and E. G. de Alverado, *Molecular Thermodynamics of Fluid-Phase Equilibria*, 2nd ed., Chapters 4 and 5. Prentice Hall, Englewood Cliffs, NJ, 1986.
7. E. A. Moelwyn-Hughes, *Physical Chemistry*, 2nd ed., Chapters VII and XIV. Pergamon Press, London, 1961.
8. J. E. Mayer and M. G. Mayer, *Statistical Mechanics*, Chapter 8. Wiley-Interscience, New York, 1977.

SYMBOLS: CHAPTER 6

$\phi, \phi_r, \phi_a, \phi_{a_{JK}}$ = potential energy of interaction (repulsion or attraction), for two particles
r, r_{JK} = distance between molecule mass centers
A, B = repulsive and attractive constants in an expression for interaction potential
$\sigma\phi_e, r_e$ = parameters in an expression for interaction potential
A, B, C = various molecule species, one time use in Figure 6.2
$\mathscr{H}(p, q)$ = the Hamiltonian
$Q(q_1 \cdots q_{3N})$ = potential energy of a system of particles
$p, q, q_1 \cdots q_N, p_1 \cdots p_N$ = particle momentum and positional coordinates
$U(r_{JK})$ = pair potential energy of interaction
Λ = notation for $(h^2/2\pi m k T)^{1/2}$
τ = collection of positional coordinates to reflect a volume
Q_N = configurational integral
V_f, v_f = free volume in a molecule system; per molecule
a, b = van der Waals constants, molecular (molar in Table 6.2)
v = molecule volume
g, c = subscript designation for gas or condensed phase
T_c, P_c, V_c = critical parameters
T_r, P_r, V_r = reduced parameters
Z = compressibility factor
$B(T), C(T), B'(T), C'(T)$ = second and third virial coefficients
$a_0, b_0, c_0, \alpha, \beta, \gamma$ = empirical constants in the Beattie–Bridgeman equation of state, one time use in Table 6.1
$E_t, H_t, C_{V,t}$ = translational components of a thermodynamic parameter
$\gamma_1, \gamma_2 \cdots \gamma_K$ = activity coefficient; general, component specific
$\mathfrak{f}, \mathfrak{f}_1 \cdots \mathfrak{f}_K$ = fugacity; general and component specific
α_P = isobaric compressibility
κ_T = isothermal compressibility
κ_S = adiabatic compressibility
μ_{JT} = Joule–Thompson coefficient

Carryover Symbols

$V, P, N, N_A, k, T, A, A_t, \mathscr{Z}, E, E_t, H, S, C_P, C_V, \gamma_c, \mu, \omega, \Theta_r, \Theta_v$.

7

INTERACTING PARTICLE SYSTEMS II: SOLID AND LIQUID PHASES

Our consideration of intermolecular interactions in the van der Waals framework demonstrated that the PVT properties of the gas phase and any general condensed phase will be quite different. The difference will be most obviously reflected by the nearly incompressible character of the condensed phase. Although the equation of state concept has general validity it will not fulfill the dominant role in solid and liquid phase considerations that it displayed in gas phase considerations. The thermodynamics that we developed in the nonideal gas chapter in terms of various compressibilities is general and applicable to all substances. Although equations of state, such as van der Waals, do extend to the liquid condensed phase they are not exact and they do not predict the solid–liquid transition.

The statistical approach to condensed phases, while it will become increasingly involved, will still serve as the most appropriate basis for developing intuitive understanding. The thermodynamic functions (A, G, H, S) that we developed in the ideal gas and real gas treatments have, as previously noted, complete generality in that they all derive in some manner from the partition function, i.e., from that particular form of the sum over available energy states. All general arguments relating to statistical equilibrium and the distribution of systems among energy states as reflected by the maximization of Ω (or S) in a supersystem including the interacting environment and the minimization of the free energy expressions for the thermodynamic system of interest under their particular constraints must therefore apply to these condensed phases. Clearly then, we must address the formulation of effective models from which available energy states can be obtained to produce the partition function.

All experimental evidence supports the thesis that effective models for the solid and liquid condensed phases will fit the particles into a structured, rigid framework for solids and into a very loosely structured and nonrigid framework for liquids. The X-ray diffraction patterns for many solids are extremely well defined for reasonably pure materials. Such solids are referred to as crystalline. Amorphous solids that do not evidence such extended order cannot be included in the arguments that follow. The X-ray diffraction patterns for liquids indicate that the long-range order characteristic of their solid phase is a very short-range and transient characteristic in the liquid phase. We assume that the translational energy of gas phase molecules moving randomly in volumes that are large compared to the molecule volumes has degraded into other motions, which we take to be vibrations, around fixed lattice points when the gas condenses into a solid. In the case of liquids there are "lattice" vacancies, which either do not exist in the solid phase or which are isolated by potential energy barriers that normally prevent translation between occupied and unoccupied sites. The presence of significant numbers of such vacancies coupled with low potential energy barriers compared to the average energy of liquid phase molecules makes site to site translation a dominant feature of that phase to the detriment of maintaining any long-range order.

Order, or traces of order, in addition to the feature of incompressibility, are particular condensed phase features. Complete order, as would characterize the crystalline solid phase, produces as a corollary the feature of distinguishability for the particles making up the system. Just as in the gas phase, the identical particles cannot be meaningfully identified but the sites that they occupy in the solid can be. The combination of particle and site makes up a distinguishable system. In a distinguishable particle system the division of the N-fold independent particle partition function product or the combinatorial term Ω by N!, which was required for the gas phase treatment, will not be required. Incomplete order would suggest that some correction between 1 and $(N!)^{-1}$ would be required. Also we would imagine that the independent particle framework is not applicable in either case since the consideration of any single particle should include its interaction with all other particles in the system, just as with interacting particles in the nonideal gas phase. Both reservations are valid. The distinguishability feature can be addressed. The question relating to the independence of the particles can be made irrelevant for solids by the simplicity of the model chosen for consideration.

We assume that each particle in a solid is subject to a force field provided by all other molecules in the system to produce a potential well in which the central particle vibrates harmonically. In analogy to the polyatomic molecule we would regard these vibrations as normal modes of oscillation and represent them as 3N harmonic oscillators. For these modes of motion to be independent the displacements must be phase coherent so that potential wells remain defined. Such normal modes are represented crudely in Figure 7.1 where the shortest

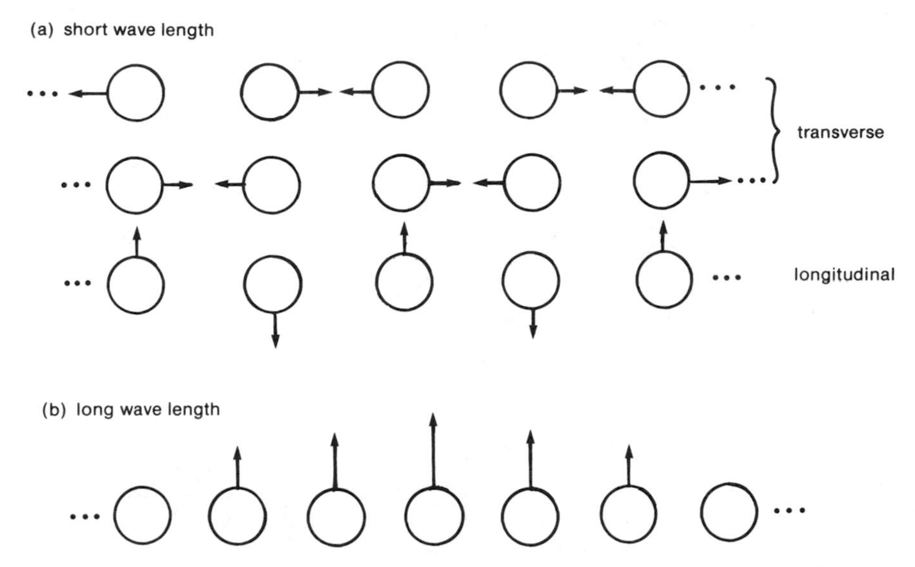

Figure 7.1. Illustration of normal vibrational modes in a crystalline solid.

wavelength associated with phase coherent motion is determined by the lattice spacing and the longest wavelength associated with phase coherent motion is determined by the crystal dimensions. As indicated in Figure 7.1, the normal modes for a monatomic solid can be resolved into two kinds of motion. These are known as transverse waves, of which there are two, and longitudinal waves. The partition function for the crystal would be constructed as if the crystal were a giant molecule consisting of 3N oscillators

$$\mathscr{Z} = \prod^{3N} \left[\frac{e^{h\nu_i/2kT}}{e^{h\nu_i/kT} - 1} \right]. \tag{7.1}$$

For any given solid there will be a particular spectrum of the normal mode frequencies that, given the number of such modes, can be regarded as a continuum of frequencies and represented by a distribution function $g(\nu)$ where $g(\nu)d\nu$ is the number of oscillator states in the interval $d\nu$. We can express the partition function in logarithmic form as

$$\ln \mathscr{Z} = 3N \int_0^\infty \ln\left[\frac{e^{-h\nu/2kT}}{1 - e^{-h\nu/kT}} \right] g(\nu) d\nu. \tag{7.2}$$

Figure 7.2 is indicative of the form that an actual distribution function assumes. It is complex for even simple materials. It is also temperature dependent. At high temperatures the short wavelength motion is dominant and at low temperatures the long wavelength motion is dominant. As a first gross

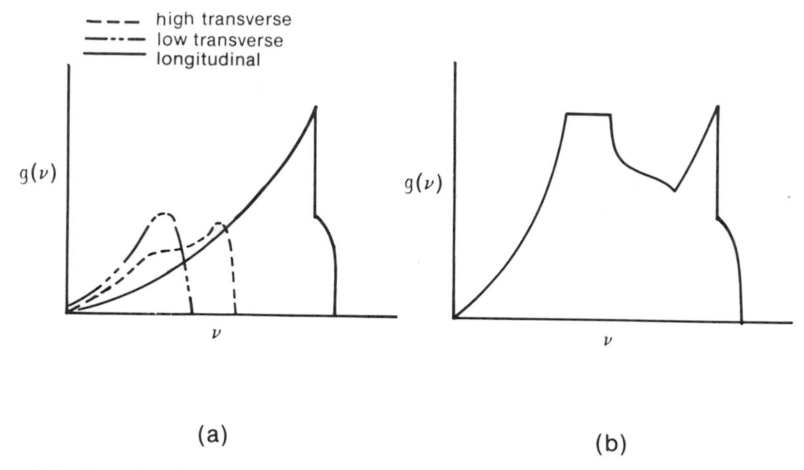

Figure 7.2. Distribution of oscillator states indicating (a) the form of distributions between various normal modes and (b) the summed frequency distribution. The general form is universal ranging from Ar to Al.

approximation introduced by Einstein, the 3N normal modes are replaced by a single vibration. The model would be that of a particle moving in a parabolic potential well as shown in Figure 7.3 where A and C occupy fixed positions and the solid lines reflect the potential energy of B as that particle moves along the line of centers. The curves "a" are a Figure 6.1 type potential function representing B moving toward A and away from C. The curves "c" reflect B moving toward C and away from A. The dotted line is the sum of curves "a" and "c" and reflects the potential well in which B oscillates. The well is assumed to be spherically symmetric for isotropic crystals and the inclusion of next nearest neighbors does not distort that symmetry. If $\phi(r)$ is the potential energy of the central molecule with r being the displacement from the cell center then

$$\phi(r) = \phi_0 + \tfrac{1}{2}\mathrm{f}r^2 + \cdots$$

where ϕ_0 is the hypothetical potential energy of a particle with all particles at rest at their lattice points and f is a force constant. Classically, the vibration

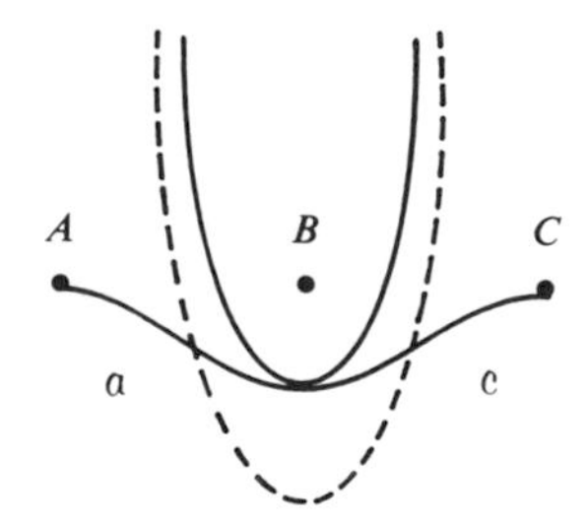

Figure 7.3. Illustration of the Einstein model for a particle oscillating in a potential well defined by at rest neighbors.

frequency would be

$$\nu = \frac{1}{2\pi}\left(\frac{f}{m}\right)^{1/2}$$

with f, and therefore ν, being dependent on the form of the potential well, therefore on the lattice spacing, therefore on V/N.

When the vibrational spectrum is represented by a single frequency ν_E

$$g(\nu)d\nu = 1 \qquad \nu = \nu_E$$
$$g(\nu)d\nu = 0 \qquad \nu \neq \nu_E$$

and the integral in Eq. (7.2) is nonzero only at ν_E. The problem has been reduced to a system of 3N identical, independent oscillators for which the partition function would be

$$\mathscr{Z}(N, T, V) = e^{-N\phi_0/2kT} z^{3N} \tag{7.3}$$

where $N\phi_0/2$ is the potential energy of the particle system with respect to an energy zero in which the particles are at rest and infinitely separated and the independent particle partition function for a single vibrational mode is given by

$$z = \frac{e^{-h\nu_E/2kT}}{1 - e^{-h\nu_E/kT}}. \tag{7.4}$$

We shall rewrite Eq. (7.4) in the form

$$z = \frac{e^{\Theta_E/2T}}{e^{\Theta_E/T} - 1}, \qquad \Theta_E = \frac{h\nu_E}{k} \tag{7.5}$$

as a reminder that ν_E is a single average frequency assigned to each of the independent oscillators and designating the characteristic temperature Θ_E associated with that vibration as the Einstein characteristic temperature. We will also refer to this model as the Einstein model recognizing its origin.

For the assembly of N particles

$$\ln \mathscr{Z} = -\frac{N\phi_0}{2kT} - \frac{3Nh\nu_E}{2kT} - 3N \ln(1 - e^{-h\nu_E/kT}). \tag{7.6}$$

As always

$$E = kT^2 \frac{\partial \ln \mathscr{Z}}{\partial T} = \frac{N\phi_0}{2} + \frac{3Nh\nu_E}{2} + \frac{3Nh\nu_E e^{-h\nu_E/kT}}{1 - e^{-h\nu_E/kT}} \tag{7.7a}$$

$$E - E_0 = \frac{3NkT\Theta_E/T}{e^{\Theta_E/T} - 1} \tag{7.7b}$$

where $E_0 = N\phi_0/2 + 3Nh\nu_E/2$. The zero point of energy is the zero point vibrational average plus the energy required to separate the particle system at rest to positions of zero potential energy with respect to each other. The heat capacity is

$$C_V = \left(\frac{\partial E}{\partial T}\right)_v = 3Nh\nu_E \frac{\partial}{\partial T}(e^{h\nu_E/kT} - 1)^{-1} \tag{7.8a}$$

$$= -3Nh\nu_E(e^{h\nu_E/kT} - 1)^{-2}\left(-\frac{h\nu_E}{kT^2}\right)e^{h\nu_E/kT}$$

$$= \frac{3Nk(h\nu_E/kT)^2 e^{h\nu_E/kT}}{(e^{h\nu_E/kT} - 1)^2} \tag{7.8b}$$

$$= \frac{3(\Theta_E/T)^2 e^{\Theta_E/T}}{(e^{\Theta_E/T} - 1)^2}. \tag{7.9}$$

Other thermodynamic functions are:

$$A = -kT \ln \mathscr{Z} = \frac{N\phi_0}{2} + \frac{3}{2}Nh\nu_0 + 3NkT\ln(1 - e^{-h\nu_E/kT})$$

$$= E_0 + 3NkT\ln(1 - e^{-\Theta_E/T}) \tag{7.10}$$

$$S = \frac{E}{T} - \frac{A}{T} = \frac{1}{T}\left[E_0 + \frac{3NkT\Theta_E/T}{(e^{\Theta_E/T} - 1)}\right] - \frac{1}{T}[E_0 + 3NkT\ln(1 - e^{-\Theta_E/T})]$$

$$= 3Nk\left[\frac{\Theta_E/T}{e^{-\Theta_E/T} - 1} - \ln(1 - e^{-\Theta_E/T})\right]. \tag{7.11}$$

Since, for a condensed phase in general, the PV product will be a very small term, it can be ignored for such systems making $A \simeq G$ and $E \simeq H$. This then completes the set of important thermodynamic functions, except for chemical potential. The chemical potential follows from $G = A = N\mu$ and

$$\mu = \frac{E_0}{N} + 3kT\ln(1 - e^{-\Theta_E/T}). \tag{7.12}$$

The table of Einstein functions (Table 5.4), which were originally introduced in the context of their application to calculations involving the Einstein model for solids, will simplify the arithmetic operations when applying Eqs. (7.7), (7.9), (7.10), and (7.11).

We note that in the high temperature limit, $\Theta_E/T \to 0$,

$$\lim_{T\to\infty} C_V = 3Nk \tag{7.13}$$

since

$$(e^{\Theta_E/T} - 1)^2 \to \left(1 + \frac{\Theta_E}{T} - 1\right)^2 = \left(\frac{\Theta_E}{T}\right)^2$$

and

$$C_V \to 3Nk \frac{(\Theta_E/T)^2}{(\Theta_E/T)^2}.$$

In the low temperature limit, $\Theta_E/T \to \infty$,

$$(e^{\Theta_E/T} - 1)^2 \to e^{2\Theta_E/T}, \qquad C_V \to Nk\left(\frac{\Theta_E}{T}\right)^2 e^{-\Theta_E/T}$$

and

$$\lim_{T \to 0} C_V \to 0. \tag{7.14}$$

Also $E \to E_0$ and $A \to E_0$ as $T \to 0$.

The limitations of the model are immediately apparent. Although in the high temperature limit the model does predict classical behavior, i.e., a $kT/2$ contribution to the heat capacity for each of the $2 \times 3N$ squared terms in the energy expression for 3N harmonic oscillators (this has been long recognized experimentally in the empirical observations of Dulong and Petit, which suggested that the heat capacity of all crystalline solids should be 6.3 cal mole^{-1} K^{-1}) and the model does predict a zero heat capacity at 0 K, the approach to this limiting value does not follow the Eq. (7.13) dependence. Also, C_V/Nk is only dependent on Θ_E/T and there should be a law of corresponding states in which the different Θ_E values for different solids produce the same C_V/Nk value when divided by the appropriate value of T. Experimental heat capacities do in fact follow such a law of corresponding states although, even at moderate temperatures, it does not follow the Einstein curve. Given the limitations of the model no more should be expected.

Since the Einstein approximation produces a reasonable high temperature heat capacity but fails at low temperature it is clear that the high-frequency region of the vibrational spectrum does not have to be accurately represented but the low-frequency region requires a closer address. It is in this region that the normal modes are least affected by the detailed structure. The Debye treatment focuses on this feature in assuming that the solid can be regarded as a continuous medium and that the vibrational spectrum is determined by the elastic constants of the solid. Resolution of the distribution function $g(\nu)$ from the elastic constants requires much more attention to the mechanical properties of solids than we can develop here. The relationship that emerges from the complete treatment is

$$g(\nu)d\nu = V\left(\frac{1}{C_l^3} + \frac{2}{C_t^3}\right)4\pi\nu^2 d\nu \tag{7.15}$$

where C_l and C_t are the longitudinal and transverse velocities of sound in the solid. Together with the requirement limiting the number of possible vibrations

$$\int g(\nu)d\nu = 3N \tag{7.16}$$

we impose a limiting frequency ν_m and obtain

$$\begin{aligned} g(\nu) &= \frac{9N\nu^2}{\nu_m^3} \qquad 0 \leqslant \nu \leqslant \nu_m \\ g(\nu) &= 0 \qquad\qquad \nu > \nu_m. \end{aligned} \tag{7.17}$$

With this result Eq. (7.2), including the zero point energy term and new limits, becomes

$$\begin{aligned} \ln \mathscr{Z} &= -\frac{N\phi_0}{2kT} + \int_0^{\nu_m} -\left[\frac{h\nu}{2kT} + \ln(1 - e^{-h\nu/kT})^{-1}\right]\frac{9N\nu^2}{\nu_m^3}d\nu \\ &= \frac{N\phi_0}{2kT} - \frac{9Nh}{kT\nu_m^3}\int_0^{\nu_m}\nu^3 d\nu - \frac{9N}{\nu_m^3}\int_0^{\nu_m}\frac{\nu^2}{\ln(1 - e^{-h\nu/kT})}d\nu \\ &= \frac{N\phi_0}{2kT} - \frac{9Nh\nu_m}{8kT} - \frac{9N}{\nu_m^3}\int_0^{\nu_m}\frac{\nu^2}{\ln(1 - e^{-h\nu/kT})}d\nu. \end{aligned} \tag{7.18}$$

Thermodynamic functions for the solid follow from routine relationships but they will all contain an integral form deriving from Eq. (7.18). The energy relationship

$$E = kT^2\frac{\partial \ln \mathscr{Z}}{\partial T}$$

leads to the expression

$$E = \frac{N\phi_0}{2} + \frac{9Nh\nu_m}{8} + \frac{9NkT^2}{\nu_m^3}\frac{\partial}{\partial T}\left[\left[\int_0^{\nu_m}\nu^2 \ln(1 - e^{-h\nu/kT})d\nu\right]\right.$$

and differentiation of the logarithmic term gives

$$\frac{e^{-h\nu/kT}}{1 - e^{-h\nu/kT}}\left(\frac{h\nu}{kT^2}\right)$$

so that the Debye energy is

$$E = E_0 + \frac{9Nh}{\nu_m^3}\int_0^{\nu_m}\frac{\nu^3}{e^{h\nu/kT} - 1}d\nu. \tag{7.19}$$

The constant volume heat capacity of the solid is

$$C_V = \left(\frac{\partial E}{\partial T}\right)_{V,N} = \frac{9N}{\nu_m^3}\int_0^{\nu_m} \nu^3 d\nu \frac{\partial}{\partial T}(e^{-h\nu/kT} - 1)^{-1}$$

$$= \frac{9N}{\nu_m^3}\frac{h}{kT^2}\int_0^{\nu_m} \frac{\nu^4 e^{h\nu/kT}}{(e^{-h\nu/kT} - 1)^2} d\nu \tag{7.20}$$

the Helmholz free energy is, from Eq. (7.18),

$$A = -kT \ln \mathscr{Z} = E_0 + \frac{9NT}{\nu_m^3}\int_0^{\nu_m} \frac{\nu^2}{\ln(1 - e^{-h\nu/kT})} d\nu \tag{7.21}$$

and the entropy is

$$S = \frac{E - A}{T} = \frac{9Nk}{\nu_m^3}\int_0^{\nu_m}\left[\frac{\nu/kT}{e^{h\nu/kT} + 1} - \ln(1 - e^{-h\nu/kT})\right]\nu^3 d\nu. \tag{7.22}$$

It will be convenient to simplify the expressions by replacing $h\nu/kT$ with x. In this form with $\nu = (kT/h)x$ and $d\nu = (kT/h)dx$ the upper limit of integration becomes $h\nu_m/kT$ which we represent as u

$$E = E_0 + \frac{9NkT}{u_m^3}\int_0^u \frac{x^3 dx}{e^x - 1} \tag{7.23a}$$

$$C_V = \frac{9N}{u_m^3}\int_0^u \frac{x^4 e^x dx}{(e^x - 1)^2} \tag{7.24a}$$

$$A = E_0 + \frac{9NkT}{u_m^3}\int_0^u \frac{x^2 dx}{\ln(1 - e^{-x})} \tag{7.25a}$$

$$S = \frac{9Nk}{u^3}\left[\int_0^u \frac{x^3 dx}{(e^x - 1)} + \int_0^u \frac{x^2 dx}{\ln(1 - e^{-x})}\right. \tag{7.26a}$$

In calculation of the thermodynamic functions for solids much effort is saved through use of tabulated numerical calculations for the Debye function defined as

$$D(u) = \frac{3}{u^3}\int_0^u \frac{x^3 du}{e^x - 1} = D\left(\frac{\Theta_D}{T}\right). \tag{7.27}$$

Rewriting the expressions for energy, free energy, entropy, and heat capacity in terms of the Debye function we have (without arithmetic)

$$E = E_0 + 3NkTD\left(\frac{\Theta_D}{T}\right) \tag{7.23b}$$

$$C_V = 2Nk\left[4D\left(\frac{\Theta_D}{T}\right) - \frac{3(\Theta_D/T)}{e^{\Theta_D/T} - 1}\right] \tag{7.24b}$$

$$A = \frac{9NkT}{8}\left(\frac{\Theta_D}{T}\right) + 3NkT\ln(1 - e^{-\Theta_D/T}) + NkTD\left(\frac{\Theta_D}{T}\right) \tag{7.25b}$$

$$S = 3Nk\left[\frac{4}{3}D\left(\frac{\Theta_D}{T}\right) - \ln(1 - e^{-\Theta_D/T})\right]. \tag{7.26b}$$

Table 7.1 is an adequate listing of Debye function values for most use.

We would construct an equation of state from the free energy expression [Eq. (7.24)]. We begin with the thermodynamic relationship

$$(\partial A/\partial V)_{T,N} = -P$$

where A is volume dependent, through Θ_D/T. We express this dependence as $A/T = f(\Theta_D/T) = f(u)$ and

$$-P = \left(\frac{\partial A}{\partial \Theta_D}\right)_{T,N}\left(\frac{\partial \Theta_D}{\partial V}\right)_{T,N}. \tag{7.27}$$

Then from

$$\left[\frac{\partial f(u)}{\partial \Theta_D}\right]_T = \frac{df(u)}{du}\left(\frac{\partial u}{\partial \Theta_D}\right)_T = \frac{df(u)}{du}\left(\frac{1}{T}\right)$$

$$\left[\frac{\partial f(u)}{\partial T}\right]_{\Theta_D} = \frac{df(u)}{du}\left(\frac{\partial u}{\partial T}\right)_{\Theta_D} = -\frac{df(u)}{du}\left(\frac{\Theta_D}{T^2}\right)$$

$$T\left[\frac{\partial f(u)}{\partial T}\right]_{\Theta_D} = \Theta_D\left(\frac{\partial f(u)}{\partial \Theta_D}\right)_T \tag{7.28}$$

and from

$$A = -kT\ln \mathscr{Z}$$

$$\left(\frac{\partial A/T}{\partial T}\right)_{V,N} = -k\left(\frac{\partial \ln \mathscr{Z}}{\partial T}\right)_{V,N} = -\frac{E}{T^2}$$

so that

$$E = -T^2\left(\frac{\partial A/T}{\partial T}\right)_{V,N} = -T^2\left(\frac{\partial A/T}{\partial T}\right)_{\Theta_D,N} \tag{7.29}$$

TABLE 7.1. The Debye Function $D(u) = \dfrac{3}{u^3}\displaystyle\int_0^u \dfrac{x^3dx}{e^x - 1}$

u	0.0	0.1	0.2	0.3	0.4	0.5	0.6	0.7	0.8	0.9	1.0
0.0	1.0000	0.9630	0.9270	0.8920	0.8580	0.8250	0.7929	0.7619	0.7318	0.7026	0.6744
1.0	0.6744	0.6471	0.6208	0.5954	0.5708	0.5471	0.5243	0.5023	0.4811	0.4607	0.4411
2.0	0.4411	0.4223	0.4042	0.3868	0.3701	0.3541	0.3388	0.3241	0.3100	0.2965	0.2836
3.0	0.2836	0.2712	0.2594	0.2481	0.2373	0.2269	0.2170	0.2076	0.1986	0.1900	0.1817
4.0	0.1817	0.1739	0.1664	0.1592	0.1524	0.1459	0.1397	0.1338	0.1281	0.1227	0.1176
5.0	0.1176	0.1127	0.1080	0.1036	0.09930	0.09524	0.09137	0.08768	0.08415	0.08079	0.07758
6.0	0.07758	0.07452	0.07160	0.06881	0.06615	0.06360	0.06118	0.05886	0.05664	0.05453	0.05251
7.0	0.05251	0.05057	0.04873	0.04696	0.04527	0.04366	0.04211	0.04063	0.03921	0.03786	0.03656
8.0	0.03656	0.03532	0.03413	0.03298	0.03189	0.03084	0.02983	0.02887	0.02794	0.02705	0.02620
9.0	0.02620	0.02538	0.02459	0.02384	0.02311	0.02241	0.02174	0.02109	0.02047	0.01987	0.01930
10.0	0.01930	0.01874	0.01821	0.01769	0.01720	0.01672	0.01626	0.01581	0.01538	0.01497	0.01457
11.0	0.01457	0.01418	0.01381	0.01345	0.01311	0.01277	0.01245	0.01213	0.01183	0.01153	0.01125
12.0	0.01125	0.01098	0.01071	0.01045	0.01020	0.00996	0.00973	0.00950	0.00928	0.00907	0.00886
13.0	0.00886	0.00866	0.00846	0.00827	0.00809	0.00791	0.00774	0.00757	0.00741	0.00725	0.00710
14.0	0.00710	0.00695	0.00680	0.00666	0.00652	0.00639	0.00626	0.00613	0.00601	0.00589	0.00577
15.0	0.00577	0.00566	0.00555	0.00544	0.00533	0.00523	0.00513	0.00503	0.00494	0.00485	0.00476

Taken from S. M. Blinder, *Advanced Physical Chemistry*, MacMillan, Toronto, 1969.

since, if V is constant Θ_D is constant. Then, from Eq. (7.42),

$$E = \Theta_D\left(\frac{\partial A}{\partial \Theta_D}\right)_{T,N} \tag{7.30}$$

and from Eq. (7.27)

$$P = \frac{E}{\Theta_D}\left(\frac{\partial \Theta_D}{\partial V}\right)_T = -\frac{E}{V}\left(\frac{\partial \ln \Theta_D}{\partial V}\right)_{T,N} = -\frac{E}{V}\gamma_G \tag{7.31}$$

with γ_G [the Grüniesen function $\equiv (\partial \ln \Theta_D/\partial V)_{T,N}$] further defined as follows:

$$\left(\frac{\partial P}{\partial T}\right)_V = -\left(\frac{\partial E}{\partial T}\right)_V \frac{\sigma_G}{V} = \frac{C_V\gamma_G}{V} \tag{7.32}$$

which, with the identity [Eq. (6.58)],

$$\left(\frac{\partial P}{\partial T}\right)_V = -\left(\frac{\partial P}{\partial V}\right)_T\left(\frac{\partial V}{\partial T}\right)_P = \frac{\alpha_P}{\kappa_T}$$

produces $\gamma_G = \alpha_P V/\kappa_T C_V$, so that the equation of state can be reconstructed as

$$P = -E\frac{\alpha_P}{\kappa_T}. \tag{7.33}$$

TABLE 7.2. The Grüneisen Function $\gamma_G = (\partial \ln \Theta_D/\partial V)_{T,N}$ for Selected Solids

Crystal	Temperature Range (κ)	γ_G Range
Ar	0–50	2.8 –3.1
Kr	0–50	2.75–3.05
Xe	0–50	2.75–3.0
Cu	0–600	1.7 –1.95
Ag	0–300	2.7 –2.8
Au	0–300	2.3 ± 0.05
Al	0–400	2.3 –2.7
Na	0–200	0.9 –1.1
K	0–150	1.05–1.2

From J. A. Reissland, *The Physics of Phonons*, Wiley, New York, 1973, as reprinted in R. S. Berry, S. A. Rice, and J. Ross, *Physical Chemistry*, Wiley, New York, 1960. Reprinted with permission by John Wiley and Sons Inc.

Of course, γ_G is a constant only insofar as α_P, κ_T, and C_V are constants. To a certain extent, the variations would be expected to cancel but, even for the simplest systems, the temperature range over which constancy is a feature is both intuitively unpredictable and often quite limited as indicated by the Table 7.2 values.

The Eq. (7.26) expression for heat capacity assumes large importance since it is the experimentally accessible quantity and $E_T = E_0 + \int_0^T C_V dT$. Also some very important conclusions derive from limiting values. As $T \to 0$ and the integration limit $\to \infty$

$$D\left(\frac{\Theta_D}{T}\right) \to \frac{3}{(\Theta_D/T)^3} \int_0^u \frac{x^4 e^x\,dx}{(e^x - 1)^2} = \frac{4\pi^4}{5}\left(\frac{T}{\Theta_D}\right)^3$$

$$C_V \to \frac{12 N k \pi^4}{5}\left(\frac{T}{\Theta_D}\right)^3. \tag{7.34}$$

As $T \to \infty$ expansion of the integral provides

$$D\left(\frac{\Theta_D}{T}\right) = 1 - \frac{u_m^2}{2D} + \frac{u_m^4}{56D} - \frac{u_m^6}{18144} \to 1$$

and

$$C_V \to 3Nk \tag{7.35}$$

which is the classical expectation. It is also important that, from Eq. (7.26), the heat capacity is a universal function of Θ/T. All monatomic crystalline solids should provide the same lattice contribution to the molar heat capacity at common values of Θ_D/T. A wide variety of monatomic solids conform closely to this law of corresponding states as demonstrated in Figure 7.4. The T^3

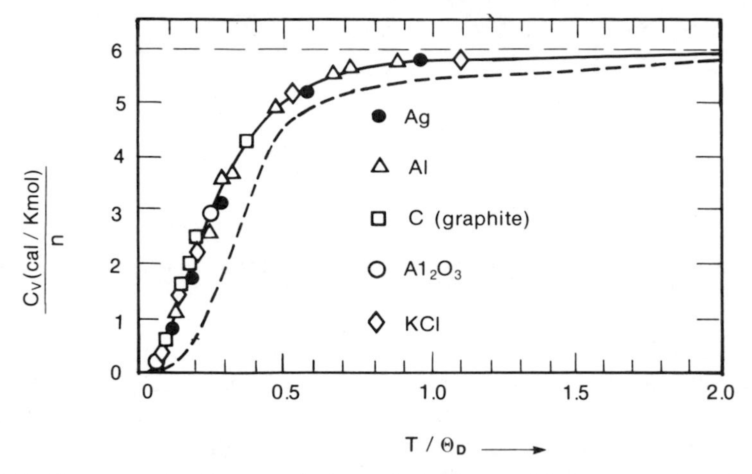

Figure 7.4. Heat capacity C_V according to the Debye and Einstein models (solid and dashed lines) with experimental data for various solids. Following F. Seitz, *Modern Theory of Solids*, 1st ed., McGraw Hill, New York, 1940.

dependence of heat capacity [Eq. (7.48)] at temperatures approaching 0 K is confirmed by experiment. Heat capacity measurements can be extended routinely down to a few degrees K and can be linearly extrapolated in a T^3 plot to the absolute zero to provide one measure of Θ_D. We should, however, recognize that the Θ_D value obtained in this manner will not compare exactly with that obtained from fitting higher temperature heat capacity data to the Eq. (11.26) form or with the values obtained from the elastic constants.[1] This reflects limitations of the model in that the assumption $g(\nu) \sim \nu^2$ is increasingly inadequate as the frequency increases. Also, we expect Θ_D/T to be temperature dependent in that the term is volume dependent. The heat capacity approach to $3N\mathscr{k}$ as a limiting value, however, suggests that these deficiencies are not of large consequence. The Einstein model with the single vibrational frequency, which comparison of Eqs. (7.7) and (7.19) demonstrates is $3\nu_m/4$, exhibits the same high temperature limit [Eqn. (7.13)].

If the substance of interest is not a monatomic solid the vibrational spectrum will be even more complex than the Figure 7.2 form. If there is more than one particle in the unit structure (e.g., face or body centered cubic structure) and/or the particles making up the structure are dissimilar (e.g., NaCl) the restoring force deriving from displacement of adjacent particles in one cell will be different from the restoring force between adjacent atoms in a neighboring cell. This will give rise to vibrations within the unit cell. It is common to refer to the Figure 7.1 modes as acoustic vibrations. The vibrations internal to the cell are called optic

[1] The bracketed term $[(1/C_l^3) + (2/C_t^3)]$ in Eq. (7.15) can be shown to be given by $\rho^{3/2}\kappa^{3/2}f(\sigma)$ where ρ is density, κ is compressibility, and $f(\sigma)$ is a complex function of Poisson's ratio: the ratio of relative radial compression to relative axial elongation when the solid is subjected to axial stress.

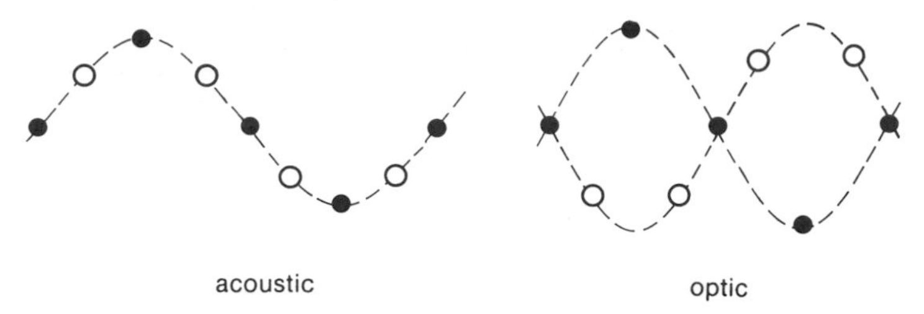

Figure 7.5. Acoustic and optic modes of vibration within a two particle species unit cell.

modes. Figure 7.5 illustrates the long wavelength acoustic and optic modes for a linear diatomic lattice. On the basis of 1 mol of particles the new modes of motion are contained within the limit imposed by Eq. (7.16). Data for KCl fit the Figure 7.4 plot as well as the monatomic data do. When the lattice sites are occupied by polyatomic particles there are internal rotational and vibrational modes of motion that are, in principle, coupled throughout the system through molecular interactions.

Again, however, recognizing that forces related to bonding are very large compared to intermolecular interactions, it is a reasonable approximation to regard the intermolecular and internal vibrations as independent with the latter only slightly perturbed by the former. We would therefore superimpose the internal vibration spectrum on the Debye spectrum. Rotational motion of the molecule will be largely perturbed by the close proximity of other molecules and their interactions. Normally the free rotation characteristic of the gas phase will be observed as a hindered rotation and represented as a torsional oscillation.

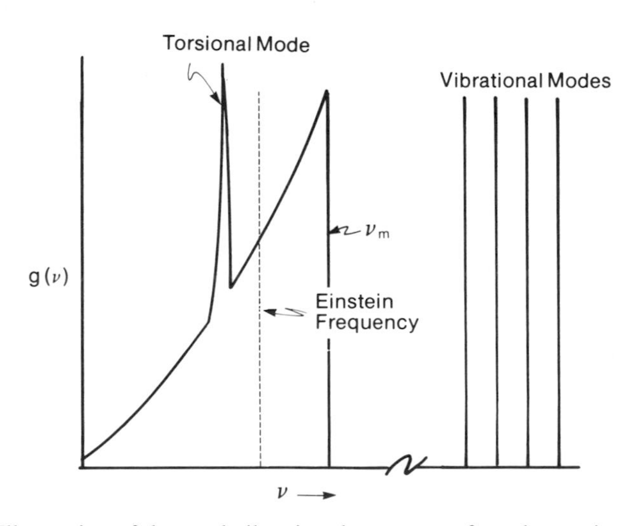

Figure 7.6. Illustration of the total vibrational spectrum of a polyatomic solid according to the Debye and Einstein models.

The frequency spectrum of a polyatomic solid, in the absence of strong intermolecular interaction (e.g., hydrogen bonding or extended covalency in the crystal) can be represented by Figure 7.6. The torsional mode (modes) associated with hindered rotation of the molecule would normally be found within the Debye spectrum of frequencies. Vibrational modes including hindered internal rotations, would normally be far outside the Debye spectrum and are unperturbed from their gas phase values. Commonly the vibrational modes will be inactive at temperatures in the range of thermal stability of the solid. When this is not true the internal vibrational modes will produce the normal additive contribution to thermodynamic functions for the solid.

In treating monatomic solids as a collection of 3N vibrators we have ignored a complication in the case of metals which, on first consideration, should produce a substantially higher experimental limiting heat capacity. One conventional concept of metallic conductors is that valence electrons are freely mobile in a framework of positive ions. In fact we commonly refer to the electron gas and we should anticipate that, in addition to the $3Nk$ heat capacity contribution from the 3N modes of positive ion vibrations, there should be a $3Nk/2$ contribution from the translational electron motion. This is known not to be true; there is no such contribution. The explanation lies in the nature of the electron, the small mass, and the spin.

The application of Boltzmann statistics to a system of indistinguishable particles leads us to the distribution numbers expression

$$N_i = g_i e^{-\alpha} e^{-\beta \epsilon_i}. \tag{1.5}$$

In arriving at the statistical argument we imposed no constraints on the number of particles that could be assigned to an energy level. However, when the particles are electrons there is a constraint imposed by the Pauli exclusion principle; only two electrons, spin paired, can occupy the same energy level. This constraint leads to a different statistical expression[2] and

$$N_i = \frac{g_i e^{-\alpha} e^{-\beta \epsilon_i}}{1 + e^{-\alpha} e^{-\beta \epsilon_i}}.$$

As always $\alpha = -\mu/kT$ and $\beta = 1/kT$ so that

$$\frac{N_i}{g_i} = \frac{1}{e^{-(\epsilon_i - \mu_i)/kT} + 1}. \tag{7.36}$$

As $T \to 0$ we recognize that, if $\epsilon_i > \mu_0$ then $N_i/g_i \to 0$, if $\epsilon_i < \mu_0$ then $N_i/g_i \to 1$ but

[2] Quantum statistics, deriving from quantum constraints on energy state occupancy, fall into two classes which are known as Fermi–Dirac and Bose–Einstein statistics. Electrons follow Fermi–Dirac statistics. See Appendix C.

if $\epsilon_i = \mu_0$ then $N_i/g_i = 1/2$. That is, at low temperatures all energy levels are occupied by two electrons up to $\epsilon_i = \mu_0$. All energy levels above μ_0, the chemical potential at $T = 0$, are empty.[3] To evaluate μ_0 for the N electrons as $T \to 0$ we recall the translational quantum energy level expression

$$\epsilon = \frac{h^2}{8m_e} \frac{n^2}{V^{2/3}} \tag{7.37}$$

where m_e is now the electron mass. As we have noted in prior argument (see the discussion related to Figure 3.1), the volume of the octant of the sphere of radius n represents all positive number sets leading to energies ϵ_n or less. This is the number of energy levels and, for $\epsilon_n = \mu_0$, equal to N/2, where N is the number of electrons available to doubly occupy as spin pairs that number of energy levels. For $\mathscr{V}_{\text{octant}} = \frac{1}{8}(4\pi n^3/3)$,

$$\mathscr{V}_{\text{octant}} = \frac{N}{2} = \frac{\pi}{6}\left(\frac{8m_e\mu_0}{h^2}\right)^{3/2} V$$

and

$$N = \frac{\pi}{3}\left(\frac{8m_e\mu_0}{h^2}\right)^{3/2} V \tag{7.38}$$

with the highest occupied energy level being

$$\mu_0 = \left(\frac{h^2}{8m_e}\right)\left(\frac{3N}{\pi V}\right)^{2/3}. \tag{7.39}$$

The distribution of energy states $\omega(\epsilon)$ is $2(\partial\mathscr{V}_{\text{octant}}/\partial\epsilon)$

$$\omega(\epsilon)d\epsilon = 2\left(\frac{\pi}{4}\right)\left(\frac{8m_e}{h^2}\right)^{3/2} V\epsilon^{1/2} d\epsilon. \tag{7.40}$$

From $E = \sum n_i \epsilon_i$ replaced by an integral since we recognize the close spacing of translational energy levels and with N_i given by Eq. (7.36), we write

$$\begin{aligned} E &= \int \omega(\epsilon) \frac{1}{e^{-(\epsilon_i - \mu_i)/kT} + 1} d\epsilon \\ &= 4\pi\left(\frac{2m_e}{h^2}\right)^{3/2} V \int_0^{\mu_0} \frac{\epsilon^{3/2}}{e^{-(\epsilon_i - \mu_i)/kT} + 1} d\epsilon. \end{aligned} \tag{7.41}$$

[3] μ_0 is referred to as the Fermi level.

The energy at T = 0 is

$$E_0 = \frac{4}{5}(4\pi V)\left(\frac{2m_e}{h^2}\right)^{3/2} \mu_0^{5/2} = \frac{3}{5} N\mu_0. \tag{7.42}$$

The heat capacity C_v at T = 0 is zero.

The entropy of the Fermi gas is zero at T = 0 since there is only one way in which the (indistinguishable) electrons can be put into the same number of energy states ($\Omega = 1$). The pressure will satisfy the ideal gas law

$$PV = \tfrac{2}{3}E$$

so that, at 0 K

$$P = \left(\frac{h^2}{5m_e}\right)\left(\frac{3}{8\pi}\right)^{2/3}\left(\frac{N}{V}\right)^{5/3}. \tag{7.43}$$

The large PV product, deriving from a pressure term (0) 10^5 atm, is balanced by the potential energy of interactions in the metal. We can regard this $(2E_0/3)$ as the cohesive energy of the metal.

To evaluate the integral appearing in Eq. (7.41) over a range of temperatures involves extended arithmetic which does not add to our understanding so we set down only the results

$$E = \frac{3}{5} N\mu_0\left[1 + \frac{5\pi^2}{12}\left(\frac{kT}{\mu_0}\right)^2 + \cdots\right] \tag{7.44}$$

$$C_v = \left(\frac{\partial E}{\partial T}\right)_V = \frac{\pi^2}{2}\frac{N}{\mu_0}k^2T + \cdots. \tag{7.45}$$

Rephrasing Eq. (7.45), we see that

$$C_v = \frac{\pi^2}{3}\left(\frac{kT}{\mu_0}\right)\frac{3}{2}Nk \tag{7.46}$$

so that the molar contribution to the heat capacity arising from the mobile electron gas is not 3R/2 but a smaller term by the factor $\pi^2 kT/3\mu_0$. The contribution is (0) 10^{-2} of the vibrational heat capacity of the solid and can be ignored.

Both the Einstein treatment and the Debye treatment predict a zero value for the entropy at T = 0. If Eq. (7.22) is rewritten in the form

$$S = \frac{9Nk}{u_m^3}\int_0^u \left[\frac{x}{(e^x - 1)} - \ln(1 - e^{-x})\right]x^2\,dx$$

and the limit of integration again taken as $u = \infty$, then

$$S = \frac{9N\mathscr{k}}{u_m^3}\left(\frac{\pi^4}{15} + \frac{\pi^4}{45}\right) = \frac{4\pi^4 N\mathscr{k}}{5}\left(\frac{T}{\Theta_0}\right)^3 \tag{7.47}$$

so that $S \to 0$ as $T \to 0$. This is an anticipated result since there is only one way in which N particles all in their lowest energy state can be arranged on the N sites, $S = \mathscr{k} \ln \Omega = 0$. Since the E_0 terms cancel in the expression $S = [(E - E_0) - (A - E_0)]$, $E \to A$ as $T \to 0$. If we exclude possible ground state nuclear degeneracy and entropic effects related to isotope mixing, as we would commonly do in addressing chemical processes, the conclusion that $S = 0$ at $T = 0$ is a general one that would apply to any perfectly ordered solid. This is a statistical statement of the third law of thermodynamics[4], which permits the construction of tables of standard absolute entropies. Such values immediately follow from Eq. (7.26) for solids at any temperature (298 K for tabulation purposes) given an appropriate value of Θ_D for the subject species. There is negligible pressure dependence for solids so that the constraint of one atmosphere pressure, commonly a feature of the standard state conditions, is trivial. It also follows from the third law statement that the Sakur–Tetrode equation [Eq. (3.15)] for monatomic species or its analog [Eq. (5.53)] for polyatomic species provide absolute entropy values for gases under standard state conditions.

Given the third law condition and therefore the assurance that there is no intrinsic residual entropy for the solid at $T = 0$, Eq. (7.47) provides for an independent experimental verification of the absolute entropy when heat capacity measurements are extended to appropriately low temperatures. Then for

$$S(T) - S(0) = \int_0^T \frac{C_V}{T}\, dT = S(T) \tag{7.48}$$

where it is assumed that no solid phase transitions are encountered over the region $0 - T$. We defer complete consideration of the statistical vs calorimetric comparison until the thermodynamic features of the phase equilibrium have been discussed but at this point we can usefully expand on the third law statement constraint of "perfect order." Any real solid will contain a multitude of lattice defects of various kinds. Ordinarily such imperfections involve a trivial fraction of the N sites and can be ignored but there are circumstances in which defects in order are of importance.

Consider as an ordered example the molecule CO. Perfect order and lowest potential energy would require alignment ···CO–CO–CO··· but the molecule

[4] The classical origin of the third law lies in the observation that $\Delta H \to \Delta G$ as $T \to 0$ for general chemical processes. Therefore, from $\Delta S = (\Delta H - \Delta G)/T$ it follows that $\Delta S \to 0$ as $T \to 0$. The conclusion that $S = 0$ for every involved species is suggested by the universality of this observation.

has a small dipole moment and, at low temperatures, tends to be "frozen" into a random alignment $\cdots$CO–OC–CO$\cdots$. Since there are two orientations of the molecule on any given site the molecular entropy will approach $k \ln 2^N$ rather than zero as $T \rightarrow 0$. This is, of course, the reflection of a metastable state and not the true equilibrium state, but calorimetric heat capacities will recognize the limiting value of the experimental system. Entropies calculated from the calorimetric data would be expected to be low by $R \ln 2 = 1.4$ eu mole^{-1}. The actual discrepancy is 1.1 eu mole^{-1}. The difference is within experimental error.

The liquid phase has neither complete order nor complete disorder and occupies, over its range of stability, a position intermediate between solid and gas. It approaches the solid at one extreme of this range and the gas at the other with, we should imagine, some larger assumption of the character of the ordered or random particle phases at the extremes of this range. It will be very difficult to treat liquid phases in any exact manner. We must be content with constructing an intuitively acceptable model that will permit us to visualize the motion of particles and appreciate the form of the resulting statistical constructs for thermodynamic functions of interest even if we are not impressed by their accuracy or capability for extension.

A major feature of the solid–liquid phase transition is the discontinuity in volume. The solid volume is normally smaller than that of the liquid.[5] In the simplest approach to consideration of the liquid state we would assume that the solid–liquid transition for monatomic particles just involves the conversion of characteristic solid vibrational degrees of freedom into other degrees of freedom in a potential well deriving from the same forces as those which limit motion in the solid but with a broader minimum due to increased volume. We can picture such a transition as resulting from moving the particles pictured in Figure 7.3 further apart as shown in Figure 7.7.

The motion of particles in the liquid are then much like those in the solid phase in the sense that the condensed phase volume is considered to be the sum of cell volumes which in the simplest case would be defined as $v_{cell} = V/N$. The liquid differs from the solid, however, in the sense that the particles cannot be completely localized in view of observed self diffusion and other fluid properties that liquids and gases share. If the particles are free to move throughout

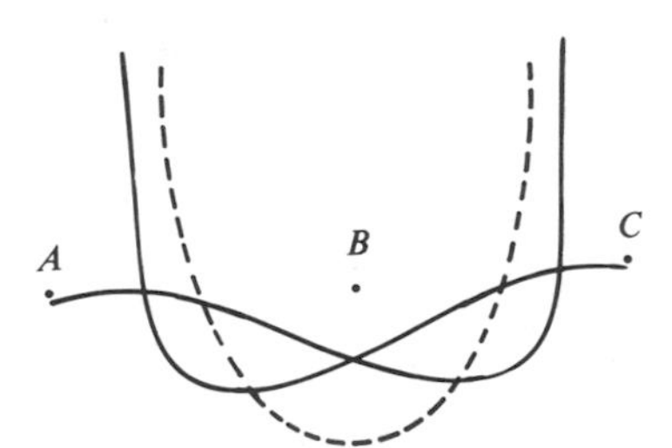

Figure 7.7. Illustration of the form of the potential well for a particle oscillating in an expanded lattice.

[5] Exceptions are bismuth, gallium, and ice.

the entire volume we would construct the partition function as

$$\mathscr{Z} = \left(\frac{2\pi m k T}{h^2}\right)^{3N/2} \frac{V^N}{N!} e^{-\phi(N)/kT} \tag{7.49}$$

where $\phi(N)$ is the potential energy of the particle system that results from intermolecular forces. If the particle is confined to the cell volume it will be necessary to remove the indistinguishability factor $(N!)^{-1}$ since the cells are distinguishable and the partition function would be

$$\mathscr{Z} = \left(\frac{2\pi m k T}{h^2}\right)^{3N/2} \left(\frac{V}{N}\right)^N e^{-\phi(N)/kT}. \tag{7.50}$$

Clearly, with $\ln N! = N \ln N - N$, and $\ln N^N = N \ln N$ Eq. (7.49) differs from Eq. (7.50) by the factor e^N. This feature has long between recognized with no clear resolution of a proper address. It has become the practice to retain the cell volume concept and insert the factor e into the partition function for the liquid which would become

$$\mathscr{Z} = \left(\frac{2\pi m k T}{h^2}\right)^{3N/2} (V_{cell} e)^N e^{-\phi(N)/kT}.$$

In individual particle terms we can express the potential energy of the particle system as $\phi(N) = (N/2)[\phi(r) - \phi_0]$ where ϕ_0 is the at rest potential energy and $\phi(r)$ is the displacement energy relating to particle motion away from the rest position. We would imagine that it is this displacement energy term that defines the somewhat smaller volume (the free volume v_f) associated with particle motion in the cell. The final partition function expression then is

$$\mathscr{Z} = \left(\frac{2\pi m k T}{h^2}\right)^{3N/2} (v_f e)^N e^{-N\phi_0/2kT} \tag{7.51}$$

with

$$v_f = \int_{cell} e^{-(\phi(r) - \phi_0)/kT} dr \tag{7.52}$$

and ϕ_0 representing the potential energy of a particle interacting with all other particles in the system of particles at rest. It is clear that v_f is dependent on system volume and temperature and $\phi(0)$ is dependent on system volume. The thermodynamic functions of common interest can be set down as

$$A = -NkT \ln\left[\left(\frac{2\pi m k T}{h^2}\right)^{3/2} v_f\right] - NkT + \frac{N\phi_0}{2} \tag{7.53}$$

$$E = \frac{3}{2} NkT + NkT^2 \left(\frac{\partial \ln v_f}{\partial T}\right)_V + \frac{N\phi_0}{2} \tag{7.54}$$

$$P = N\mathscr{k}T\left(\frac{\partial \ln v_f}{\partial V}\right)_T - \frac{N}{2}\left(\frac{\partial \phi_0}{\partial V}\right)_T \tag{7.55}$$

$$C_V = N\mathscr{k}\left[\frac{3}{2} + 2T\left(\frac{\partial \ln v_f}{\partial T}\right)_V + T^2\left(\frac{\partial^2 \ln v_f}{\partial T^2}\right)_V\right] \tag{7.56}$$

$$S = N\mathscr{k}\left[\frac{5}{2} + \ln\left(\frac{2\pi m\mathscr{k}T}{h^2}\right)^{3/2} v_f\right] + T\left(\frac{\partial \ln v_f}{\partial T}\right)_V. \tag{7.57}$$

The intuitive introduction of the factor e^N into the liquid partition function is not important in the consideration of E, P, and C_V (or H and C_p) where the partition function appears as a derivative. It is important in the consideration of A (or G) and S where the partition function appears. The additional $N\mathscr{k}$ contribution to the entropy has lead to the common practice of referring to the entropic effect of introducing mobility into the partition function as "communal entropy." Support for the argument of introducing the e^N factor to apply over the entire range of liquid stability rather than apportioning it in some manner derives from the experimental observation that the entropy of fusion is about $N\mathscr{k}$ for monatomic fluids. We would recognize that the mobility introduced by inserting the factor e^N into the partition function is an over large response to the limited ability of a particle to exchange cells with other particles.

Since the potential in which a particle moves in a cell is determined by a large number of particle–particle interactions (extending over several particle diameters) it will be much more difficult to address the free volume term through the defining Eq. (7.52) than it was to address the configuration integral for interacting particle gaseous systems where the pair interaction approximation provided a second virial coefficient sufficient for most purposes. Until recent years liquid state theory, however, did concentrate on a variety of modelistic treatments that are largely unsuccessful and tedious to follow in complete detail but the approaches should be a part of our historical awareness even if they are inadequate representations. In recent years computational methods have expanded to the point that many modelistic simplifications are not required. Even when the computational results are impressive there is little in the way of intuitive simplification that we can extract from them. But again, the present state of such studies should be a part of our awareness. We begin with the modelistic approaches.

In the earliest treatments $\phi(r)$ is taken to derive from a classical oscillator model[6]

$$\phi(r) = \frac{2\pi v^2}{2m} r^2 dr$$

[6] See reference 8 (Moelwyn-Hughes, 1961), p. 729ff, for closer detail and original references.

and the partition function becomes

$$\mathscr{Z} = \left(\frac{2\pi m \ell T}{h^2}\right)^{3N/2} \left[\frac{1}{\nu}\left(\frac{\ell T}{2\pi m}\right)^{1/2}\right]^{3N} e^{-\phi(N)/\ell T}. \tag{7.58}$$

The bracketed term is the product of a velocity and a reciprocal frequency and its cube can be regarded as a volume so that

$$V_f = \left[\frac{1}{\nu}\left(\frac{\ell T}{2\pi m}\right)^{1/2}\right]^3. \tag{7.59}$$

The partition function is just that of classical three-dimensional oscillators

$$\mathscr{Z} = \left(\frac{\ell T}{h\nu}\right)^{3N} \tag{7.60}$$

and leads to $C_v = 3N\ell$. The experimental heat capacity of monatomic liquids actually follows a somewhat complex curve beginning at $3R$ at low temperatures and decreasing to about $9/4R$ at the critical temperature. The introduction of anharmonic correction terms to the Hooke's law expression, e.g.,

$$\phi_D = k_2 r^2 + k_3 r^3 + k_4 r^4 + \cdots \tag{7.61}$$

provides a basis for empirical fit to experimental data but it is clear that the model is more appropriate to an expanded solid that to a liquid.

In a direct address to the free volume term we can approximate the potential well shown in Figure 7.7 according to Figure 7.8 and imagine that particles adjacent to the central particle are fixed in their cell centers. If each particle has an excluded volume as defined for a van der Walls gas, the dimension over which the central particle can wander is $2(V^{1/3} - \sigma)$. In van der Waals terms

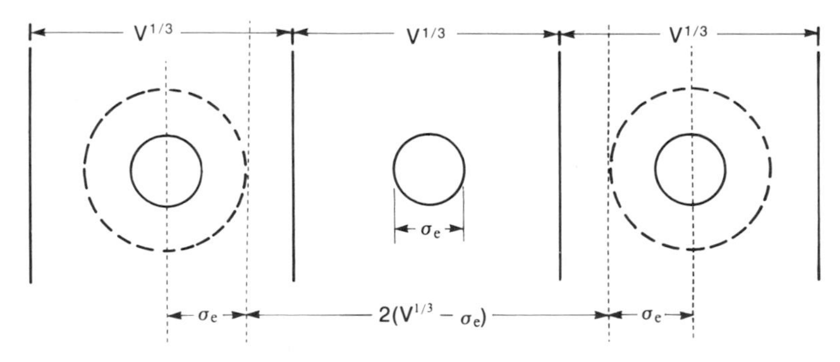

Figure 7.8. Illustration of the free volume concept for a particle occupying cell space defined by at rest neighbors.

$b = 2\pi N\sigma_e^3/3$ and σ_e is therefore $(3b/2\pi N)^{1/3}$. The free volume is

$$v_f = \left[2v^{1/3} - \left(\frac{12b}{\pi N}\right)^{1/3}\right]^3 \tag{7.62}$$

for $v = V/N$. The v_f so defined is a function of V and of T through that dependence. Inserting Eq. (7.62) into Eq. (7.51) and replacing ϕ_0 with the van der Waals term $-a^2/V$ with temperature dependence through V, an expression for the Helmholz free energy follows as

$$A = -N\ell T\left\{\ln\left(\frac{2\pi m\ell T}{h^2}\right)^{3/2} + 3\ln\left[2v^{1/3} - 2\left(\frac{3b}{2\pi N}\right)^{1/3}\right]\right\} - N\ell T + \frac{Na^2}{V} \tag{7.63}$$

and the pressure is

$$-(\partial A/\partial V)_T = -\frac{1}{N}\left(\frac{\partial A}{\partial v}\right)_T$$

so that

$$\begin{aligned} P &= \frac{3\ell T}{v^{1/3} - (3b/2\pi N)^{1/3}}\left(\frac{\partial v^{1/3}}{\partial v}\right)_T - \frac{a^2 \partial v^{-1}}{\partial v} \\ &= \frac{N\ell T}{V - (3b/2\pi)^{1/3}V^{2/3}} + \frac{N^2a^2}{V^2}. \end{aligned} \tag{7.64}$$

Equation (7.64) is known as the Eyring equation of state.[7] We can refer to it as the equation of state for a van der Waals liquid. Compressibilities for this simplified model follow from $(\partial P/\partial T)_V = \alpha_P/\kappa_T$ [Eq. (6.58)]

$$\frac{\alpha_P}{\kappa_T} = \frac{2\ell}{v_f^{1/3}v^{2/3}}$$

and from the defining expression $\kappa_T = -V^{-1}(\partial V/\partial P)_T$ [Eq. (6.57)]

$$\begin{aligned} (\partial P/\partial V)_T &= (\partial P/\partial v)_T(1/N) = -1/v\kappa_T \\ -\frac{1}{v}\left(\frac{1}{\kappa_T}\right) &= 2\ell T\frac{\partial}{\partial v}\left\{\left[2v - 2\left(\frac{3b}{2\pi N}\right)^{1/3} v^{2/3}\right]^{-1} + \frac{a}{v^2}\right. \\ \frac{1}{\kappa_T} &\cong \frac{2\ell T}{v_f^{1/3}v^{2/3}} + \frac{2a^2}{v^2}. \end{aligned} \tag{7.65}$$

[7] See reference 5 (Hirschfelder et al., 1954), p. 281ff.

TABLE 7.3. Coefficients of Compressibility $\kappa_T = -(\partial V/\partial P)_T/V$ and Thermal Expansivity $\alpha_P = (\partial V/\partial T)_P/V$

	$\kappa_T \times 10^4 \, atm^{-1}$		α_P	
Substance	Calculated[a]	Observed	Calculated[a]	Observed
$(C_2H_5)_2O$	2.12	1.29	1.68	1.58
CCl_4	1.07	1.05	1.14	1.23
$CHCl_3$	1.03	1.00	1.31	1.27
C_6H_6	0.85	0.95	1.12	1.24

Taken from J. O. Hirschfelder, C. F. Curtiss, and R. B. Bird, *The Molecular Theory of Gases and Liquids*. New York, Wiley, 1954.

[a]Calculated from the Eyring equation.

This will be the general route for comparison of modelistic approaches with experimental reality.

Illustrative data are shown in Table 7.3. In view of the crudity of the model the results for spherical molecules are reasonable. More realistic potential functions should improve things but the arithmetic quickly becomes too involved for our purposes even when unrealistic structural features are imposed on the liquid.

If we start with some (transient) structure such as face-centered cubic (a lattice model), each molecule has 12 nearest neighbors (a realistic coordination number for simple fluids) that are the adjacent molecules in the face-centered plane and the face-centered surrounding molecules (as in Figure 7.9). The free volume might assume the form of a dodecahedron with face to center distance of $a/2$ and $v = (V/N)$ is $a^3\sqrt{2}$ where a is the nearest neighbor distance. When the neighbors are fixed the central molecule is free to move so long as $v > \sigma^3/\sqrt{2}$ where σ is the molecule diameter. The faces of the dodecahedron will be replaced by concave spherical segments of decreasing radius as the v/σ^3 ratio diminishes

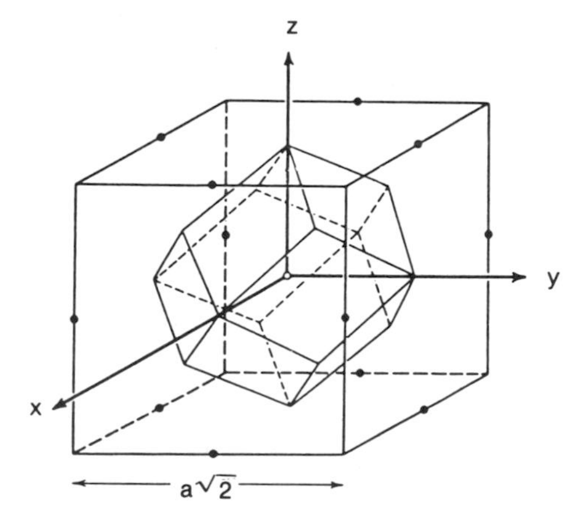

Figure 7.9. Illustration of the ideal free volume for a body centered particle in a face centered lattice of at rest particles. Taken from J. D. Hirschfelder, C. F. Curtiss, and R. D. Bird, *Molecular Theory of Gases and Liquids*, Wiley, New York, 1954. Reprinted with permission by John Wiley and Sons Inc.

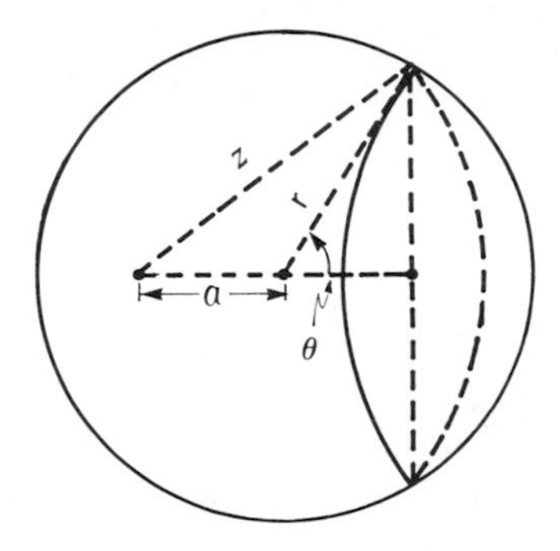

Figure 7.10. Illustration of the spherical volume approximation for the free volume Taken from T. L. Hill, *An Introduction to Statistical Thermodynamics*, Addison-Wesley, Reading, MA, 1960.

and the geometries will become quite complex. In the Leonard–Jones and Devonshire model[8] the free volume is replaced by a sphere that will fit inside the exact free volume, in effect smearing out interactions with nearest neighbors over the surface of the sphere. The interactions are averaged by supposing a molecule to be at rest at some point along a diameter while another moves over the spherical surface which has maximum radius $a\sqrt{2}$. Then, as illustrated in Figure 7.10, the distance z between the 2 atoms is

$$z = (a^2 + r^2 + 2ar\cos\theta)^{1/2} \tag{7.66}$$

and $dz/d\theta$ is $-ar\sin\theta\, d\theta$. The area of an infinitesimal ring at r and θ is $2\pi r^2 \sin\theta\, d\theta$ and the density of smeared out neighbors is $C/4\pi r^2$ where C is the number of nearest neighbors (12 in this case) so that

$$\phi(a, r) = 6\int_0^{\pi} \phi(z)\sin\theta\, d\theta.$$

For the Leonard–Jones potential

$$\phi(z) = \epsilon\left(\frac{r_e}{z}\right)^{12} - 2\epsilon\left(\frac{r_e}{z}\right)^{6}$$

with z given by Eq. (7.66) the integration will produce $\phi(\mathrm{r})$ and from this $\mathrm{v_f}$ but only in terms of integral forms which must be evaluated numerically.[9] The agreement with experiment is not good in terms of providing satisfactory critical constants and other thermodynamic features of the liquid state. It is recognized that the deficiencies are the result of building in too much order with N sites each singly occupied. The real increase in liquid volume over that of the solid

[8] See reference 4 (Hill, 1960), p. 292ff for original reference and closer detail.

[9] A more detailed treatment including original references, tables of the integrals, and calculated thermodynamic functions can be found in reference 5 (Hirschfelder et al., 1954), p. 295ff.

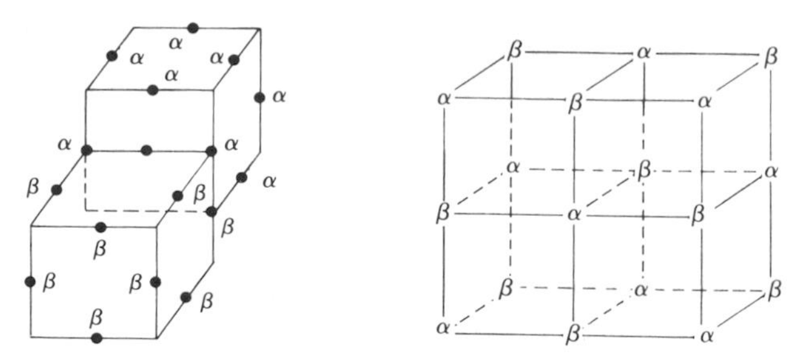

Figure 7.11. Illustration of two interpenetrating lattices forming a two site geometry for vacancy assignments.

would suggest that a lattice model will require empty sites as well as expansion of the lattice. One approach, known as order–disorder theory, develops as follows.[10]

If we imagine that there are not N but 2N sites for the N particle system where the sites occupy lattice points in interpenetrating cubic structures we can define a completely ordered structure as one in which all of one set of sites (call them α sites) is occupied while all of the other set of sites (call them β sites) are empty. A particular state of disorder exists when both site sets are partially occupied. In the ordered state the configurational part of the partition function is

$$Q = v_f^N e^{-\phi_O/kT} \tag{7.67}$$

while in the disordered state the configurational part of the partition function is

$$Q = v_f^N \frac{N!}{N_\alpha!(N - N_\alpha)!} \sum e^{-(\phi_D - \phi_O)/kT}. \tag{7.68}$$

For a lattice structure such as sketched in Figure 7.11 that may be pictured as a lattice of α sites with interpenetration by a cubic lattice of β sites, each α site has 6 nearest neighbor β sites and 12 next nearest neighbor α sites. To reduce the arithmetic to manageable dimensions the interaction energies are written as average values rather than instantaneous displacement distance values so that the displacement energy is

$$\phi_D = \tfrac{1}{2}N_\alpha(6\chi_\beta\epsilon_{\alpha\beta} + 12\chi_\alpha\epsilon_{\alpha\alpha}) + \tfrac{1}{2}N_\beta(6\chi_\alpha\epsilon_{\alpha\beta} + 12\chi_\beta\epsilon_{\alpha\alpha}) \tag{7.69}$$

where χ_α, χ_β are fractions of the respective sites occupied and the average interaction energies are $\epsilon_{\alpha\alpha}(= \epsilon_{\beta\beta})$ and $\epsilon_{\alpha\beta}(= \epsilon_{\beta\alpha})$. In the ordered state, taking

[10] See reference 8 (Moelwyn–Hughes, 1961), p. 742ff.

all α sites as occupied (occupation of all β sites is also complete order)

$$\phi_0 = \frac{N}{2}\epsilon_{\alpha\alpha}$$

and in the disordered state

$$\phi_D - \phi_0 = 6N(\epsilon_{\alpha\beta} - 2\epsilon_{\alpha\alpha})\chi_\alpha(1 - \chi_\alpha). \tag{7.70}$$

The maximum term in the Eq. (7.68) sum will result when

$$(\epsilon_{\alpha\beta} - 2\epsilon_{\alpha\alpha}) = \frac{kT}{3}\frac{\ln \chi_\alpha(1 - \chi_\alpha)}{(2\chi_\alpha - 1)}. \tag{7.71}$$

When $(\epsilon_{\alpha\beta} - 2\epsilon_{\alpha\alpha}) < kT/3$ there is one maximum for this expression at $\chi_\alpha = 0.5$; for $(\epsilon_{\alpha\beta} - 2\epsilon_{\alpha\alpha}) > kT_3$ there are two maxima, symmetrical around $\chi_\alpha = 0.5$. Setting $\chi_\alpha = \chi_0$ for the maximum the Helmholz free energy will be minimum valued for

$$A = -kT\ln\left(\frac{2\pi mkT}{h^2}\right)^{3N/2} - 6N(\epsilon_{\alpha\beta} - 2\epsilon_{\alpha\alpha})(1 - \chi_0)$$
$$- 2NkT[\chi_0 \ln \chi_0 + (1 - \chi_0)\ln(1 - \chi_0)] \tag{7.72}$$

and the pressure will follow when $(\epsilon_{\alpha\beta} - 2\epsilon_{\alpha\alpha})$ is related to the volume of the system. We have

$$\frac{\partial A}{\partial V} = \frac{\partial A}{\partial \omega}\frac{d\omega}{dV} = 6N(1 - \chi_0)\frac{d\omega}{dV} = -P \tag{7.73}$$

where we have written $(\epsilon_{\alpha\alpha} - 2\epsilon_{\alpha\beta}) = \omega$. It is not practical to attempt further calculation but the theory could be a useful empirical framework in a form such as

$$\omega = \omega_0(V_0/V)^v$$

$$\frac{d\omega}{dV} = -v(1 - \chi_0)\omega_0 V_0^v V^{-(v-1)}$$

$$\frac{PV_0}{NkT} = \frac{6v\omega_0}{kT}(1 - \chi_0)\left(\frac{V_0}{V}\right)^{v+1}. \tag{7.74}$$

From the remarks preceding Eq. (7.72), it is clear that certain values of the pressure will correspond to two volume (density) values, one partially ordered ($\chi_0 \neq 0.5$) and the other completely disordered ($\chi_0 = 0.5$). We can return to this point in the discussion of phase equilibrium.

Although lattice theories represent an important concept in liquid state theory, it is clear that lattice expansion alone will not accommodate the very large volume changes that are observed over the range of liquid stability (in terms of linear dimension, up to 50%). It will be necessary to introduce holes. Significant structure theory (Eyring) represents a simple and (empirically) successful framework for introducing holes. The holes are empty cells that increase in number to accommodate volume expansion. They are mobile and confer gas-like degrees of freedom on adjacent molecules that exchange sites with the empty cell.

If there are N occupied cells and N_0 vacant cells, the fraction of occupied cells is $N/(N + N_0) = 1/(1 + N_0/N) = 1/(1 + x)$ where x is the ratio of vacant cells to occupied cells and $x/(1 + x) = N_0/(N + N_0)$ is the fraction of total cells that is vacant. If we assume that v_{cell} (for both occupied cells and holes) is the same as that for the solid $(= V_s/N)$ then $x = (V_\ell - V_s)/V_s$ where V_ℓ and V_s are the liquid and solid total volumes and $x/(1 + x) = (V_\ell - V_s)/V_\ell$. If any occupied cell in the liquid is surrounded by C_ℓ other cells (the liquid phase coordination number) there are $n_h = C_\ell x/(1 + x)$ holes adjacent to an occupied cell and there are $[1 + n_h \exp(-\epsilon_h/kT)]$ ways in which the occupied cell and the holes can be arranged where $\exp(-\epsilon_h/kT)$ is the probability factor that a given molecule has the requisite energy to occupy a hole. The requisite energy associated with replacement of a hole by an occupied site is taken to be the traslational energy component of the hole in one of the directions in which the hole must move in order for the occupancy exchange to occur. There are $1/n_h$ occupied sites adjacent to each vacancy and the number of directions in which the hole can move is $2C_\ell/n_h = 2(1 + x)/x$. If the total translational energy of the hole is assumed to be $\Delta E_{sub}/N$ the average energy taken over all directions is $\epsilon_h = \frac{1}{2}(\Delta E_{sub}/N)x/(1 + x)$. A molecule must possess at least this energy to provide mobility to the hole through exchange. From these arguments we can construct the general partition function in terms of the product of $N[1/(1 + x)]$ solid-like particles and $N[x/(1 + x)]$ gas-like particles

$$\mathscr{Z} = z_s^{N/(1+x)} z_g^{Nx/(1+x)} \tag{7.75}$$

where

$$z_s = [1 + n_h \exp(-\epsilon_h/kT)][1 - \exp(-\Theta_E/T)]^{-1}[\exp(-\epsilon_0/kT)] \tag{7.76}$$

and

$$z_g = \left(\frac{2\pi m k T}{h^2}\right)^{3/2} \left\{\frac{(V_\ell - V_s)}{[Nx/(1 + x)]!}\right\}^{[Nx/(1+x)]^{-1}}. \tag{7.77}$$

There are, of course, alternative choices in formulating the partition function but purely modelistic address will not improve the calculated thermodynamic functions over other approaches. Quite spectacular results, however, follow in the empirical framework suggesting that the general model is an acceptable picture of liquid phase behavior. We will return to this point in the discussion of aqueous ionic solutions.

To this point the various approaches to liquid state theory that we have examined introduce structure largely as a computational convenience. Even in simple fluids (liquid argon, etc.) we would imagine that geometric features will introduce some structure and that with more complex molecules internal structure (linear chains, dipole and quadrupole moments) and pseudochemical interactions (hydrogen bonding) will impose preferred orientations that lead to structure in the liquid. Although these same features are important in determining crystal structure and provide a real justification for lattice approaches to liquid state theory, a different framework is more compatible with computer simulation studies that now dominate the field and with experimental structure studies that are becoming increasingly refined. The required framework features various forms of distribution functions that measure the probabilities associated with microscopic nonrandom density fluctuations.

In simple fluids (monatomic or spherical polyatomic nonpolar molecules) repulsive forces will dominate packing features. The repulsive forces will derive from modified (by the presence of adjacent molecules) Leonard–Jones type pair interaction energies. In general we expect any molecule to be surrounded by a coordination sphere of other molecules at a center-to-center distance of about σ_e. This argument can be extended to a second and third coordination sphere of decreasing regularity. Thermal disorder will partially override any longer range order. Such structure would be reflected, as in Figure 7.12, by local density variation around a given molecule. The presence of such structure is reflected by X-ray and neutron diffraction, but, as would be anticipated, the patterns are diffuse as compared with those for solid structures. In part this is because the structures themselves are transient and in part because they lack the full effect of phase coherent interference between successive layers in patterns of extended

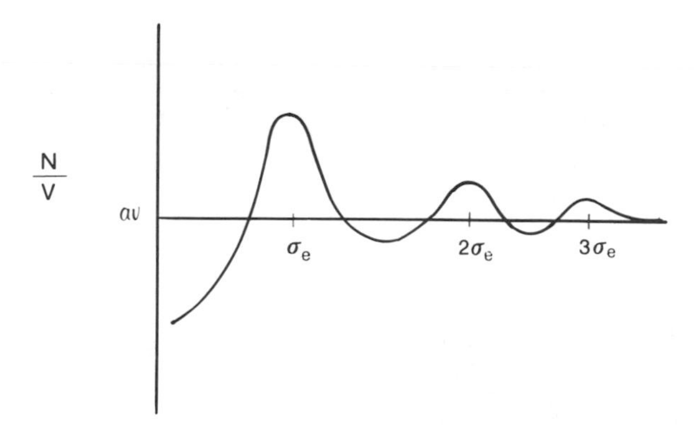

Figure 7.12. Illustration of the local density variation around a central particle due to extended pair interactions.

order. Polyatomic molecules pose special problems because bond distances are comparable to internuclear distances and because different nuclei scatter radiation in slightly different ways. However, since X-rays are scattered by electrons and neutrons by nuclei, it is possible to obtain information about the distribution of molecule centers from X-ray scattering data and information about orientation from neutron scattering data. Although, by these means, a wealth of structural data can be obtained, it will be very difficult to incorporate this into effective thermodynamic functions. We can consider here only the most simple cases.

If we designate the probability of encountering a molecule in any small volume element $d\mathbf{r}$, at the point $\mathbf{r}_1$ from some arbitrary origin as $n^{(1)}\mathbf{r}_1$, the number of molecules in the volume element is $n^{(1)}\mathbf{r}_1 d\mathbf{r}_1$. The quantity $n^{(1)}\mathbf{r}_1$ is called the singlet distribution function and in a homogeneous fluid it will be equal to N/V. Given the presence of a particle at r_1 the probability of also finding a particle at $\mathbf{r}_2$ in the volume element $d\mathbf{r}_2$ is $n^{(2)}\mathbf{r}_1\mathbf{r}_2$, which we call the pair distribution function. Since the fluid is isotropic the pair distribution function depends only on the scalar distance $r_2 - r_1 = r_{12}$. As $r \rightarrow \infty$ the pair distribution function becomes independent of the presence of a particle at $\mathbf{r}_1$ and $n^{(2)}r \rightarrow (\mathrm{N/V})^2$ but at close distances this will not be true. We introduce the pair correlation function $\mathrm{g}(r)$ to accommodate this feature

$$n^{(2)}(r) = \mathrm{g}(r)\left(\frac{\mathrm{N}}{\mathrm{V}}\right)^2 = \mathrm{g}(r)\rho^2. \tag{7.77}$$

Higher order correlation functions, $n^{(3)}(\mathbf{r}_1\mathbf{r}_2\mathbf{r}_3)$, etc. are defined in the same manner and other kinds of distribution functions can be defined. The number of particles in a spherical shell of thickness dr will not be $dn(r) = \rho 4\pi r^2 dr$ if the shell surrounds a particle at close distance but

$$dn(r) = \rho \mathrm{g}(r) 4\pi r^2 dr \tag{7.79}$$

and $dn(r)$ is referred to as the radial distribution function. The excess number of particles in the volume element $4\pi r^2 dr$ due to the presence of a particle at $\mathbf{r}_1$ is $\rho[\mathrm{g}(r) - 1]4\pi r^2$ and $h(r) = \mathrm{g}(r) - 1$ is called the total correlation function.

For very low density systems (gases) we would recognize that

$$\lim_{\rho \to 0} \mathrm{g}(r) = e^{-U_{12}(r)/kT} \tag{7.80}$$

where U_{12} is the pair potential. In dense fluids we recognize that each particle is within a force field defined by many adjacent particles and it is common to introduce the concept of the potential of mean force $U(r)$ for which

$$\mathrm{g}(r) = e^{-U(r)/kT} \tag{7.81}$$

with $U(r)$ defined in terms of the mean force $f(\mathbf{r})$

$$f(\mathbf{r}_1) = \partial U(\mathbf{r}_1)/\partial \mathbf{r}_1. \tag{7.82}$$

Quantitatively, the force acting on a particle at $\mathbf{r}_1$ is taken to be the sum of the force resulting from a particle at $\mathbf{r}_2$ and the average force exerted by all other particles

$$f(\mathbf{r}_1) = \frac{\partial U(\mathbf{r}_1\mathbf{r}_2)}{\partial \mathbf{r}_1} + \int \frac{n^{(3)}(\mathbf{r}_1\mathbf{r}_2\mathbf{r}_3)}{n^{(2)}(\mathbf{r}_1\mathbf{r}_2)} \frac{\partial U(\mathbf{r}_1\mathbf{r}_3)}{\partial \mathbf{r}_1} dr \tag{7.83}$$

where $n^{(3)}(\mathbf{r}_1\mathbf{r}_2\mathbf{r}_3)$ is the triplet distribution function. From Eqs. (7.81) and (7.82)

$$f(\mathbf{r}_1) = -\frac{\partial \ln \mathrm{g}(r)}{\partial r_1} = -\frac{\partial \ln n^{(2)}(r)/\rho}{\partial r_1}. \tag{7.84}$$

Therefore, combining Eqs. (7.83) and (7.84), we see that the integral equation for $n^{(2)}$ will involve $n^{(3)}$ and, it follows, the equation for $n^{(3)}$ will involve $n^{(4)}$, etc. If $\ln \mathrm{g}(r)$ is expanded in powers of density

$$\ln \mathrm{g}(r) = -\frac{U(r)}{kT} + \ln[1 + \rho x_1(r) + \rho^2 x_2(r) + \cdots] \tag{7.85}$$

the functions $x_1(r)$, $x_2(r), \ldots$ are integrals that must be approximated in some manner. Various forms of approximate integral equations (Born–Green–Yvon and Kirkwood, to cite two) have been constructed on the basis of the so-called superposition approximation:

$$n^{(3)}(\mathbf{r}_1\mathbf{r}_2\mathbf{r}_3) = \rho^3 \mathrm{g}(\mathbf{r}_1\mathbf{r}_2)\mathrm{g}(\mathbf{r}_2\mathbf{r}_3)\mathrm{g}(\mathbf{r}_1\mathbf{r}_3). \tag{7.86}$$

The expressions are involved and not very successful as would be anticipated for the neglect of integral terms beyond $x_1(r)$. Other approaches utilize graph theory (which we cannot address here) to classify the integrals relating to higher order interactions and to prescribe how correlation functions and the potential of mean force are to be calculated. Approximations for the direct correlation function $f(r)$ defined as

$$h(\mathbf{r}_1\mathbf{r}_2) = \mathrm{g}(\mathbf{r}_1\mathbf{r}_2) - 1 = f(\mathbf{r}_1\mathbf{r}_2) + \rho \int f(\mathbf{r}_1\mathbf{r}_2)h(\mathbf{r}_2\mathbf{r}_3)dr, \tag{7.87}$$

will involve dropping particular sets of terms that contribute to x_2, etc. The so-called hypernetted chain (HNC) and Percus–Yevic (PY) approximations are the basis for most current work in computer simulation and for association of experimental scattering data [which provide $\mathrm{g}(r)$ and $f(r)$] with thermodynamic expressions. Compressibility is particularly appropriate for comparison of calculation and experiment since the compressibility equation takes on a quite simple form

$$\kappa_T = \frac{1}{\rho k T} + \frac{1}{kT}\int_0^\infty h(r) 4\pi r^2 dr. \tag{7.88}$$

It is not particularly useful to set down forms for the set of thermodynamic functions in terms of the distribution function since they can be utilized only in connection with (very large) computer calculations. In fact, this chapter after Eq. (7.74) is not intended to provide operationally useful material but rather to indicate the complexity of liquid state theory. Calculation of thermodynamic properties of gases and solids coupled with experimental heat capacities and thermodynamic data related to phase changes as developed in Chapter 9 will be our source of liquid phase thermodynamic data.

PROBLEMS

7.1. For Al the value Θ_D as obtained from elastic constants is 402 and for Ag it is 214. Using the Table of Debye functions ($u = \Theta_D/T$) and Eqs. (7.23), (7.25) and (7.26) calculate $E - E_0$, $A - E_0$, and S for both substances at 100, 400, and 600 K.

7.2. In terms of the Debye frequency $\nu_{max} - \Theta_D k/h$, the Einstein frequency Θ_E is $3\nu_{max}/4$. Using the Table of Einstein functions from Chapter 5 calculate $E - E_0$, $A - E_0$, and S for Al at 600 K and compare with the Debye values.

7.3 By a suitable graphic or computer integration determine the entropy of metallic silver at 25°C from the following data.

T(K)	C_p ($J K^{-1}$)	T(K)	C_p ($J K^{-1}$)	T(K)	C_p ($J K^{-1}$)	T(K)	C_p ($J K^{-1}$)
15	0.67	90	19.13	170	23.61	250	25.03
30	4.77	110	20.96	190	24.09	270	25.31
50	11.65	130	22.13	210	24.42	290	25.44
70	16.33	150	22.97	230	24.73	300	25.50

Assume that the heat capacity approaches absolute zero according to a T^3 relation; that is, $C_p = \text{const. } T^3$, and $C_p = 0$ at $T = 0$.

7.4. A number of relationships were developed in Chapter 6 that are general in their application. The expressions for $(C_p - C_v)$ and for C_p/C_v in terms of compressibilities are often useful in their application to condensed phases particularly when, as with liquids, we have small confidence in our models.

The value for thermal expansivity of CCl_4 at 20°C and 1 atm is $1.236 \times 10^{-2}\,\text{deg}^{-1}$ and the isothermal expansivity is $91.6 \times 10^{-6}\,\text{atm}^{-1}$ between 0 and 100 atm. Neglecting any changes in the coefficients estimate the change in volume when 200 liters of CCl_4 is heated from 20° to 30°C in an open vessel (i.e., at constant 1 atm pressure). If the 200 liters of CCl_4 were in a closed system and compressed to 50 atm at 20°C, what volume would it occupy? If the CCl_4 completely occupies a sealed vessel at 20°C and 1 atm pressure at what temperature will the pressure reach 50 atm? (From reference 9.)

CHAPTER CONTENT SOURCES AND FURTHER READING

1. S. M. Blinder, *Advanced Physical Chemistry*, Chapter 23. Macmillan, Toronto, 1969.
2. T. M. Reed and K. E. Gubbins, *Applied Statistical Mechanics*, Chapters 8, 9, 10, and 11. McGraw-Hill, New York, 1973.
3. Y. Marcus, *Introduction to Liquid State Chemistry*, Chapters 1, 2, and 3. Wiley, New York, 1977.
4. T. L. Hill, *Introduction to Statistical Thermodynamics*, Chapters 5, 16, and 19. Addison-Wesley, Reading MA, 1960.
5. J. D. Hirschfelder, C. F. Curtiss, and R. B. Bird, *Molecular Theory of Gases and Liquids*, Chapter 4. Wiley, New York, 1954.
6. G. M. Barrow, *Physical Chemistry*, Chapter 5, 4th ed., McGraw-Hill, New York, 1961.
7. R. S. Berry, S. A. Rice, and J. Ross, *Physical Chemistry*, Chapters 22 and 23. Wiley, New York, 1960.
8. E. A. Moelwyn-Hughes, *Physical Chemistry*, 2nd ed., Chapters XIII, XV, and XVI. Pergamon Press, London, 1961.
9. M. A. Paul, *Principles of Chemical Thermodynamics*, McGraw-Hill, New York, 1951.

SYMBOLS: CHAPTER 7

New Symbols

ϕ_0 = potential energy of an at rest particle in a system of at rest particles

$g(\nu)$ = distribution function (among frequencies)

f = force constant

C_l, C_t = longitudinal and transverse velocities of sound

x = representation of $h\nu/kT$

u = representation of $h\nu_m/kT$

ν_m = limiting (maximum) frequency in a range of frequencies

Θ_D/k, D(u), D(Θ_D/T) = Debye characteristic temperature, Debye function

f(Θ_D/T), f(u) = representation of A/T (one time use)

γ_G = the Grüniesen function
μ_0 = the chemical potential at 0 K
$\phi(\mathrm{N})\phi(\mathrm{r}), \phi_0$ = potential energy of a system of particles, in motion and at rest
$\mathrm{v}, \mathrm{v}_{\mathrm{cell}}, \mathrm{v}_{\mathrm{f}}$ = volume per molecule, cell volume, free volume
α, β = designations of lattice sites (one time use)
χ_α, χ_β = fraction of filled and empty sites (one time use)
$\phi_{\mathrm{D}}, \phi_{\mathrm{O}}$ = potential energies for disordered and ordered filling of α, β sites (one time use)
ω, ω_0 = exchange energy, general and in some reference state
C_ℓ = coordination number in a liquid
$\mathrm{V}_\ell, \mathrm{V}_\mathrm{s}$ = volume of liquid, of solid
$n^{(1)}, n^{(2)}, n^{(3)}$ = singlet, pair, triplet features of the distribution function
$\mathrm{g}(r)$ = distribution function for particles at distance r
$U(r)$ = potential of mean force
$x_1(\mathrm{r}), x_2(\mathrm{r}) \cdots$ = integrals in the expression for $\ln \mathrm{g}(r)$

Carryover Symbols

ϕ, N, k, z, $\mathscr{Z}$, v_i, v_e, Θ_E, $\mathrm{C_V}$, E, $\mathrm{E_0}$, T, A, S, μ, P, V, α_p, κ_T, σ_e, ρ.

8

MULTICOMPONENT SYSTEMS

We have acquired substantial computational capability and some appreciation for the role of various energetic parameters in the description of thermodynamic states for a single component system. The most obvious effect of adding components to a system of statistical concern is that, with every component added, we expand the total set of energy states subject to the constraint that the energy states for each component are to be occupied by that component exclusively. In the simplest possible case the energy states available to a single species in the mixture would be the unperturbed states available to that species in its pure state subject to the same temperature and volume constraints. For structurally simple gases in dilute systems we would anticipate that such conditions would usually be met. In condensed systems and for gases having highly structured particles at moderate to high densities, the normal energy levels would be perturbed by the presence of other species and the interaction potentials will be altered. It is not difficult to anticipate that in mixtures of such systems, where we were forced to construct simplified models of the potential energies of interaction in the one component case, two or more components present in varying proportions will present problems that will soon exhaust our abilities to compute. We do, however, anticipate that the statistical argument will provide a more adequate framework for empirical extension than that which we could construct if unguided.

As always we begin our considerations with an examination of the partition function and with the simplest possible case; the κ-component mixture of ideal gases. For this purpose we return to the isolated, nonstructured particle system at constant E, V, and N. We relax the constraint that all particles of the system are identical, retaining all other constraints. We still have a closed, dilute gaseous system of N independent structureless particles where $N = \sum^{K} N_K$. We further require that the particles are noninteracting. We exclude chemical change but do not exclude energy exchange on collision. Being structureless, the particles of a single species are characterized only by mass m_K. Being gaseous, all particles are free to move throughout the entire volume and to exchange energy with other particles on collision. Within a species the particles are indistinguishable. The energy levels available to each particle species are determined by V and m_K and, being independent, the particles of each species distribute themselves among their available energy states as if they alone occupied the entire volume. The total energy is translational for the structureless particles and additive

$$(E_K)_\Sigma = \sum^{K} N_K \sum^{j} p_{j,K} \epsilon_{j,K}. \tag{8.1}$$

Each $p^*_{j,K}$ is

$$p^*_{j,K} = \frac{e^{-\epsilon_{j,K}/kT}}{z_K} \tag{8.2}$$

where the ϵ_K depend on m_K and V. In all expressions V and T are commonly valued in the equilibrium system and z_K depends on m_K, V, and T. The p^* are such that, for each species

$$\sum p^*_{j,K} \epsilon_{j,K} = \bar{\epsilon}_K = \tfrac{3}{2} kT \tag{8.3}$$

and all $\bar{\epsilon}_K$ are identical. In the ideal gas circumstance the energy change is indifferent to the nature of the structureless particles added to or taken from the system. These intuitive extensions of the single component system are supported in full by the detailed statistical argument.

The maximization procedure, embodied in outline in Chapter 1 and detailed in the appendices,[1] requires one constraint on each of the individual particle numbers since they cannot be interchanged,

$$\sum^{j} N_{j,1} = N_1, \qquad \sum^{j} N_{i,2} = N_2 \cdots \sum^{i,K} N_{i,K} = N_K$$

and therefore one multiplier $\alpha_1, \alpha_2, \ldots, \alpha_K$ for each constraint. Since energies can be interchanged on collision without regard to identity there is only one constraint on energy; that the total be fixed,

$$\sum^{i} N_{j,1}\epsilon_{j,1} + \sum^{j} N_{j,2}\epsilon_{j,2} + \cdots \sum^{i} N_{i,K}\epsilon_{j,K} = E$$

[1] See Appendix C on Statistical Considerations.

and there is therefore only one value of β and one value of T. Each of the undetermined multipliers $\alpha_1, \alpha_2, \ldots, \alpha_K$ is related to the species-specific partition function

$$\alpha_K = \ln \frac{z_K}{N_K} = -\frac{\mu_K}{kT}.$$

The detailed analysis also shows that

$$\Omega = \prod^{K} \frac{(g_{i,K})^{N_{i,K}}}{N_{i,K}!} \tag{8.4}$$

$$E_\Sigma = -\left(\frac{\partial \ln \mathscr{Z}_\Sigma}{\partial \beta}\right)_{V,N} = kT^2\left(\frac{\partial \ln \mathscr{Z}_\Sigma}{\partial T}\right)_{V,N} \tag{8.5}$$

where

$$\mathscr{Z}_\Sigma = \prod^{K} \frac{z_K^{N_K}}{N_K!}. \tag{8.6}$$

These results would be anticipated on the basis that the energies of the individual (independent) particles are additive by species to provide the total energy and the individual $\ln z$ terms are additive only if the partition functions combine as product terms to provide $\mathscr{Z}_\Sigma$. The individual entropies and free energies are also additive by species to provide the total entropy on the basis of Eqs. (8.4) and (8.6) by the same argument,

$$S_\Sigma = k \ln \Omega = k \sum^{i,K} N_{i,K} \ln g_{i,K} - k \sum^{i,K} \ln N_{i,K} \tag{8.7}$$

$$A_\Sigma = -\sum^{K} N_K kT \ln \frac{z_K e}{N_K}. \tag{8.8}$$

Finally, in the ideal gas with all particle species having the same average energy, the average momenta are also equal, heavier particles having a lower average velocity. The total pressure, reflecting the summed momentum exchanges in the ideal gas system, is dependent only on the number of particles not on their nature. It is, however, often useful to sum the momentum exchanges with bounding walls over each species separately providing

$$P_\Sigma = \frac{kT}{V} \sum^{K} N_K = \sum^{K} P_K \tag{8.9}$$

where each P_K is referred to as the κth component partial pressure. Each partial pressure reflects the number of particles of that species present.

We summarize these multicomponent system considerations in the ideal gas as follows. The pressure is the sum of the individual species partial pressures. The energy, entropy, free energy, and heat capacity of the multicomponent system of independent particles are obtained by summing over the individual species present. Each individual term involves the species specific partition function arising from the species specific energy level set [Eq. (1.1)]. Since the energies are all translational, the partition functions are of the same form and show the same dependence on T. Both E_K and $C_{V,K}$, which are partition function derivatives with respect to T, have the same per particle value for all species

$$\frac{E_K}{N_K} = \tfrac{3}{2}kT \tag{8.10}$$

$$\frac{C_{V,K}}{N_K} = \tfrac{3}{2}k. \tag{8.11}$$

Both the total energy and the total constant volume heat capacity are indifferent, in the ideal gas case, to the fact that several species are present. Together with the system pressure, only the total number of particles is important. The enthalpy is related to E and the PV product, therefore both H and C_P are also indifferent to the particle species

$$H_K = \tfrac{3}{2}N_K kT + N_K kT \tag{8.12}$$

$$C_{P,K} = \tfrac{3}{2}N_K k + N_K k. \tag{8.13}$$

The entropy involves the partition function directly so it is species specific as are the Gibbs and Helmholz free energies

$$S_K = k \ln \mathcal{Z}_K + \frac{E_K}{T} = N_K k \ln \frac{z_K}{N_K} + \frac{5}{2}N_K k \tag{8.14}$$

$$A_K = -N_K kT \ln \frac{z_K}{N_K} - N_K kT \tag{8.15}$$

$$G_K = -N_K kT \ln \frac{z_K}{N_K}. \tag{8.16}$$

As always, the total thermodynamic function is the sum of species contributions

$$H_\Sigma = \tfrac{5}{2}kT\sum N_K, \qquad C_{P,\Sigma} = \tfrac{5}{2}k\sum N_K, \qquad G_\Sigma = -\sum N_K kT \ln \frac{z_K}{N_K}, \text{ etc.}$$

When these arguments are extended to real gas mixtures in PVT regions where intermolecular interactions are trivial we would regard the real gas mixture as behaving ideally. If the real gas mixture is a dilute mixture of nonatomic gases the above relationships are complete. If the real gas mixture is a dilute mixture of structured particle gases, then all energy states that would be available to the molecules of a single species if that species alone occupied the entire system volume will be available to that species in the mixture. In the absence of interactions the energy states will be the set of normal unperturbed states.

In such dilute real gas mixtures

$$\mathscr{Z}_{\Sigma} = \prod^{\mathrm{K}} \frac{z_{\mathrm{t,K}}^{\mathrm{N_K}}}{\mathrm{N_K}!} \prod^{\mathrm{K}} (z_{\mathrm{r,K}}, z_{\mathrm{v,K}}, z_{\mathrm{c,K}})^{\mathrm{N_K}} \tag{8.17}$$

and all remarks developed for the ideal gas system apply to the first product term in Eq. (8.15). There is no volume dependence in any of the partition function contributions included in the second product term and

$$\mathrm{P} = k\mathrm{T}\left(\frac{\partial \ln \mathscr{Z}_{\mathrm{K}}}{\partial \mathrm{V}}\right)_{\mathrm{T,N}} = k\mathrm{T}\left[\sum^{\mathrm{K}} \frac{\partial}{\partial \mathrm{V}} \frac{(\ln z_{\mathrm{t,K}}^{\mathrm{N_K}})}{\mathrm{N}!}\right] = \sum^{\mathrm{K}} \mathrm{P_K}. \tag{8.18}$$

Equation (8.18) produces the general ideal gas argument that the PV product is structure independent

$$\mathrm{PV} = k\mathrm{T}\sum^{\mathrm{K}} \mathrm{N_K} = \mathrm{N}_{\Sigma} k\mathrm{T}$$

and will make the same contribution to any thermodynamic function in which it appears whether or not structure is a feature of other contributions to that function. Therefore, we obtain for the total energy in a structured particle ideal gas mixture of K components

$$\mathrm{E}_{\Sigma} = k\mathrm{T}^2 \sum^{\mathrm{K}} \left(\frac{\partial \ln \mathscr{Z}_{\mathrm{K}}}{\partial \mathrm{T}}\right)_{\mathrm{V,N}} \tag{8.19}$$

and

$$\mathrm{H}_{\Sigma} = k\mathrm{T}^2 \sum^{\mathrm{K}} \left(\frac{\partial \ln \mathscr{Z}_{\mathrm{K}}}{\partial \mathrm{T}}\right)_{\mathrm{V,N}} + k\mathrm{T}\sum^{\mathrm{K}} \mathrm{N_K} \tag{8.20}$$

similarly

$$\mathrm{A}_{\Sigma} = -k\mathrm{T}\sum^{\mathrm{K}} \ln \mathscr{Z}_{\mathrm{K}} \tag{8.21}$$

$$\mathrm{G}_{\Sigma} = -k\mathrm{T}\sum^{\mathrm{K}} \ln \mathscr{Z}_{\mathrm{K}} + k\mathrm{T}\sum \mathrm{N_K}. \tag{8.22}$$

Clearly, even though the PV product is structure, therefore species, independent for the ideal gas none of the other thermodynamic functions of interest is. There is, however, no difficulty in constructing the appropriate contribution for each species from Eq. (8.17) only noting that V appearing in the system partition function must be the volume occupied by the mixed system. This precludes, generally, obtaining the partition function for individual species each occupying an initial volume $V_{i,K}$ and multiplying these to obtain $\mathscr{Z}_{\Sigma}$. The problem extends beyond the mixing of gaseous systems but the analysis is best addressed in the ideal gas framework that can, as in prior considerations, always be extended to more complex systems.

We consider a particular ideal gas mixing process. Our mixing conditions will be to allow two equal volumes of different monatomic gases under the same pressure to combine into a single volume that is the sum of the unmixed volumes. Such a process is quite general in that it is parallel to the mixing process when two liquids are mixed where the final volume is (ideally) additive (but in practice almost never quite so). Also, there are common statistical features associated with allowing particles access to larger volumes that are most easily recognized when the number density of particles is not changed in the mixing process. Consider two closed systems $\mathscr{A}$ and $\mathscr{A}'$ each of constant volume and $V_{\mathscr{A}} = V_{\mathscr{A}'}$. They contain the same number of identical effectively unstructured particles ($N_{\mathscr{A}} = N_{\mathscr{A}'}$) and each is in contact with the surroundings (a large thermostat at T) through closed, rigid diathermal boundaries. These surroundings and systems $\mathscr{A}$ and $\mathscr{A}'$ are supersystem $\mathscr{A}\mathscr{A}'\mathscr{E}$. The supersystem equilibrium state is that state where $T_{\mathscr{E}} = T_{\mathscr{A}} = T_{\mathscr{A}'}$. The systems $\mathscr{A}$ and $\mathscr{A}'$ can now be isolated by changing their diathermal boundaries to adiabatic boundaries to produce supersystem $\mathscr{A}\mathscr{A}'$. If the physical state is gaseous and the gas is ideal, E is proportional to T and $E_{\mathscr{A}} = E_{\mathscr{A}'}$; the systems are individual isolated systems with the same constant E, V, and N values. If the boundary partitioning the two systems is now made material permeable, the identical particle systems mix with no change in total energy, total volume, or total number of particles. No change whatever in any thermodynamic property has occurred. If the boundary were again made nonpermeable each system is indistinguishable from the original state. If, however, starting as before with the exception that the particles, while the same in number, are of different species ($N_{\mathscr{A}} = N_{\mathscr{B}}$) repetition of the mixing process produces a different result.

To examine the process in statistical detail we clearly, even with structured particle gaseous systems, need consider only the translational component of the partition function since only volume is changed in the isothermal mixing process. Then, for the like particle original systems

$$\mathscr{Z}_{\mathscr{A}} = \left[\left(\frac{2\pi m_{\mathscr{A}} kT}{h^2}\right)^{3/2}\left(\frac{V_{\mathscr{A}} e}{N_{\mathscr{A}}}\right)\right]^{N_{\mathscr{A}}} \tag{8.23a}$$

$$\mathscr{Z}_{\mathscr{A}'} = \left[\left(\frac{2\pi m_{\mathscr{A}'} kT}{h^2}\right)^{3/2}\left(\frac{V_{\mathscr{A}'} e}{N_{\mathscr{A}'}}\right)\right]^{N_{\mathscr{A}'}} \tag{8.23b}$$

and for the $\mathscr{A}\mathscr{A}'$ supersystem before mixing is permitted the total Helmholz free energy is

$$A_\Sigma = A_{\mathscr{A}} + A_{\mathscr{A}'} = -kT \ln \mathscr{Z}_{\mathscr{A}} \mathscr{Z}_{\mathscr{A}'}. \tag{8.24}$$

For the $\mathscr{A}\mathscr{A}'$ supersystem after mixing

$$\mathscr{Z}_{\mathscr{A}\mathscr{A}'} = \left[\left(\frac{2\pi m_{\mathscr{A}} m_{\mathscr{A}'} kT}{h^2} \right)^{3/2} \left(\frac{V_{\mathscr{A}\mathscr{A}'}}{N_{\mathscr{A}} + N_{\mathscr{A}'}} \right) \right]^{N_{\mathscr{A}} + N_{\mathscr{A}'}} \tag{8.25}$$

with

$$V_{\mathscr{A}\mathscr{A}'} = 2V_{\mathscr{A}} = 2V_{\mathscr{A}'} \qquad \text{and} \qquad N_{\mathscr{A}} + N_{\mathscr{A}'} = 2N_{\mathscr{A}} = 2N_{\mathscr{A}'}$$

$$\ln \mathscr{Z}_{\mathscr{A}\mathscr{A}'} = \ln \mathscr{Z}_{\mathscr{A}} \mathscr{Z}_{\mathscr{A}'}$$

$$(A_{\mathscr{A}} + A_{\mathscr{A}'})_{\text{unmixed}} = (A_{\mathscr{A}\mathscr{A}'})_{\text{mixed}}. \tag{8.26}$$

For the mixing of unlike particles however for a species $\mathscr{A}$ with the partition function given by Eq. (8.23a) and a species $\mathscr{B}$ with the partition function given by

$$\mathscr{Z}_{\mathscr{B}} = \left[\left(\frac{2\pi m_{\mathscr{B}} kT}{h^2} \right)^{3/2} \left(\frac{V_{\mathscr{B}} e}{N_{\mathscr{B}}} \right) \right]^{N_{\mathscr{B}}} \tag{8.27}$$

and

$$A_\Sigma = -kT \ln \mathscr{Z}_{\mathscr{A}} \mathscr{Z}_{\mathscr{B}}$$

before mixing. After mixing the partition function is

$$\mathscr{Z}_{\mathscr{A}\mathscr{B}} = \left(\frac{2\pi kT}{h^2} \right)^{3/2} \left[\left(\frac{m_{\mathscr{A}}^{3/2} V_{\mathscr{A}\mathscr{B}}}{N_{\mathscr{A}}} \right)^{N_{\mathscr{A}}} \left(\frac{m_{\mathscr{B}}^{3/2} V_{\mathscr{A}\mathscr{B}}}{N_{\mathscr{B}}} \right)^{N_{\mathscr{B}}} \right] \tag{8.28}$$

since each species will occupy the entire volume and energy states will be changed for only the number of particles for which the set of states is available. Even without arithmetic it is clear that $\mathscr{Z}_{\mathscr{A}\mathscr{B}} \neq \mathscr{Z}_{\mathscr{A}} \mathscr{Z}_{\mathscr{B}}$ and A_Σ unmixed $\neq A_\Sigma$ mixed. This result will be a feature of any multicomponent system prepared by mixing pure species and will be reflected by any thermodynamic property that is volume dependent. It will be convenient, particularly when we consider condensed phase systems, to focus on the unique properties of the mixture by defining mixing terms as follows. For example, the Helmholz free energy is

$$(\Delta A_{\text{mix}})_T = (A_{\text{mixed system}})_T - \left(\sum^{K} (A_K^0)_T \right) \tag{8.29}$$

where the $(A_K^0)_T$ are the free energies of the pure substances κ in amounts appropriate to their presence in the mixed system.

We would obtain free energies of mixing from

$$(\Delta A_{mix})_T = -kT\Delta \ln \mathscr{Z} = -kT[\ln \mathscr{Z}_{\mathscr{A}\mathscr{B}} - \ln \mathscr{Z}_{\mathscr{A}}\mathscr{Z}_{\mathscr{B}}]. \tag{8.30}$$

The term $(\Delta A_{mix})_T$ is zero for the mixing of like particles under isothermal conditions where the volumes are additive and molecule density N/V is unchanged in the mixing process. Under these same conditions the free energy change on mixing unlike particles is

$$\Delta A_{mix} = -kT\left[N_{\mathscr{A}} \ln m_{\mathscr{A}}^{3/2}\frac{V_{\mathscr{A}\mathscr{B}}}{N_{\mathscr{A}}} + N_{\mathscr{B}} \ln m_{\mathscr{B}}^{3/2}\frac{V_{\mathscr{A}\mathscr{B}}}{N_{\mathscr{B}}}\right]$$

$$+\left[(N_{\mathscr{A}} + N_{\mathscr{B}}) - N_{\mathscr{A}} \ln m_{\mathscr{A}}^{3/2}\frac{V_{\mathscr{A}}}{N_{\mathscr{A}}} - N_{\mathscr{B}} \ln m_{\mathscr{B}}^{3/2}\frac{V_{\mathscr{B}}}{N_{\mathscr{B}}} - (N_{\mathscr{A}} + N_{\mathscr{B}})\right]$$

$$\Delta A_{mix} = -kT\left[N_{\mathscr{A}} \ln \frac{V_{\mathscr{A}\mathscr{B}}}{V_{\mathscr{A}}} + N_{\mathscr{B}} \ln \frac{V_{\mathscr{A}\mathscr{B}}}{V_{\mathscr{B}}}\right]. \tag{8.31}$$

Since the final volume is larger than either $V_{\mathscr{A}}$ or $V_{\mathscr{B}}$, the free energy change is negative as we would anticipate for the spontaneous mixing process.

There is also an entropy of mixing,

$$\Delta S_{mix} = \frac{\Delta E_{mix} - \Delta A_{mix}}{T} \tag{8.32}$$

which, since $\Delta E_{mix} = 0$ (arising from $E = \frac{3}{2}NkT$, independent of V for the ideal gas), is

$$\Delta S_{mix} = k\left[N_{\mathscr{A}} \ln \frac{V_{\mathscr{A}\mathscr{B}}}{V_{\mathscr{A}}} + N_{\mathscr{B}} \ln \frac{V_{\mathscr{A}\mathscr{B}}}{V_{\mathscr{B}}}\right] \tag{8.33}$$

and the change is positive as anticipated. Since $G = A + PV$ and $\Delta G = \Delta A + \Delta(PV)$ with $\Delta(PV)$ being zero for the ideal gas, $\Delta G_{mix} = \Delta A_{mix}$. Just as $\Delta E_{mix} = 0$ in the ideal gas system there is no $\Delta H_{mix}(= \Delta E_{mix} + \Delta(PV)_{mix})$. Equations (8.28) and (8.30) can be rephrased in a form that will be more amenable to comparison with nonideal gas results which follow if the volumes are replaced by NkT/P giving, since P is unchanged in the process,

$$\Delta A_{mix} = \Delta G_{mix} = -kT\left[N_{\mathscr{A}} \ln \frac{(N_{\mathscr{A}} + N_{\mathscr{B}})kT/P}{N_{\mathscr{A}}(kT/P)} + N_{\mathscr{B}} \ln \frac{(N_{\mathscr{A}} + N_{\mathscr{B}})kT/P}{N_{\mathscr{B}}(kT/P)}\right]$$

$$= kT\left[N_{\mathscr{A}} \ln \frac{N_{\mathscr{A}}}{N_{\mathscr{A}} + N_{\mathscr{B}}} + N_{\mathscr{B}} \ln \frac{N_{\mathscr{B}}}{N_{\mathscr{A}} + N_{\mathscr{B}}}\right] \tag{8.34}$$

$$\Delta S_{mix} = k\left[N_{\mathscr{A}} \ln \frac{N_{\mathscr{A}}}{N_{\mathscr{A}} + N_{\mathscr{B}}} + N_{\mathscr{B}} \ln \frac{N_{\mathscr{B}}}{N_{\mathscr{A}} + N_{\mathscr{B}}}\right]. \tag{8.35}$$

Equations (8.34) and (8.35) can be extended to any number of species, e.g.,

$$\Delta A_{mix} = \mathscr{k}T\sum^{K}\left(N_K \ln \frac{N_K}{\sum N_K}\right). \tag{8.36}$$

Also, a mixture of any overall composition can be constructed in that the original V_K are not required to be equal. In our development we required that $N_{\mathscr{A}}/V_{\mathscr{A}} = N_{\mathscr{B}}/V_{\mathscr{B}}$ so that the gas phase pressure can remain constant. To meet this requirement in a several component system we require that all N/V values are the same. We also equilibrated each subsystem with a common thermostat so that (ideal gas) T is unchanged during the mixing.

Mixing processes at constant T and P will be those of normal concern because the results are not confused by other terms but, in the event that the criteria regarding T and P are not met by prior construct, the mixed system will find a common value of P ($= P_{f,\mathscr{A}\mathscr{B}}$) and T ($= T_{f,\mathscr{A}\mathscr{B}}$) and the changes in free energy and entropy associated with the change in thermodynamic state are those that would have occurred had both subsystems been brought independently to the final thermodynamic state $T_{f,\mathscr{A}\mathscr{B}}$, $P_{f,\mathscr{A}\mathscr{B}}$ with the mixing term unchanged except that $T = T_{f,\mathscr{A}\mathscr{B}}$.

We can summarize the statistical origin of mixing terms in binary particle systems as follows. For both like and unlike particle gaseous systems, increase in system volume results in a compression of energy levels with no change in their degeneracy for ideal gases. In an energy interval there is an increased number of energy states and, therefore, an increase in the particle partition function $\mathscr{z}$. Also, populations must change in both cases to maintain E constant (when there is no change in T) but the changes are subject to quite different constraints for like and unlike particle systems. For like particle systems the single enlarged set of energy states is available to all particles, for unlike particle systems two enlarged sets of energy states are each available to only one species. The important term is $\mathscr{z}/N$, which is unchanged when both $\mathscr{z}$ and N are increased proportionately but increases when $\mathscr{z}$ is increased and N is unchanged.

The above arguments are consistent with the commonly encountered and intuitively useful concept that entropy measures randomness. For our purposes randomness increases with increase in the number of ways in which a fixed number of particles can be distributed among available energy states. The combinatorial term Ω and its analog entropy increase as randomness, so defined, increases. The combinatorial term recognizes, after correction for gas phase indistinguishability, all distribution consequences. When mixing unlike species every distribution of species $\mathscr{A}$ particles must be coupled with every distribution of species $\mathscr{B}$ particles to provide terms in the combinatorial term sum. Since the two species are independent

$$\Omega = \Omega_{\mathscr{A}}\Omega_{\mathscr{B}}.$$

This product is immensely larger than either $\Omega_{\mathscr{A}}$ or $\Omega_{\mathscr{B}}$. Of course, S is

(O) $10^{-16} \ln \Omega_{\mathscr{A}} \Omega_{\mathscr{B}}$ and is only substantially larger than $S_{\mathscr{A}}$, (O) $10^{-16} \ln \Omega_{\mathscr{A}}$ or $S_{\mathscr{B}}$, (O) $10^{-16} \ln \Omega_{\mathscr{B}}$.

In randomness terms, anything that increases the number of available energy states per particle, increases entropy. This can include increase in temperature, which makes available states that were not available at lower temperature, and, for gases, increase of volume, which compresses the energy levels bringing nonaccessible levels down into the range of T. In condensed phases where the particles must be considered as localized or partially localized there will be combinatorial term contributions arising from spatial arrangements in mixed systems that increase Ω and therefore S. Of course such system related processes do not always result in a spontaneous process. It is $\ln \mathscr{Z}$ which must increase ($\Delta A = -kT \ln \Delta \mathscr{Z}$) and $\mathscr{Z}$ is not a simple function in complex systems. In the arguments that follow we will concentrate on the terms ΔE, ΔS, and ΔA for the mixing process since these are the terms from which all others follow by familiar routines.

The extension of multicomponent system considerations to nonideal gases and to condensed systems offers some problems as would be anticipated. We first consider mixing in nonideal gases. There are a number of obvious simplifications that we can apply to the gaseous system. The single component partition function can be written as

$$\mathscr{Z} = \frac{1}{N!}\left(\frac{1}{\Lambda}\right)^{3N} V_f^N e^{-N\phi_a/2kT} \tag{6.17}$$

where z_t is the translational component for the gas occupying volume V and behaving ideally. The free energy is

$$A = -kT\left[-N \ln N + N - \frac{3N}{2} \ln \Lambda + N \ln V\right] - kT\left[N \ln \frac{V_f}{V} - \frac{N\phi_a}{2kT}\right]. \tag{8.37}$$

All arguments developed for mixing in the ideal gas apply to the first bracketed term in Eq. (8.37). This will produce a $(\Delta A_{mix})_{ideal}$. Contributions from mixing which arise from considerations of the second bracketed term we will refer to as an excess mixing property deriving from nonideality in the system. Following convention, we define ${}^E(\Delta A_{mix}) = \Delta A_{mix} - (\Delta A_{mix})_{ideal}$ and other excess properties by analogy. For a van der Waals gas $V_f = V - (2\pi/3)Nr_e^3$ and $\phi_{\mathscr{A}} = (4\pi N\phi_e r_e^3/3V)$. The exact form of the potential function determines ϕ_e and r_e. It will be different for interactions involving unlike molecules than for those involving like molecules. In a two species system where the species are identified as $N_{\mathscr{A}}$ and $N_{\mathscr{B}}$ molecules of types $\mathscr{A}$ and $\mathscr{B}$ we take $r_{e\mathscr{A}\mathscr{B}} = \frac{1}{2}(r_{e,\mathscr{A}} + r_{e,\mathscr{B}})$. This will be appropriate for the hard sphere model that we are using as can be seen by constructing a Figure 6.3 like model. We take $\phi_{\mathscr{A}\mathscr{B}} = (\phi_{e,\mathscr{A}}\phi_{e,\mathscr{B}})^{1/2}$, which is not obvious. This relationship is obtained as a result in the consideration of dispersion forces between unlike molecules and is therefore appropriate to the 6th power attractive potential that we are using.

In either the species $\mathscr{A}$ or the species $\mathscr{B}$ subsystems before mixing there will be only like molecule interactions, which we identify by number as $N_{\mathscr{A}\mathscr{A}}$ and $N_{\mathscr{B}\mathscr{B}}$. In the mixed system there will also be $N_{\mathscr{A}\mathscr{B}}$ interactions. We can assume that, since we recognize only pair interactions, the number of $\mathscr{A}\mathscr{A}(=N_{\mathscr{A}}^2/2)$ and $\mathscr{B}\mathscr{B}(=N_{\mathscr{B}}^2/2)$ interactions will be unchanged in the mixed system and there will be $N_{\mathscr{A}}N_{\mathscr{B}}/2$ of the $\mathscr{A}\mathscr{B}$ interactions. We can drop the one half factors, inserted to prevent double counting, since they have already been inserted in Eq. (8.37).

We note from Eq. (8.37) that the term important to excess properties is really two terms of opposite sign, one relating to the free volume and one relating to attractive interactions. The free volume term is really not important in the hard sphere approximation, the total free volume is the sum of the free volumes of the two systems and, for component $\mathscr{A}$, $N_{\mathscr{A}}\ln(V_f/V)_{\mathscr{A}}$ in system $\mathscr{A}$ goes to $N_{\mathscr{A}}\ln(V_f/V)_{\mathscr{A}\mathscr{B}}$ with the same order of change for component $\mathscr{B}$, $N_2\ln(V_f/V)_{\mathscr{B}}$ going to $N_{\mathscr{B}}\ln(V_f/V)_{\mathscr{A}\mathscr{B}}$.

Focusing on the energetic terms, for the unmixed systems,

$$A_{\mathscr{A}} + A_{\mathscr{B}} = \frac{2\pi kT}{3}\left[N_{\mathscr{A}}^2\frac{\phi_{e,\mathscr{A}}r_{e,\mathscr{A}}^3}{V_{\mathscr{A}}} + N_{\mathscr{B}}^2\frac{\phi_{e,\mathscr{B}}r_{e,\mathscr{B}}^3}{V_{\mathscr{B}}}\right] \tag{8.38a}$$

and for the mixed system

$$A_{\mathscr{A}\mathscr{B}} = \frac{2\pi kT}{3}\left[\frac{N_{\mathscr{A}}^2\phi_{e,\mathscr{A}}r_{e,\mathscr{A}}^3}{V_{\mathscr{A}\mathscr{B}}} + \frac{N_{\mathscr{B}}^2\phi_{e,\mathscr{B}}r_{e,\mathscr{B}}^3}{V_{\mathscr{A}\mathscr{B}}} + N_{\mathscr{A}}N_{\mathscr{B}}(\phi_{e,\mathscr{A}}\phi_{e,\mathscr{B}})^{1/2}(r_{e,\mathscr{A}} + r_{e,\mathscr{B}})^3\right]. \tag{8.38b}$$

If $N_{\mathscr{A}} = N_{\mathscr{B}}$ and $V_{\mathscr{A}} = V_{\mathscr{B}} = V_{\mathscr{A}\mathscr{B}}/2$, the excess free energy of mixing assumes the relatively simple form

$$^{E}(\Delta A_{\text{mix}}) = \frac{\pi kT}{3V_{\mathscr{A}}}\left[\frac{(\phi_{e,\mathscr{A}}\phi_{e,\mathscr{B}})^{1/2}(r_{e,\mathscr{A}} + r_{e,\mathscr{B}})^3}{8} - 2(\phi_{e,\mathscr{A}}r_{e,\mathscr{A}}^3 + \phi_{e,\mathscr{B}}r_{e,\mathscr{B}}^3)\right]. \tag{8.39}$$

Extending mixing considerations to a solid phase that is the next order of complexity in the statistical sense requires a somewhat different line of thought since the concept of mixing is not as real as with gases. When we speak of a mixed solid we will, of course, be referring to a solid solution rather than an assemblage of various solid phases to which no special considerations apply.[2] In the mixed solid the components are distributed over the lattice sites and the

[2] In all of our discussions relating to condensed phase mixtures the terms mixture and solution are interchangeable.

general concept of a potential well at each site is still valid although the description of the potential will be changed in accordance with the character of near neighbors if there are differences in the interaction energies between the species making up the solid phase. If there are large differences in the masses of substances present in the solid the force constants associated with vibrations will be quite different and yet the normal modes must represent phase coherent motions as in the single component solids. There may be substantial changes in the vibrational spectrum from that of any of the components of the mixture. Also the structure of the vibrational spectrum may favor a nonrandom distribution of components on the lattice sites. The lattice structure in the mixed solid may not be that of either of the substances making up the solid although we would imagine that the similarity of lattice structure would be a practical precondition for forming the solid solution.

Effectively none of the above concerns can be reasonably reflected in any simple model for the solid solution, but the concept of an ideal free energy or entropy of mixing is readily constructed. Consider that there are $N_{\mathscr{A}}$ and $N_{\mathscr{B}}$ molecules occupying lattice sites. Then the combinatorial term which reflects normal species $\mathscr{A}$ and $\mathscr{B}$ interchanges within each site set is supplemented by a term reflecting the exchange of sites.

$$\Omega_{\text{mix}} = \Omega_{\mathscr{A}\mathscr{B}}(\Omega_{\mathscr{A}}\Omega_{\mathscr{B}})_{\text{mix}} = \frac{(N_{\mathscr{A}} + N_{\mathscr{B}})!}{N_{\mathscr{A}}!N_{\mathscr{B}}!}(\Omega_{\mathscr{A}}\Omega_{\mathscr{B}})_{\text{mix}}$$

and

$$\ln \Omega_{\text{mix}} - (\ln \Omega_{\mathscr{A}} - \ln \Omega_{\mathscr{B}})_{\text{mix}} = \ln \frac{(N_{\mathscr{A}} + N_{\mathscr{B}})!}{N_{\mathscr{A}}!N_{\mathscr{B}}!}.$$

The additional combinatorial term produces directly the ideal entropy of mixing if the $\Omega_{\mathscr{A}}$, and $\Omega_{\mathscr{B}}$ terms are unchanged on mixing, i.e., if the energy states of the two components are effectively the same. Expansion of the $\ln N!$ terms in Ω produces

$$\Delta S_{\text{mix}} = -N_{\mathscr{A}}k \ln \frac{N_{\mathscr{A}}}{N_{\mathscr{A}} + N_{\mathscr{B}}} - N_{\mathscr{B}}k \ln \frac{N_{\mathscr{B}}}{N_{\mathscr{A}} + N_{\mathscr{B}}} \tag{8.40}$$

as in the ideal gas case. Complete arithmetic would produce

$$\Delta A_{\text{mix}} = \Delta G_{\text{mix}} = T\Delta S_{\text{mix}} \qquad \text{for} \qquad \Delta E_{\text{mix}} = 0, \qquad \Delta H_{\text{mix}} = 0, \qquad \Delta V_{\text{mix}} = 0.$$

A different combinatorial term would be required if the distribution of components were not random (as would arise from association of like species or particular site occupancies) and ΔS_{mix} would not be given by the Eq. (8.40) form even if ΔE_{mix} and ΔV_{mix} were zero (unlikely under these circumstances.)

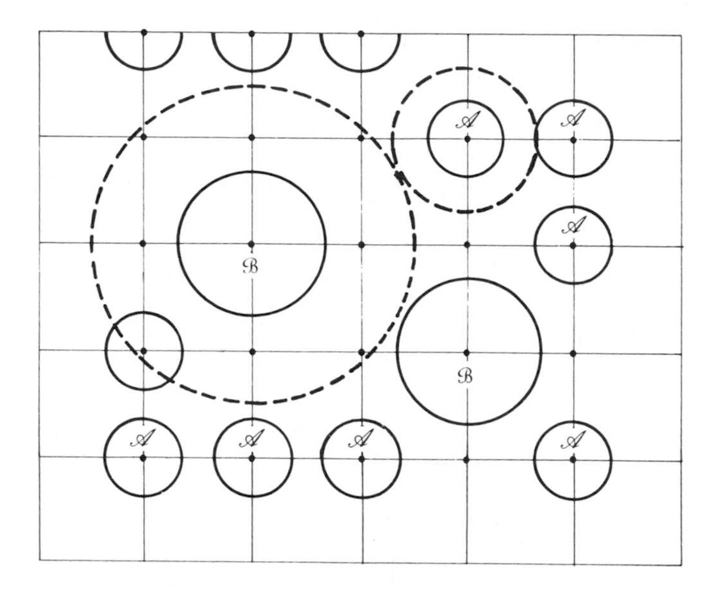

Figure 8.1. Illustration of lattice sites excluded from occupancy by large molecules dispersed in a small molecule lattice structure.

If the two components are of substantially different volumes it is still possible to produce a simple treatment if the distribution is random, first recognizing that even if the attractive part of the potential function is nearly the same for the two components the repulsive (free volume) part, which is controlling in condensed systems, will be different. Consider a mixture of large molecules $\mathcal{B}$ and small molecules $\mathcal{A}$ occupying N lattice sites in a solid scaled to accommodate $\mathcal{A}$-type molecules in minimum energy configuration. As indicated in Figure 8.1, occupancy of a site by a molecule $\mathcal{B}$ can, depending on excluded volumes and lattice spacing, also deny a number of adjacent sites to other $\mathcal{B}$-type molecules. We examine the most simple case in which sites adjacent to $\mathcal{B}$ will still physically accommodate $\mathcal{A}$-type molecules. If we place the $\mathcal{B}$-type molecules into the lattice one at a time and if a $\mathcal{B}$-type molecule denies z total sites to another $\mathcal{B}$-type molecule,[3] we have N choices for the first, $N - z$ choices for the second, $N - 2z$ choices for the third,... or generally $N - (N_{\mathcal{B}} - 1)z$ for placing a molecule $\mathcal{B}$ into the lattice structure where $N_{\mathcal{B}}$ $\mathcal{B}$-type molecules are already present. The remaining sites are occupied by $\mathcal{A}$-type molecules and the total number of configurations is

$$\Omega = N_{\mathcal{A}}! \prod_{N_{\mathcal{B}}=1}^{N_{\mathcal{B}}} [N - (N_{\mathcal{B}} - 1)z]/N_{\mathcal{A}}!N_{\mathcal{B}}! \qquad (8.41)$$

[3] If $\mathcal{B}$ type molecules occupy next nearest neighbor sites the total number of excluded sites is not $2z$ but in solutions dilute in $\mathcal{B}$ adjacent occupancies will not often be encountered.

where the division by product $N_{\mathscr{A}}!N_{\mathscr{B}}!$ corrects for indistinguishability. To relate Eq. (8.41) to system volume (V) and excluded volume b per molecule we note that $N = V/v_{cell}$ and that the total excluded volume b is zv_{cell} so that

$$\Omega = \frac{1}{N_{\mathscr{B}}!}\prod_{N_{\mathscr{B}}=1}^{N_{\mathscr{B}}}\frac{V-(N_{\mathscr{B}}-1)b}{v_{cell}} = \frac{1}{N_{\mathscr{B}}!}\left(\frac{1}{v_{cell}}\right)^{N_{\mathscr{B}}}\prod_{N_{\mathscr{B}}=1}^{N_{\mathscr{B}}} V\left[1-\left(\frac{N_{\mathscr{B}}-1}{V}\right)b\right]. \qquad (8.42)$$

The entropy of the mixed system is

$$k\ln\Omega = k\left\{\ln N_{\mathscr{B}}! - N_{\mathscr{B}}\ln\frac{N}{V} + N_{\mathscr{B}}\ln V + \sum_{N_{\mathscr{B}}=1}^{N_{\mathscr{B}}}\ln\left[1-\left(\frac{N_{\mathscr{B}}-1}{V}\right)b\right]\right\}.$$

For small $N_{\mathscr{B}}$ (dilute mixtures) the last term can be approximated by

$$\sum_{N_{\mathscr{B}}=1}^{N_{\mathscr{B}}}\ln\left[1-\left(\frac{N_{\mathscr{B}}-1}{V}\right)b\right] = \sum_{N_{\mathscr{B}}=1}^{N_{\mathscr{B}}}\left(\frac{N_{\mathscr{B}}-1}{V}\right)b \cong \frac{N_{\mathscr{B}}^2 b}{2V} \qquad (8.43)$$

and the entropy is

$$S_{\Sigma} = -k\left\{N_{\mathscr{B}}\ln N_{\mathscr{B}} + N_{\mathscr{B}} - N_{\mathscr{B}}\ln N + 2N_{\mathscr{B}}\ln V + \frac{N_{\mathscr{B}}^2 b}{2V}\right\}. \qquad (8.44)$$

If the total volume is represented in terms of molecule volumes $v_{\mathscr{A}}$ and $v_{\mathscr{B}}$,

$$V = N_{\mathscr{A}}v_{\mathscr{A}} + N_{\mathscr{B}}v_{\mathscr{B}}$$

$$S_{\Sigma} = k\{-N_{\mathscr{B}}\ln N_{\mathscr{B}} + N_{\mathscr{B}} + N_{\mathscr{B}}\ln N + 2N_{\mathscr{B}}\ln(N_{\mathscr{A}}v_{\mathscr{A}} + N_{\mathscr{B}}v_{\mathscr{B}}) + N_{\mathscr{B}}^2 b/2(N_{\mathscr{A}}v_{\mathscr{A}} + N_{\mathscr{B}}v_{\mathscr{B}}) \qquad (8.45)$$

and $\Delta S_{mix} = S_{\Sigma} - (S_{\mathscr{A}} - S_{\mathscr{B}})$. The pure component entropies are just Eq. (8.45) with $N_{\mathscr{A}}$ and $N_{\mathscr{B}}$ separately set equal to zero and

$$\Delta S_{mix} = -N_{\mathscr{B}}k\ln\upsilon_{\mathscr{B}} + \frac{N_{\mathscr{B}}kb}{2v_{\mathscr{B}}}\upsilon_{\mathscr{A}} \qquad (8.46)$$

where $\upsilon_{\mathscr{B}}$, $\upsilon_{\mathscr{A}}$ are the volume fractions. It is unlikely that the interaction energies would be unchanged for the greatly dissimilar forms of the potential wells at and around a $\mathscr{B}$-type molecule so that we would anticipate $\Delta E_{mix} \neq 0$ and excess mixing values for all thermodynamic functions. Our assumption would be that, given the constraints that control normal vibrational modes in solids, appropriate mechanical data for the mixed solid would provide an effective Θ_D value and the general statistical framework for solids would apply. Except for unknown changes in the E_0 values (which we related to the energy of sublimation at 0 K) this is true so, in principle and in conjunction with experimental data, calculation would be accomplished. Our interest in mixed solid solutions does not extend to the pursuit of such calculations but the general features of mixed species interactions in a lattice structure that we have developed for solids are of wide interest as they can be applied to liquids and liquid mixtures to the extent that lattice theory is applicable.

We have a general form for the single component liquid phase partition function in terms of the free volume v_f associated with a cell volume. If we are not interested in a precise address to v_f we can write

$$\mathscr{Z} = \prod^{i} z_i^{N} (ve)^{N} e^{-\phi_\Sigma/kT}$$

where the z_i represent all motion related components of the partition function, v is the cell volume, and ϕ_Σ the total potential energy of the N particle system. If we now make up a mixture of $N_{\mathscr{A}} + N_{\mathscr{B}}$ molecules of species $\mathscr{A}$ and $\mathscr{B}$ distributed among $N_{\mathscr{A}} + N_{\mathscr{B}}$ cells in a lattice model liquid, the combinatorial term will take the same form as the combinatorial term for lattice sites in the solid (after we drop out the factor e^{N}) and the partition function for the mixture is

$$\mathscr{Z} = \prod^{i} z_{\mathscr{A},i}^{N_{\mathscr{A}}} \prod^{i} z_{\mathscr{B},i}^{N_{\mathscr{B}}} \frac{(N_{\mathscr{A}} + N_{\mathscr{B}})!}{N_{\mathscr{A}}! N_{\mathscr{B}}!} v_{\mathscr{A}}^{N_{\mathscr{A}}} v_{\mathscr{B}}^{N_{\mathscr{B}}} e^{-\phi_\Sigma/(N_{\mathscr{A}}+N_{\mathscr{B}})kT} \tag{8.47}$$

and the free energy of mixing is

$$\Delta A_{mix} = (N_{\mathscr{A}} A_{\mathscr{A}} + N_{\mathscr{B}} A_{\mathscr{B}})_{mixture} - (N_{\mathscr{A}} A_{\mathscr{A}}^0 + N_{\mathscr{B}} A_{\mathscr{B}}^0)$$

where A is the per molecule free energy and A^0 refers to the pure component. Using Eqs. (8.45) and (8.46), ignoring the motion-related terms that are unchanged in the mixing process,

$$\begin{aligned}\Delta A_{mix} = & -kT[N_{\mathscr{A}} \ln v_{\mathscr{A}} + N_{\mathscr{B}} \ln v_{\mathscr{B}}] + \phi_\Sigma/(N_{\mathscr{A}} + N_{\mathscr{B}}) \\ & + kT[(N_{\mathscr{A}} + N_{\mathscr{B}}) \ln(N_{\mathscr{A}} + N_{\mathscr{B}}) - (N_{\mathscr{A}} + N_{\mathscr{B}}) \\ & - N_{\mathscr{A}} \ln N_{\mathscr{A}} + N_{\mathscr{A}} - N_{\mathscr{B}} \ln N_{\mathscr{B}} + N_{\mathscr{B}}] \\ & + kT[N_{\mathscr{A}} \ln v_{\mathscr{A}}^0 + N_2 \ln v_{\mathscr{B}}^0] - \phi_{\mathscr{A}}^0/N_{\mathscr{A}} - \phi_{\mathscr{B}}^0/N_{\mathscr{B}}\end{aligned}$$

and collecting terms

$$\begin{aligned}\Delta A_{mix} = & kT\left[N_{\mathscr{A}} \ln \frac{v_{\mathscr{A}}^0}{v_{\mathscr{A}}} + N_{\mathscr{B}} \ln \frac{v_{\mathscr{B}}^0}{v_{\mathscr{B}}}\right] + \frac{\phi_\Sigma}{N_{\mathscr{A}} + N_{\mathscr{B}}} - \frac{\phi_{\mathscr{A}}^0}{N_{\mathscr{A}}} - \frac{\phi_{\mathscr{B}}^0}{N_{\mathscr{B}}} \\ & + kT\left[N_{\mathscr{A}} \ln \frac{N_{\mathscr{A}}}{N_{\mathscr{A}} + N_{\mathscr{B}}} + N_{\mathscr{B}} \ln \frac{N_{\mathscr{B}}}{N_{\mathscr{A}} + N_{\mathscr{B}}}\right].\end{aligned} \tag{8.48}$$

In the event that the cell volumes are unchanged in the mixing process ($\Delta V_{mix} = 0$), the volume ratio terms vanish. In the event that $\phi_{\mathscr{A}\mathscr{B}} = \phi_{\mathscr{A}\mathscr{A}} =$

$\phi_{\mathscr{B}\mathscr{B}}$ ($\Delta E_{mix} = 0$) the terms involving these energies vanish and the ideal free energy of mixing is

$$(\Delta A_{mix})_{ideal} = \ell T\left(N_{\mathscr{A}} \ln \frac{N_{\mathscr{A}}}{N_{\mathscr{A}} + N_{\mathscr{B}}} + N_{\mathscr{B}} \ln \frac{N_{\mathscr{B}}}{N_{\mathscr{A}} + N_{\mathscr{B}}} \right).$$

In the event that the ideality criteria are not met the excess free energy of mixing is just those terms which were dropped from Eq. (8.48)

$${}^{E}(\Delta A_{mix}) = \ell T\left[N_{\mathscr{A}} \ln \frac{v^0_{\mathscr{A}}}{v_{\mathscr{A}}} + N_{\mathscr{B}} \ln \frac{v^0_{\mathscr{B}}}{v_{\mathscr{B}}} \right] + \Delta E_{mix}. \tag{8.49}$$

The ideal entropy of mixing is, from $(\partial A/\partial T)_{V,N} = -S$,

$$(\Delta S_{mix})_{ideal} = \ell \left[N_{\mathscr{A}} \ln \frac{N_{\mathscr{A}} + N_{\mathscr{B}}}{N_{\mathscr{A}}} + N_{\mathscr{B}} \ln \frac{N_{\mathscr{A}} + N_{\mathscr{B}}}{N_{\mathscr{B}}} \right]$$

and the excess entropy of mixing is

$${}^{E}(\Delta S_{mix}) = -\ell \left[N_{\mathscr{A}} \left(\frac{\partial \ln v^0_{\mathscr{A}}/v_{\mathscr{A}}}{\partial T} \right)_{P,N} + N_{\mathscr{B}} \left(\frac{\partial \ln v^0_{\mathscr{B}}/v_{\mathscr{B}}}{\partial T} \right)_{P,N} + \left(\frac{\partial \Delta E_{mix}}{\partial T} \right)_{P,N} \right]. \tag{8.50}$$

The volume terms in the excess functions ${}^{E}(\Delta A_{mix})$ and ${}^{E}(\Delta S_{mix})$ do not require further consideration. The undefined interaction energy terms represent the lattice energy of the mixed system and can be addressed if we introduce some simplifying assumptions and specify the distribution of the pair interactions that sum to ϕ_{Σ}.

The $(N_{\mathscr{A}} + N_{\mathscr{B}})$ molecules that make up the lattice are each surrounded by c neighbors and there will be interactions between $\mathscr{A}$-type molecules, between $\mathscr{B}$-type molecules and between $\mathscr{A}$- and $\mathscr{B}$-type molecules. We designate the number of such interactions as $N_{\mathscr{A}\mathscr{A}}$, $N_{\mathscr{B}\mathscr{B}}$, and $N_{\mathscr{A}\mathscr{B}}$ and the interaction energies as $\phi_{\mathscr{A}\mathscr{A}}$, $\phi_{\mathscr{B}\mathscr{B}}$, and $\phi_{\mathscr{A}\mathscr{B}}$ so that the lattice energy is

$$\phi_{lattice} = N_{\mathscr{A}\mathscr{A}}\phi_{\mathscr{A}\mathscr{A}} + N_{\mathscr{B}\mathscr{B}}\phi_{\mathscr{B}\mathscr{B}} + N_{\mathscr{A}\mathscr{B}}\phi_{\mathscr{A}\mathscr{B}}. \tag{8.51}$$

The $cN_{\mathscr{A}}$ total interactions consist of $2N_{\mathscr{A}\mathscr{A}} + N_{\mathscr{A}\mathscr{B}}$ pair interactions and the $cN_{\mathscr{B}}$ total interactions consist of $2N_{\mathscr{B}\mathscr{B}} + N_{\mathscr{A}\mathscr{B}}$ pair interactions so

$$\begin{aligned} \phi_{lattice} &= \left(\frac{cN_{\mathscr{A}} - N_{\mathscr{A}\mathscr{B}}}{2} \right)\phi_{\mathscr{A}\mathscr{A}} + \left(\frac{cN_{\mathscr{B}} - N_{\mathscr{A}\mathscr{B}}}{2} \right)\phi_{\mathscr{A}\mathscr{B}} + N_{\mathscr{A}\mathscr{B}}\phi_{\mathscr{A}\mathscr{B}} \\ &= \frac{c}{2}N_{\mathscr{A}}\phi_{\mathscr{A}\mathscr{A}} + \frac{c}{2}N_{\mathscr{B}}\phi_{\mathscr{B}\mathscr{B}} + \omega N_{\mathscr{A}\mathscr{B}} \end{aligned} \tag{8.52}$$

$$\omega_{\mathscr{A}\mathscr{B}} = \phi_{\mathscr{A}\mathscr{B}} - \tfrac{1}{2}(\phi_{\mathscr{A}\mathscr{A}} + \phi_{\mathscr{B}\mathscr{B}}). \tag{8.53}$$

If the distribution of $\mathscr{A}$-type and $\mathscr{B}$-type molecules is completely random the probability of an $\mathscr{A}$-type molecule being on any site is $N_{\mathscr{A}}/(N_{\mathscr{A}} + N_{\mathscr{B}})$ and that of a $\mathscr{B}$-type molecule being on any site is $N_{\mathscr{B}}/(N_{\mathscr{A}} + N_{\mathscr{B}})$ so the probability of types $\mathscr{A}$ and $\mathscr{B}$ being on adjacent sites is $2N_{\mathscr{A}}N_{\mathscr{B}}/(N_{\mathscr{A}} + N_{\mathscr{B}})^2$ (since $\mathscr{A}$–$\mathscr{B}$ and $\mathscr{B}$–$\mathscr{A}$ are equally probable and either arrangement leads to an $\mathscr{A}\mathscr{B}$ interaction). There are $c(N_{\mathscr{A}} + N_{\mathscr{B}})/2$ pairs of neighboring sites so the total number of $\mathscr{A}\mathscr{B}$ interactions for the random distribution is

$$N_{\mathscr{A}\mathscr{B}} = \frac{c}{2}(N_{\mathscr{A}} + N_{\mathscr{B}})\frac{2N_{\mathscr{A}}N_{\mathscr{B}}}{(N_{\mathscr{A}} + N_{\mathscr{B}})^2} = \frac{cN_{\mathscr{A}}N_{\mathscr{B}}}{(N_{\mathscr{A}} + N_{\mathscr{B}})} \tag{8.54}$$

and

$$\Delta E_{mix} = c\left(\frac{N_{\mathscr{A}}N_{\mathscr{B}}}{N_{\mathscr{A}} + N_{\mathscr{B}}}\right)\omega_{\mathscr{A}\mathscr{B}}. \tag{8.55}$$

The excess free energy of mixing is

$$^{E}(\Delta A_{mix}) = kT\left[N_{\mathscr{A}} \ln \frac{v^0_{\mathscr{A}}}{v_{\mathscr{A}}} + N_{\mathscr{B}} \ln \frac{v^0_{\mathscr{B}}}{v_{\mathscr{B}}}\right] + c\left(\frac{N_{\mathscr{A}}N_{\mathscr{B}}}{N_{\mathscr{A}} + N_{\mathscr{B}}}\right)\omega_{\mathscr{A}\mathscr{B}} \tag{8.56}$$

and, in the event that ΔV_{mix} is zero or negligible

$$^{E}(\Delta S_{mix}) = 0 \tag{8.57}$$

since there is no temperature dependence in the interchange energy term as we have constructed it. Mixtures that conform to these constraints $(\Delta V_{mix}) = 0$, $^{E}(\Delta S_{mix}) = 0$ are called regular.

We can summarize functions of mixing at constant T as follows. The ideal free energy or entropy of mixing is a statistical phenomenon related to increase in the combinatorial term Ω when two or more species are transferred from their pure state to a random mixed system. The ideal free energy of mixing, for solids, liquids, and gases, can be reduced to

$$(\Delta A_{mix})_{ideal} = kT\sum^{K} N_K \ln \frac{N_K}{\sum N_K} \tag{8.58}$$

$$(\Delta S_{mix})_{ideal} = k\sum^{K} N_K \ln \frac{\sum N_K}{N_K}. \tag{8.59}$$

The ideal mixed system involving condensed phases is one in which there is no change in total volume on mixing and the interaction between unlike particles is negligibly different from the interaction between like particles. For all ideal systems the form of the mixing functions is the same and as indicated in Figure 8.2 for a two-component ($\mathscr{A}$ and $\mathscr{B}$) system. Nonideal mixed systems will exhibit positive or negative deviations from the Figure 8.2 curves, the differences being dictated by the excess function values. We can expect to encounter ideal mixtures, particularly when the mixed species are similar in character and size. There will also be circumstances in which ΔV_{mix} and $^E(\Delta S_{mix})$ are zero (regular mixtures).

We fit the above statistical considerations into our general thermodynamic framework as follows. For a single component system we have

$$dE = N\sum^{j} \epsilon_j d p_j + \sum^{j} p_j d\epsilon_j + \sum^{j} p_j \epsilon_j dN. \tag{4.3}$$

The statistical terms identify process terms through

$$dE = TdS - PdV + \mu dN$$

where $\mu = kT\ln(z/N)$ is identified as $\sum p_j\epsilon_j$ or as $(\partial E/\partial N)_{S,V}$ [Eq. (4.11)]. This was expanded to $\mu = (\partial A/\partial N)_{T,V}$ [Eq. (4.17)] and to $G_{T,P} = N\mu$ [Eq. (4.34)] with μ further defined as $\mu^0(T) + kT\ln P_{atm}$ where $\mu^0(T)$ contains all temperature-dependent components of the partition function and the conversion factor atm $(\text{dyne cm}^2)^{-1}$. Since most of the mixed systems of concern will be condensed phase mixtures constructed at ambient (1 atm) pressure, the Gibbs free energy function [G(T, P, N)] will be the most appropriate basis for phrasing thermodynamic argument.

In a mixture we recognize the additive contribution for each component by constructing

$$G_\Sigma(T, P) = \sum^{K} N_K\mu_K \tag{8.60}$$

$$dE_\Sigma = TdS_\Sigma - PdV_\Sigma + \sum^{K} \mu_K dN_K \tag{8.61}$$

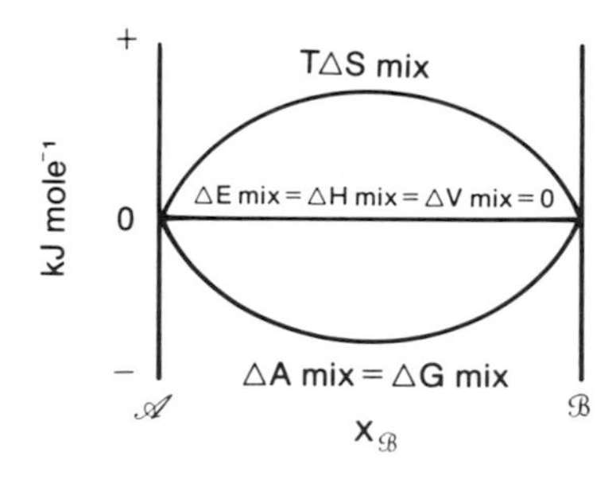

Figure 8.2. Illustration of the form of thermodynamic functions for an ideal mixing process.

$$dG_\Sigma = d(E_\Sigma + PV_\Sigma - TS_\Sigma)$$
$$= S_\Sigma dT + V_\Sigma dP + \sum^{K} \mu_K dN_K. \tag{8.62}$$

Since differentiation of Eq. (8.60) produces

$$(dG_\Sigma)_{T,P} = \left(\sum^{K} N_K d\mu_K + \sum^{K} \mu_K dN_K \right)_{T,P} \tag{8.63}$$

comparison with Eq. (8.62) when T and P are constant produces the general result

$$\left(\sum^{K} N_K d\mu_K \right)_{T,P} = 0 \tag{8.64}$$

which is referred to as the Gibbs–Duham equation. It will be useful in defining constraints on the independent variations in chemical potentials in mixed systems. In a single component system $(\partial G/\partial N)_{T,P}$ is a thermodynamic definition of chemical potential as a molecular feature of the system. In molar terms

$$N_A \left(\frac{\partial G}{\partial N} \right)_{T,P} = \left(\frac{\partial G}{\partial n} \right)_{T,P} = N_A \mu(T, P) = \tilde{G}(T, P). \tag{8.65}$$

In a mixture at constant temperature and pressure from Eq (8.60)

$$(G^*_\Sigma)_{T,P,N} = \left(\sum^{K} \mu_K N_K \right)_{T,P,N} \tag{8.66}$$

where G^*_Σ is the total Gibbs free energy of the mixed system and

$$\left(\frac{\partial G^*_\Sigma}{\partial N_K} \right)_{T,P,N_J} = (\mu_K)_{T,P,N_J} \tag{8.67}$$

defines each chemical potential in the mixture. The notation N_J is a statement that N is constant for all components except for that one involved in the infinitisimal variation. Since G_Σ is the Gibbs free energy of the mixture all mixing effects are included in the term and μ_K for any component in the mixture will differ from the pure component values

$$(\Delta G_\Sigma)_{mix} = (G^*_\Sigma)_{T,P} - \sum^{K} (G^0_K)_{T,P} \tag{8.68}$$

$$(\Delta \mu_K)_{mix} = \mu^*_K(T, P) - \mu^0_K(T, P) \tag{8.69}$$

where G_Σ^* and μ_K^* are the total free energy and Kth component chemical potential in the mixture of specified N_K at fixed T and P. If T and P are the standard state values (298 K; 1 atm) the values $\sum^K (G_K^0)_{T,P}$ and $(\mu_K^0)_{T,P}$ would be written as G_Σ° and μ_K°. For the ideal mixture with $(\Delta G_\Sigma)_{mix} = (\Delta A_\Sigma)_{mix}$, Eqs. (8.58) and (8.67) provide

$$(\Delta\mu_K)_{mix} = \left(\frac{\partial(\Delta G_\Sigma)_{mix}}{\partial N_K}\right)_{T,P,N_J} = kT\left[\sum^K \frac{\partial N_K \ln \dfrac{N_K}{\Sigma N_K}}{\partial N_K}\right]_{T,P,N_J}$$

$$= kT \ln \frac{N_K}{\Sigma N_K} = kT \ln X_K \tag{8.70}$$

where the X_K are component mole fractions.
For an ideal mixture under standard state conditions the, we write for any component

$$\mu_K^* = \mu_K^\circ + kT \ln X_K \tag{8.71}$$

and for the mixture

$$\sum^K N_K \mu_K^* = \sum^K N_K \mu_K^\circ + kT \sum^K N_K \ln X_K. \tag{8.72}$$

Alternatively, multiplying all terms by $N_A | \sum_K N_K$, we can write

$$\sum^K X_K \bar{G}_K = \sum^K X_K \tilde{G}_K^0 + N_A kT \sum^K X_K \ln X_K \tag{8.73}$$

where $\bar{G}(= N_A\mu_K)$, $\tilde{G}_K^0$, and $N_A k (= R)$ are molar terms. The overbar notation represents the molar property of a component in a mixture. We would refer to $\bar{G}_K$ as the partial molar free energy of component K in a mixture.

The framework represented by Eqs. (8.67)–(8.72) is applicable to any ideal mixture on the basis of statistical argument. As applied to gases, however, the μ° terms do not have exactly the same meaning as they do for condensed phases since we defined μ and μ° for the gas phase in a manner that incorporated the ideal gas concept

$$\mu = \mu^\circ + kT \ln P.$$

In the ideal mixed gas system

$$\mu_K^* = \mu_K^\circ + kT \ln P + kT \ln X_K$$

or in a mixture of nonideal gases that nevertheless exhibited ideal mixing behavior (intermolecular interactions effectively the same between all species)

$$\mu_K^* = \mu_K^\circ + kT \ln f_K^* \tag{8.74}$$

where the fugacity term reflects the nonideal behavior of each component [Eq. (6.50)] and for the imagined circumstance would be unchanged from the pure gases. If the mixing process is a nonideal process there would be additional terms deriving from Eq. (8.39). We only note here that the excess (standard) free energy of mixing over that of ideal mixing is

$$^{E}(\mu_K)_{mix} = kT \ln \frac{f_K^*}{(f_K^*)_{ideal}} \tag{8.75}$$

where it is assumed that the energy levels as reflected in the μ_K° values are unchanged in the mixing process but the fugacity f_K may be different from that of the pure component f_K. Normally fugacity values for the mixture would be obtained from compressibility data for the mixture [Eq. (6.54)] or from generalized fugacity tables for the various components of the mixture making various assumptions about critical states appropriate to the mixture. Since our strongest interest will not lie with gas mixtures but in condensed phase mixtures (solutions) of various kinds (liquids and solids in liquids predominantely) we will not pursue the detail.[4]

Ideal solutions can be defined as those for which Eqs. (8.70)–(8.73) apply. As we have defined the μ° terms for pure condensed phases, their calculation (from free energies) is difficult for solids and not at all satisfactory for liquids. Each such value would include cell volumes, interaction energies, and molecule motion terms. In constructing the statistical arguments for ideal solutions we have assumed that all of these features are unchanged for all components of the solution; then μ_K° in the pure component phase and the solution is the same. In nonideal solutions, even though we would anticipate that internal motions of the molecules (such as vibration) are not perturbed, effectively all other features of the partition function for unlike molecule interactions will be different for those between like interactions and μ_K° for pure components will not be μ_K° for those components in solution. However, if we modify Eq. (8.72) by introducing the concept of activity (just as we introduced the fugacity concept for deviation from ideal gas behavior) we can write

$$\mu_K = \mu_K^\circ + kT \ln a_K \tag{8.76}$$

where each component activity is $a_K = \gamma_K X_K$. Then γ_K, which we will refer to as the activity coefficient for the Kth solution component, will represent all deviation from the behavior of that component in an ideal solution. We require that $\gamma_K \to 1$ as $X_K \to 1$ so that Eq. (8.76) reduces to the pure component expression in an appropriate manner.[5] The activity coefficient will reflect

[4] See reference 2, Prausnitz, et al., 1986, p 151 ff.

[5] Operationally, other conventions are required in some cases where solubilities are limited and other conventions are sometimes just more convenient. Any other convention, however, requires a redefinition of μ_K^0. Discussion is deferred to later chapters.

nonideality through the excess functions as we defined them. Then, for condensed phase solutions

$$
\begin{aligned}
{}^{E}(\Delta G_\Sigma)_{mix} &= \sum^{K}(N_K\mu_K)_{nonideal} - \sum^{K}(N_K\mu_K)_{ideal} \\
&= \ell T \sum^{K} N_K \ln \gamma_K
\end{aligned}
$$

or

$$
{}^{E}(\Delta G_m)_{mix} = N_A \ell T \sum X_K \ln \gamma_K \tag{8.77}
$$

where we use (ΔG_m) to express the free energy change per mole of mixture (the tilde we reserve for molar properties of pure substances). Most commonly discussions of excess thermodynamic functions will feature the partial molar notation in which [extending Eq. (8.73)]

$$
\sum X_K {}^{E}(\bar{G}_K) = N_A \ell T \sum X_K \ln \gamma_K
$$

and for any component

$$
{}^{E}(\bar{G}_K) = \ell T \ln \gamma_K. \tag{8.78}
$$

The partial molar property concept is a general one, not just confined to the Gibbs free energy. For any state function F we can define[6]

$$
N_A \left(\frac{\partial F_\Sigma}{\partial N_K} \right)_{T,P,N_J} = \bar{F}_K = \left(\frac{\partial F_K^*}{\partial n_K} \right)_{T,P,n_J} \tag{8.79}
$$

$$
(X_m)_{mix} = \sum X_K \bar{F}_K. \tag{8.80}
$$

It also follows that in a nonideal solution we can define excess thermodynamic functions and express them in terms of partial molar properties

$$
{}^{E}(\Delta X_m)_{mix} = \sum X_K {}^{E}(\bar{F}_K). \tag{8.81}
$$

For a two-component system a useful graphical construct can produce the partial molar property for any mixed system. Figure 8.3 illustrates such a general construct where the components are designated as 1 and 2. The slope at any overall composition χ, when extended to the pure component axes, produces

[6] It is important to remember that partial molar terms are by definition given by Eq. (8.80) but only $\bar{G}_K = \mu_K$. Terms such as $(\partial E_\Sigma/\partial N_K)_{S,V,N}$ [Eqs. (4.9) and (4.10)] also define μ_K but such terms are not partial molar properties.

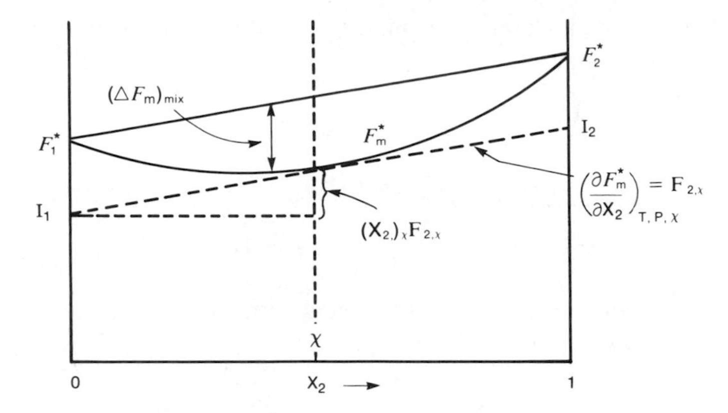

Figure 8.3. Illustration of the intercept method for resolving partial molar properties from the mixed system property.

intercepts designated as I_1 and I_2. Since

$$\begin{aligned}(\bar{F}_m)_\chi &= X_{1,\chi}\bar{F}_{1,\chi} + X_{2,\chi}\bar{F}_{2,\chi} \\ &= I_1 + X_{2,\chi}\bar{F}_{2,\chi}\end{aligned} \tag{8.82}$$

$$I_1 = \bar{F}_{1,\chi} @ \chi_1 = 1. \tag{8.83}$$

The reverse argument relates I_2 to $\bar{F}_{2,\chi}$. Such a construct can be accomplished when ΔF_{mix} can be obtained either by calculation or experiment. Both volume and energy changes on mixing are available experimental quantities.

Since $(\Delta E)_{mix} = (\Delta H)_{mix}$ and ΔV_{mix} are zero in the ideal solution, the excess quantities in nonideal solutions are (for 1 mol of mixture)

$$^E(\Delta H_m)_{mix} = (\Delta H_m)_{mix} = \sum^{K} X_K \bar{H}_K \tag{8.84}$$

$$^E(\Delta V_m)_{mix} = (\Delta V_m)_{mix} = \sum^{K} X_K \bar{V}. \tag{8.85}$$

As applied to volume changes on mixing to provide a composition χ, $V_{\Sigma,\chi}$ is a directly measured quantity as is $\tilde{V}_1^0$ and $\tilde{V}_2^0$ and

$$V_{\Sigma,\chi} = n_1\tilde{V}_1^0 + n_2\tilde{V}_2^0 + (\Delta V_{mix})_\chi \tag{8.86}$$

$$\frac{V_{\Sigma,\chi}}{n_1 + n_2} = V_m^* = X_1\tilde{V}_1^0 + X_2\tilde{V}_2^0 + \frac{(\Delta V_{mix})_\chi}{n_1 + n_2}. \tag{8.87}$$

As applied to energy changes on mixing, the directly measured term is the integral heat of mixing

$$(\Delta H_{mix})_\chi = H_{\Sigma,\chi} - (n_1 H^0_{m,1} + n_2 H^0_{m,2}) \tag{8.88}$$

$$H_{\Sigma,\chi} = n_1 \bar{H}_{1,\chi} + n_2 \bar{H}_{2,\chi} \tag{8.89}$$

$$(\Delta H_{mix})_\chi = n_1(\bar{H}_{1,\chi} - H^0_1) + n_2(\bar{H}_{2,\chi} - H^0_2). \tag{8.90}$$

To avoid absolute enthalpies the bracketed terms in Eq. (8.90) are normally reported as differential heats of solution

$$\left[\frac{\partial(\Delta H_{mix})_\chi}{\partial n_1}\right]_{T,P} = (\bar{H}_{1,\chi} - H^0_1)$$

$$\left[\frac{\partial(\Delta H_{mix})_\chi}{\partial n_2}\right]_{T,P} = (\bar{H}_2 - H^0_2) \tag{8.91}$$

and obtained graphically from the integral heat data. These are particularly useful terms since the calorimetric heat ($= -\Delta H_{mix}$) of mixing and the volume change on mixing are directly related to properties of the activity coefficient through relationships developed in earlier chapters and Eq. (8.79)

$$\left[\frac{\partial^E(\bar{G}_K/T)}{\partial T}\right]_{P,N_J} = \frac{^E(\bar{H}_K)}{T^2} = R\left(\frac{\partial \ln \gamma_K}{\partial T}\right)_{P,N_J} \tag{8.92}$$

$$\left[\frac{\partial(\bar{G}_K/T)}{\partial P}\right]_{T,N_J} = {}^E(\bar{V}_K) = R\left(\frac{\partial \ln \gamma_K}{\partial P}\right)_{T,N_J}. \tag{8.93}$$

For solutions which are not ideal but regular, $\Delta V_{mix} = 0$, the measured $\Delta H_{mix} = {}^E(\Delta H_{mix})$ will provide ${}^E(\Delta E_{mix})$. In this context Eq. (8.55) is the basis for useful empirical approaches to activity coefficients in nonideal solutions. From Eqs. (8.55) and (8.79)

$${}^E(\bar{G}_K) = RT \ln \gamma_K = \frac{\partial \sum n\, {}^E(\Delta G_m)_{mix}}{\partial n_K} \tag{8.94}$$

$${}^E(\Delta E)_{mix} = c\omega_{12} X_1 X_2 = {}^E(\Delta G_m)_{mix} \tag{8.95}$$

since ${}^E(\Delta S_{mix})$ is also zero. Substituting $X_1 = n_1/\Sigma n$, $X_2 = n_2/\Sigma n$, and differentiating produces

$$\ln \gamma_1 = \frac{c\omega_{12}}{kT} X_2^2 \tag{8.96}$$

$$\ln \gamma_2 = \frac{c\omega_{12}}{kT} X_1^2. \qquad (8.97)$$

Even given that the term $c\omega$ is uncertain[7] and based on an oversimplified model, the intuitive association is useful and the general expression[8]

$${}^E(\Delta G_m) = AX_1X_2 \qquad (8.98)$$

is an effective framework for data treatment when expanded as

$${}^E(\Delta G)_{mix} = X_1X_2[A + B(X_1 - X_2) + C(X_1 - X_2)^2 + \cdots]. \qquad (8.99)$$

The term A produces a symmetrical parabola while the term B will provide a tilt and term C will sharpen or flatten the parabola thus providing a very flexible empirical data treatment.

As a final observation relating to the general framework of treating deviations from ideality in terms of the activity coefficient we return to the Gibbs–Duhem equation in the form

$$\sum X_K d\mu_K = 0$$

obtained by dividing each term in Eq. (8.64) by $\sum N_K$. For an infinitisimal change in concentration in a two-component system

$$X_1\left(\frac{\partial \mu_1}{\partial X_1}\right)_{T,P,N_J} dX_1 = -X_2\left(\frac{\partial \mu_2}{\partial X_2}\right)_{T,P,N_J} dX_2 \qquad (8.100)$$

and since dX_1 will be equal to $-dX_2$ and ${}^E(\mu_K) = kT \ln \gamma_K$

$$X_1\left(\frac{\partial \ln \gamma_1}{\partial X_1}\right)_{T,P} = X_2\left(\frac{\partial \ln \gamma_2}{\partial X_2}\right)_{T,P}$$

which can be written as

$$\left(\frac{\partial \ln \gamma_1/\gamma_2}{\partial X_2}\right)_{T,P} = \frac{1}{X_2}\left(\frac{\partial \ln \gamma_2}{\partial X_2}\right)_{T,P}. \qquad (8.101)$$

[7] The van Laar equations deriving from a treatment in which the mixed liquids are considered to be van der Waals fluids is not in itself adequate except as an empirical form. The Scatchard–Hildebrand theory based on a parameter $C = E_{vap}/\tilde{V}_l$ and assumptions relating C_{11}, C_{22}, and C_{12} to ω is much more successful. See reference 2 (Prausnitz et al., 1986), Chapter 7 for further detail.

[8] Equation (8.98) is known as the Margules equation and Eq. (8.99) is the Redlich–Kistler expansion. See reference 2 (Prausnitz et al., 1986), Chapters 6 and 7, for further detail.

Where component 1 is solvent it is often possible to obtain γ_1 values experimentally over a range of compositions. The data can be fitted to

$$\ln \gamma_1 = \sum a_i X_2^{b_i} \tag{8.102}$$

where the a_i and b_i are empirical constants ($b_i < 1$). Then

$$\frac{1}{X_2}\left(\frac{\partial \sum a_i X_2^{b_i}}{\partial X_2}\right)_{T,P} = \left(\frac{\partial \ln \gamma_1/\gamma_2}{\partial X_2}\right)_{T,P}$$

and

$$\ln \gamma_2 - \ln \gamma_1 = \int \sum a_i b_i X_2^{(b_i - 2)} dX_2 \tag{8.103}$$

$$\ln \gamma_2 = \sum a_i X_i^{b_i} - \sum \frac{a_i b_i}{(b_i - 1)} X_2^{(b_i - 1)} - I \tag{8.104}$$

where I is a constant of integration which is

$$I = \sum a_i + \sum \frac{a_i b_i}{(b_i - 1)} \tag{8.105}$$

since our present convention is $\gamma_2 = 1$ when $X_2 = 1$. In a final compact form

$$\ln \gamma_2 = \sum a_i X_i^{b_i} - \sum \frac{a_i}{(b_i - 1)} (b_i X_2^{(b_i - 1)} - 1). \tag{8.106}$$

Lastly we observe that for

$${}^E(\Delta G_m)_{mix} = RT(X_1 \ln \gamma_1 + X_2 \ln \gamma_2) \tag{8.107}$$

$$\frac{d^E(\Delta G_m)_{mix}/RT}{dX_1} = X_1 \frac{\partial \ln \gamma_1}{\partial X_1} + \ln \gamma_1 + X_2 \frac{\partial \ln \gamma_2}{\partial X_1} + \ln \gamma_2 \frac{dX_2}{dX_1} \tag{8.108}$$

and with $dX_1 = -dX_2$

$$\frac{d^E(\Delta G_m)_{mix}/RT}{dX_1} = \ln \frac{\gamma_1}{\gamma_2}. \tag{8.109}$$

Integrating over the range of component 1 mole fraction

$$\int_0^1 d^E(\Delta G_m)_{mix}/RT = \int_0^1 \ln \frac{\gamma_1}{\gamma_2} dX_1 \tag{8.110}$$

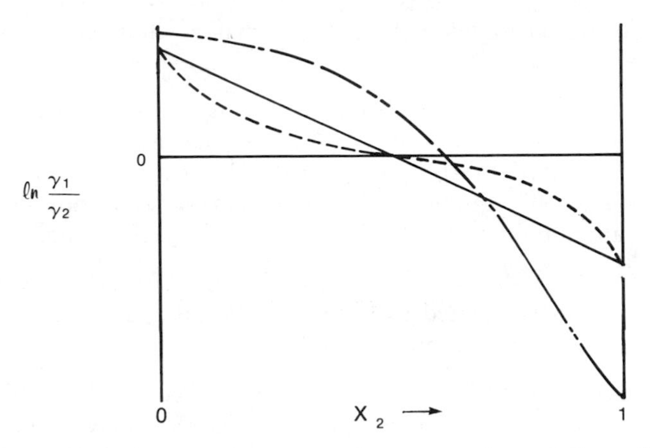

Figure 8.4. Illustration of an equal areas consistency test for activity coefficients in a two-component system.

recognizing that

$$^E(\Delta G_m)_{mix}/RT = 0 @ X = 0$$

$$^E(\Delta G_m)_{mix}/RT = 0 @ X = 1$$

$$\int_0^1 \ln \frac{\gamma_1}{\gamma_2} dX_1 = 0 \tag{8.111}$$

so that if $\ln \gamma_1/\gamma_2$ is plotted against X_1 we should observe equal areas above and below the $\ln \gamma_1/\gamma_2 = 0$ line regardless of the complexity of the solution as shown in Figure 8.4. This is a general consistency test.

In the following chapters we extend mixed system considerations to multiphase systems where we find that equilibrium constraints both complicate and simplify our considerations.

PROBLEMS

8.1. Equations (8.34) and (8.35) express the free energy and entropy changes on mixing ideal gases when the gaseous systems exist at the same temperature and pressure before mixing. Calculate the Gibbs free energy and the entropy of mixing 2 mole of N_2 with 1 mole of O_2 when each gas is initially under standard conditions of temperature and pressure assuming that both gases behave ideally under these conditions.

8.2. Assume that the mixing process is removal of a barrier between separate volumes. The 1 mole of oxygen is at 2 atmospheres pressure and 260 K while the 2 mole of nitrogen is at 298 K and 1 atmosphere pressure. Calculate the Gibbs free energy and the entropy changes that accompany removal of the barrier.

8.3. If both nitrogen and oxygen are at 10 atmospheres pressure and at 500 K before they are mixed we can assume that they are not ideal gas systems, mixed or unmixed. Starting with Eq. (8.37), show that the van der Waals correction to the equation of state volume term is not an important consideration in obtaining the excess free energy of mixing.

8.4. Starting with Eq. (8.39), assuming that N_2 and O_2 under the problem 8.3 conditions can be treated as a van der Waals gas, calculate the excess Helmholz free energy of mixing.

8.5. In a regular solution Eq. (8.55) for the excess free energy of mixing would reduce to the energy of mixing term. Assume that $CHCl_3$ and CCl_4 form a regular mixture at mole fraction 0.5 (each component). The van der Waals a for $CHCl_3$ is 20.39 liter^2 atm mole^{-2} and for CCl_4 it is 15.17. Calculate the Helmholz free energy of mixing for this system.

8.6. The volumes of mixtures of chloroform and acetone are observed to be

mole fraction $CHCl_3$	0	0.194	0.385	0.559	0.788	0.889	1.00
molar volume mixture	73.99	75.29	76.50	77.55	79.08	79.82	80.67

Using the method of intercepts obtain the partial molar volume of each component at 0.2 and at 0.7 mole fraction $CHCl_3$.

CHAPTER CONTENT SOURCES AND FURTHER READING

1. I. Klotz, *Chemical Thermodynamics*, Chapters 13 and 14. W. A. Benjamin, New York, 1964.
2. J. M. Prausnitz, R. N. Lichtenthaler, E. G. de Azevado, *Molecular Thermodynamics of Fluid-Phase Equilibria*, 2nd ed., Chapters 5, 6, and 7 and Appendix VIII. Prentice-Hall, Englewood Cliffs, NJ, 1986.
3. O. K. Rice, *Statistical Mechanics, Thermodynamics and Kinetics*, Chapter 12. W. H. Freeman, San Francisco, 1967.
4. G. S. Rushbrooke, *Statistical Mechanics*, Chapters XI, XII, XIII, and XIV. Oxford, London 1949.
5. W. E. Acree, Jr., *Thermodynamics Properties of Nonelectrolyte Solutions*, Chapters 2 and 5. Academic Press, New York 1984.

SYMBOLS: CHAPTER 8

New Symbols

P_K, P_Σ = partial pressure of a component in a mixture, total pressure of the mixed system

$A_{\mathscr{A}}, A_{\mathscr{A}\mathscr{A}'} \cdots \mathscr{Z}_{\mathscr{A}\mathscr{A}'}$ = value of a thermodynamic function for systems $\mathscr{A}$, $\mathscr{A}'$ and mixed $\mathscr{A}\mathscr{A}'$

$\Delta A_{mix} \cdots \Delta S_{mix}$ = value of the change in a thermodynamic function on mixing

$\phi_{\mathscr{A}}$ = potential energy for a molecule $\mathscr{A}$ in a system of $\mathscr{A}$-type molecules

$\phi_{e\mathscr{A}}, r_{e,\mathscr{A}}$ = potential energy parameters for molecules $\mathscr{A}$ in a system of $\mathscr{A}$-type molecules

$\phi_{\mathscr{A}\mathscr{B}}, \gamma_{\mathscr{A}\mathscr{B}}$ = potential energy parameters for molecules in an $\mathscr{A}\mathscr{B}$ mixture

z = number of lattice sites denied to another molecule (one time use)

$N_{\mathscr{B}}(= 1 \cdots N_{\mathscr{B}})$ = a number of $\mathscr{B}$-type molecules added to a lattice structure of other molecules

$v_{\mathscr{A}}, v_{\mathscr{B}} \cdots v^0_{\mathscr{A}}, v^0_{\mathscr{B}}$ = molecule volumes in a mixture, in the pure state

$\upsilon_{\mathscr{A}}, \upsilon_{\mathscr{B}}$ = volume fractions of $\mathscr{A}$- and $\mathscr{B}$-type molecules in a mixture (one time use)

ϕ_{Σ} = total potential energy in a molecule system

$\omega_{\mathscr{A}\mathscr{B}}$ = exchange energy, pure systems to mixed system

c = number of molecule interactions

H^*, G^*, μ^* = total value of a thermodynamic function in a mixture

$\bar{H}_{\kappa}, \bar{G}_{\kappa} \cdots$ = partial molar property of a component κ in a mixture

${}^{E}(\Delta H)_{mix} \cdots {}^{E}(\Delta H_m)_{mix}$ = excess value of a thermodynamic function in a mixture, per mole of mixture

$\bar{F}, F_m \cdots$ = general partial molar thermodynamic property, property per mole of mixture (one time use)

$n_{\mathscr{A}}, n_{\mathscr{B}} \cdots n_1, n_2 \cdots$ = number of moles of a component $\mathscr{A}, \mathscr{B} \cdots 1, 2 \cdots$ in a mixture

$X_1 \cdots X_{\kappa}$ = mole fraction of a component in a mixture

$(\Delta H_m)_{mix} \cdots$ = property of a mixture per mole of mixture

A, B, C = constants in an empirical free energy expression (one time use)

a_i, b_i, I = constants in an empirical activity coefficient expression (one time use)

Carryover Symbols

E, E_{κ}, E_{Σ}, N_{κ}, $\wp_{j,\kappa}$, $\epsilon_{j,\kappa}$, m, V, T, $\mathscr{Z}$, $\mathscr{Z}_{\Sigma}$, S, S_{Σ}, A, A_{Σ}, P, $C_{V,\kappa}$, $C_{P,\kappa}$, H_{Σ}, G_{Σ}, Ω, $\Omega_{\mathscr{A}}$, $\Omega_{\mathscr{A}\mathscr{B}}$, v_f, v_{cell}

9

PHASE AND CHEMICAL EQUILIBRIUM I: STATISTICAL CONSIDERATIONS

We have introduced the concept of equilibrium as a state of the isolated thermodynamic system in which all statistical features and all observable features that we associate with them are unchanging with time. The most fundamental statistical feature of the equilibrium state is the identification of a particular distribution of particles among energy states, which leads to the maximization of entropy ($k \ln \Omega$) in the isolated system. This distribution is arrived at through, and maintained by, the constant dynamic exchange of energies through weak interactions involving all particles in the system. The constant particle exchange within the set of available quantum states, while maintaining a certain fixed distribution, is a further feature of equilibrium that can be extended to the circumstance that identical particles can exchange between energy states associated with different phases (phase equilibrium) or with different species (chemical equilibrium).

Having now developed the capacity to construct partition functions for all normal phases for mixed systems within a single phase, and for complex molecules, we are prepared to examine the statistical circumstances under which one or several components could be distributed among several coexisting phases to provide a maximum value for the total entropy within an isolated system, which includes the several components and phases. We will refer to this circumstance as phase equilibrium. Maximization of entropy for the distribution of atoms among molecular species in an isolated system also characterizes chemical equilibrium, but the arguments must necessarily be phrased separately. We will first consider phase equilibrium in a one-component system.

We imagine any isolated gas phase system of identical molecules of any structural complexity; E, V, and N are specified. We will place the system in isothermal contact with a thermal reservoir at T and further alter the constraints by introducing a displaceable boundary across which we apply an external force to reversibly decrease the system volume at constant T. From nonideal gas arguments developed in Chapter 6 we anticipate that, for all $T < T_c$, condensation will occur at some point in the isothermal process and that, if unhindered, the overall process $a \rightarrow d$ would be represented by Figure 9.1 where "c" represents a condensed phase and "g" the gas phase. Since different phases are described by different partition functions there will be a discontinuity in certain thermodynamic properties. The nature of such discontinuities may be quite complex and under some circumstances quite unique. The Figure 9.1 type process is referred to as a first-order phase transition.

Clearly we can interrupt the Figure 9.1 process at any point, say at "b" and reimpose the isolation conditions. Since the process has been conducted reversibly, isolation will not effect any change in the system. Even if we had not attempted to conduct the overall process ($a \rightarrow b$) reversibly, only the overall process parameters q and w could be different from the reversible process parameters. All state functions (S, E, G, etc.) would change by amounts determined by the initial and final isolated system values of T, V, P, and N as constrained by the appropriate equations of state. The final isolated state at point "b" is a two-phase equilibrium system characterized by maximum valued entropy ($k \ln \Omega$).

The statistical route to the description of the two-phase system is clear. A combinatorial term for the system should be constructed and then maximized. Following familiar statistical arguments we would consider an isolated system of specified E, V, and N containing a single species but present now both as gas and condensed phase. We would understand from our analysis of the equation of state behavior that E, V, and N cannot all be randomly selected

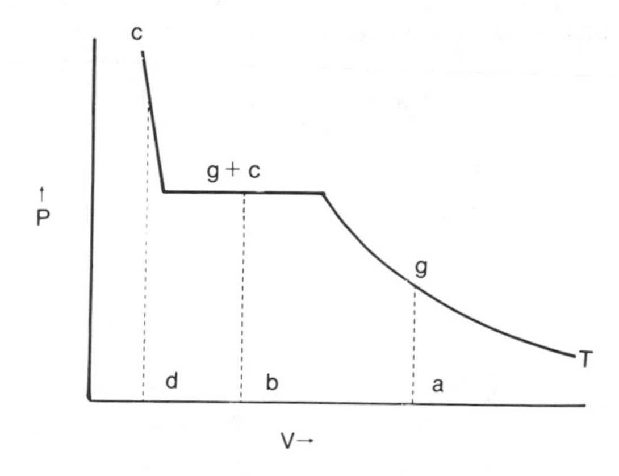

Figure 9.1. Isothermal condensation of a gas phase.

and we depend on the statistical argument to provide the criteria that dictate their values when the state of two-phase equilibrium prevails.

Any molecule can be in any of the energy states $\epsilon_1, \epsilon_2, \ldots, \epsilon_j$ of the gas phase or $\eta_1, \eta_2, \ldots, \eta_j$ of the condensed phase. We take there to be N_g molecules in the gas phase with distribution numbers $N_1, N_2, \ldots, N_j$ and M_c molecules in the condensed phase with distribution numbers $M_1, M_2, \ldots, M_j$. There are altogether $N_\Sigma = M_c + N_g$ molecules in the system. Any distribution of the N_g gas phase molecules among their energy states can be coupled with any distribution of the M_c condensed phase molecules among their energy states. The combinatorial term is

$$\Omega = \sum \frac{N_g!}{\prod N_j!} \frac{1}{N_g!} \frac{M_c!}{\prod M_j!}. \tag{9.1}$$

Since all energy states are available to all molecules and molecules can exchange between phases, there are only two equations of constraint requiring only two undetermined multipliers

$$\sum^j N_j + \sum^j M_j = N_\Sigma, \qquad -\alpha\left[\sum^j dN_j + \sum^j dM_j\right] = 0$$

$$\sum^j N_j\epsilon_j + \sum^j M_j\eta_j = E_\Sigma, \qquad -\beta\left[\sum^j \epsilon_j dN_j + \sum^j \eta_j dN_j\right] = 0$$

and the maximum term procedure involves the expression

$$d\ln t_{max} - \alpha\left[\sum^j dN_j + \sum^j dN_j\right] - \beta\left[\sum \epsilon_j dN_j + \sum \eta_j dN_j\right] = 0$$

where $d\ln t_{max} = d[\ln M! - \sum^j \ln N_j! - \sum^j \ln M_j!$. The detailed arithmetic is provided in Appendix C on Statistical Considerations and the solution is

$$N_j^* = e^{-\alpha} e^{-\beta\varepsilon_j} \tag{9.2}$$

$$\frac{M_j^*}{\sum M_j^*} = e^{-\alpha} e^{-\beta\eta_j}. \tag{9.3}$$

Then

$$\sum N_j^* = N_g^* = e^{-\alpha} \sum^j e^{-\beta\epsilon_j} = e^{-\alpha} z_g \tag{9.4}$$

$$\frac{\sum M_j^*}{\sum M_j^*} = 1 = e^{-\alpha} \sum e^{-\beta\eta_j} = e^{-\alpha} z_c \tag{9.5}$$

and

$$\alpha = \ln \frac{z_g}{N_g^*} = \ln z_c.$$

There is one value of chemical potential $\mu = kT\alpha$ for the two phases in complete accord with prior argument that chemical potential for a species is uniform in an equilibrium system. Also, $N_g^* = z_g/z_c$, providing the gas molecule density, i.e., the gas phase pressure when T and V are specified. This, of course, follows from the consideration that the gas phase partition function contains the term V, which can (over most of the two phase region) be taken as the total volume ($V_g = V - V_c \cong V$). Then $z_g = \Lambda^{-3/2} V z_g'$ and

$$\frac{N_g}{V} = \frac{\Lambda^{-3/2} z_g'}{z_c} = \frac{P^*}{kT}, \qquad P^* = kT \frac{\Lambda^{-3/2} z_g'}{z_c} \tag{9.6}$$

assuming that the gas phase behaves ideally in this (normally) low pressure circumstance.

There is a two-phase coexistence pressure (P^*), which we will call the vapor pressure, fixed by T, Λ (specific only in terms of molecule mass), z_g' representing components of the gas phase partition function other than translation (with their unique T dependence), and the condensed phase partition function z_c, whatever its particular form might be (with its unique temperature dependence). The equilibrium partition functions for particles of mass m in a one-component system are defined only when T and V (or P) are specified.

The results for single component phase equilibrium that emerge from the above treatment are important enough to warrant emphasis. The chemical potential is uniform throughout the system although the molecular species present is distributed between two very different phases that are described by very different partition functions. Clearly, only at the unique value of P and T that characterizes any particular two-phase equilibrium do the very different partition function expressions have the same value since, generally, $G = N\mu$,

$$G_g = N_g\mu_g = -kT \ln \mathscr{Z}_g - N_g kT$$

$$G_c = N_c\mu_c = -kT \ln \mathscr{Z}_c$$

for

$$(\mu_g = \mu_c)_{eq}, \qquad (\ln \mathscr{Z} - N_g)_{P^*,T} = (\ln \mathscr{Z}_c)_{P^*,T}.$$

Figure 9.1, abstracted from Figure 5.5, is a unique isotherm (P^*, T). A continuous variation in T produces a set of intersecting PV surfaces of the Figure 9.2 form, each unique specification of T providing a unique value of vapor pressure which we will identify as $P^*(T)$. Our conclusions apply to any point located in the two-phase (heavily shaded) surface and its (bounded)

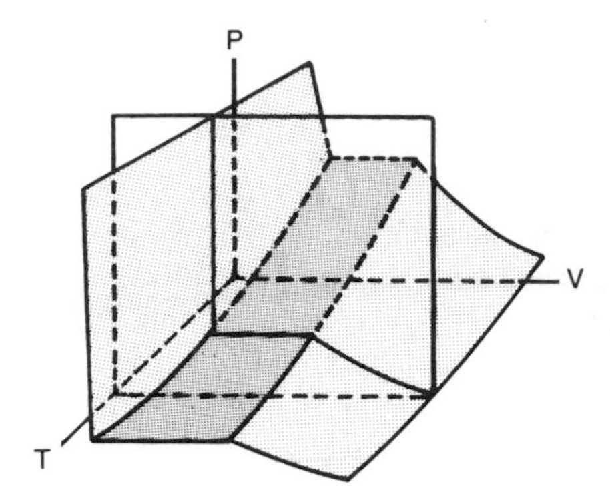

Figure 9.2. The PVT surface for a one component system.

extensions. Within this region the quite different surfaces (not shown) representing

$$\left[\left(\frac{\partial kT \ln \mathscr{Z}_g}{\partial T}\right)_{P^*N_\Sigma} - N_g k\right] \quad \text{and} \quad \left(\frac{\partial kT \ln \mathscr{Z}_c}{\partial T}\right)_{P^*N_\Sigma} \tag{9.7}$$

intersect for all P^*, N_Σ, a generalizing feature to which we will return.

The important results are quite general as can be recognized from consideration of their origin, which derives from the form of the combinatorial term and the constraints on its maximization and not from any particular detail associated with specific phase character although such detail will be reflected in the unique equilibrium distribution. The unique distribution of particles among phases follows immediately from partition function detail when the constraint of a single value for each of the undetermined multipliers α and β is applied. The constraint on α, however, applies to the availability of energy states to individual particles independently of the separability of energy state sets for separate phases. The single value of β is a feature of the equilibrium state rather than a feature of the particular phases present since no particle is confined to a region of the energy state spectrum by any constraint other than that imposed by the total energy E.

If the system were to contain not one but several components, some details of the statistical argument will be changed. A two-component, two-phase system will be sufficiently general to include all necessary extensions. Let us designate the components as $\mathscr{A}$ and $\mathscr{B}$ and assume that each component is present in each of the two phases, which we will again take to be a gas phase behaving ideally and any general condensed phase. Following the notation developed for the one-component system we now have energy states $\epsilon_{1,\mathscr{A}}, \epsilon_{2,\mathscr{A}}, \ldots, \epsilon_{j,\mathscr{A}}$ and $\epsilon_{1,\mathscr{B}}, \epsilon_{2,\mathscr{B}} \ldots, \epsilon_{j,\mathscr{B}}$ in the gas phase with distribution numbers $N_{1,\mathscr{A}}, N_{2,\mathscr{A}}, \ldots, N_{j,\mathscr{A}}$ and $N_{1,\mathscr{B}}, N_{2,\mathscr{B}}, \ldots, N_{j,\mathscr{B}}$. Corresponding energy states and distribution numbers in the condensed phase are $\eta_{1,\mathscr{A}}, \ldots, \eta_{j,\mathscr{B}}$ and $M_{1,\mathscr{A}}, \ldots, M_{j,\mathscr{B}}$. There are

$N_{\mathscr{A}} + N_{\mathscr{B}} = N_g$ total gas phase particles and $M_{\mathscr{A}} + M_{\mathscr{B}} = M_c$ condensed phase particles.

We construct the combinatorial term, again with recognition that the gas phase particles are nonlocalized and that a completely general treatment will regard the condensed phase particles as localized (lattice sites, or cell volumes, etc.). If the particles are similar they may occupy lattice sites in a single solid phase. If so, there are two components that must be assigned to the $M_{\mathscr{A}} + M_{\mathscr{B}}$ localized sites and permutations among the site assignments will lead to a term $(M_{\mathscr{A}} + M_{\mathscr{B}})!/M_{\mathscr{A}}!M_{\mathscr{B}}!$. If the particles are quite dissimilar each species will be confined to the sites associated with its normal condensed phase and the site permutation term will be absent. We consider these implications in detail in the following chapter. Following the arithmetic associated with maximizing the combinatorial term expression for the single condensed phase system we have

$$\Omega = \sum \frac{(M_{\mathscr{A}} + M_{\mathscr{B}})!}{M_{\mathscr{A}}!M_{\mathscr{B}}!} \frac{M_{\mathscr{A}}!}{\prod M_{j,\mathscr{A}}} \frac{M_{\mathscr{B}}!}{\prod M_{j,\mathscr{B}}!} \frac{1}{\prod N_{j,\mathscr{A}}!} \frac{1}{\prod N_{j,\mathscr{B}}!} \tag{9.8}$$

and $\Omega = t_{max}$ as always. The maximization constraints in the isolated system must include our recognition that (in principle) all particles in the system can exchange energy through the weak interaction mechanism, i.e., there is only one energy constraint

$$\sum(M_{j,\mathscr{A}}\eta_{j\mathscr{A}} + M_{j,\mathscr{B}}\eta_{j\mathscr{B}}) + \sum(N_{j,\mathscr{A}}\epsilon_{j\mathscr{A}} + N_{j,\mathscr{B}}\epsilon_{j,\mathscr{B}}) = E$$

requiring the assignment of one undetermined multiplier β.

$$\beta[\sum(\eta_{j,\mathscr{A}}dM_{j,\mathscr{A}}, \ldots, \epsilon_{j,\mathscr{B}}dN_{j,\mathscr{B}})] = 0.$$

Constraints on the distribution numbers must be expressed in the form of two independent equations (since, e.g., the $N_{\mathscr{A}}$ molecules must be distributed only among the $\epsilon_{j,\mathscr{A}}$ and the $\eta_{j,\mathscr{A}}$ energy states i.e.,

$$\sum M_{j,\mathscr{A}} + \sum N_{j,\mathscr{A}} = N_{\mathscr{A}}, \qquad \alpha_{\mathscr{A}}[\sum(dM_{j,\mathscr{A}} + dN_{j,\mathscr{A}})] = 0$$

$$\sum M_{j,\mathscr{B}} + \sum N_{j,\mathscr{B}} = N_{\mathscr{B}}, \qquad \alpha_{\mathscr{B}}[\sum dM_{j,\mathscr{B}} + dN_{j,\mathscr{B}}] = 0$$

and one undetermined multiplier α will be required for each equation of constraint. The arithmetic associated with the maximization is routine and confined to Appendix C on Statistical Considerations. The results are

$$N^*_{j,\mathscr{A}} = e^{-\alpha_{\mathscr{A}}} e^{-\beta\epsilon_{j,\mathscr{A}}} \tag{9.9}$$

$$N^*_{j,\mathscr{B}} = e^{-\alpha_{\mathscr{B}}} e^{-\beta\epsilon_{j,\mathscr{B}}} \tag{9.10}$$

$$M^*_{j,\mathscr{A}} = (M^*_{\mathscr{A}} + M^*_{\mathscr{B}}) e^{-\alpha_{\mathscr{A}}} e^{-\beta\eta_{j,\mathscr{A}}} \tag{9.11}$$

$$M^*_{j,\mathscr{B}} = (M^*_{\mathscr{A}} + M^*_{\mathscr{B}}) e^{-\alpha_{\mathscr{B}}} e^{-\beta\eta_{j,\mathscr{B}}}. \tag{9.12}$$

Then, following familiar arithmetic,

$$\frac{N^*_{g,\mathscr{A}}}{z_{g,\mathscr{A}}} = e^{-\alpha_{\mathscr{A}}} = \frac{M^*_{c,\mathscr{A}}}{M^*_{c,\mathscr{A}} + M^*_{c,\mathscr{B}}} \frac{1}{z_{c,\mathscr{A}}} \tag{9.13}$$

$$\frac{N_{g,\mathscr{B}}}{z_{g,\mathscr{B}}} = e^{-\alpha_{\mathscr{B}}} = \frac{M^*_{c,\mathscr{B}}}{M^*_{c,\mathscr{A}} + M^*_{c,\mathscr{B}}} \frac{1}{z_{c,\mathscr{B}}} \tag{9.14}$$

yielding

$$N^*_{g,\mathscr{A}} = \frac{M^*_{c,\mathscr{A}}}{M^*_{c,\mathscr{A}} + M^*_{c,\mathscr{B}}} \frac{z_{g,\mathscr{A}}}{z_{c,\mathscr{A}}} \tag{9.15}$$

$$N^*_{g,\mathscr{B}} = \frac{M^*_{c,\mathscr{B}}}{M^*_{c,\mathscr{A}} + M^*_{c,\mathscr{B}}} \frac{z_{g,\mathscr{B}}}{z_{c,\mathscr{B}}}. \tag{9.16}$$

Together with the constraints

$$N^*_{g,\mathscr{A}} + M^*_{c,\mathscr{A}} = N_{\mathscr{A}}$$

$$N^*_{g,\mathscr{B}} + M^*_{c,\mathscr{B}} = N_{\mathscr{B}}$$

the distributions are completely determined when the appropriate partition functions are constructed.

The statistical argument to this point is completely general and can be extended to any number of components in two-phase equilibrium. There will be one undetermined multiplier α for each component, which leads to the requirement that the chemical potential of that component in the equilibrium system is uniform throughout. The argument, demonstrated for one- and two-phase systems, is clearly general. However many phases or components are present, the chemical potential of a component is uniform throughout the equilibrium system.

It follows then that we can avoid the statistical arithmetic by formulating expressions for the chemical potential of each component in each phase and then equating such expressions. Each value of μ_K is obtained from $(\partial A/\partial N_K)_{TV}$, and each such general expression is a function of T, V, and N. For any single value of T, specified V and N (within the constraint system) fixes P through the equation of state. We are prepared to construct partition functions and therefore expressions for the Helmholz free energy for normal (solid, liquid, gas) phases and we imagine that we can extend the methods developed there to other circumstances (e.g., adsorbed phases). It is required that we concern ourselves with the constraints, if any, that control the variety of phases permitted at equilibrium as well as the manner in which the chemical potential in each such phase varies with T and P. We examine these features separately and as follows.

The circumstances that attach to the single undetermined multiplier for each component are associated with the availability of energy states. In a multicomponent system each component species is restricted to certain energy state sets and excluded from all others. The particles of a particular species must be distributed among the energy states available to that species and α is the undetermined multiplier applied to that constraint. For κ components it is required that there be $\alpha_1, \ldots, \alpha_\kappa$ undetermined multipliers, each of which defines a chemical potential $\mu_1, \ldots, \mu_\kappa$. There are κ equations of constraint relating chemical potentials for each of the $\mathfrak{p}$ phases present, e.g., for solid, liquid, and gaseous phases in coexistence

$$\kappa \begin{cases} \overbrace{\mu_{1,g} = \mu_{1,\ell} = \mu_{1,s}}^{\mathfrak{p}} \\ \vdots \\ \mu_{\kappa,g} = \mu_{\kappa,\ell} = \mu_{\kappa,s}. \end{cases}$$

It follows there are $\kappa(\mathfrak{p}-1)$ nonredundant equations relating the chemical potentials, i.e., if $\mu_{1,g} = \mu_{1,\ell}$ and $\mu_{1,\ell} = \mu_{1,s}$ then $\mu_{1,g} = \mu_{1,s}$ follows without specification. There are $1(3-1) = 2$ such equations in a single component, three-phase system. This is the total number of general constraints that must be satisfied and this number of constraints cannot exceed the number of independent variables (T, P, and phase compositions). The composition of each phase is determined by the specification of $(\kappa - 1)$ concentrations and there are $\mathfrak{p}(\kappa - 1)$ such specifications. Then, defining degrees of freedom (F) = number of variables − number of constraints,

$$F = \mathfrak{p}(\kappa - 1) + 2 - \kappa(\mathfrak{p} - 1) = \kappa - \mathfrak{p} + 2. \tag{9.17}$$

The argument is completely general and provides the constraint on the number of coexisting phases permitted in an equilibrium system of any complexity. It is known as the Gibbs Phase Rule. As applied to a one-component system it admits to two degrees of freedom (excess variables that we are free to assign independently) when a single phase is present, one degree of freedom when two phases are present, and no degrees of freedom when three phases are present. All of this is consistent with our results to this point. Over wide ranges of T and P we are free to specify both T and P independently arriving at partition function values for solid or liquid or gaseous phases from which state variables characteristic of that thermodynamic state follow. In the region of two-phase coexistence we can specify P* but in so doing we have placed the system on a particular isotherm or in specifying a particular isotherm we have fixed P* = P*(T). Without recourse to detailed statistical analysis, the Phase Rule alone tells us that when three phases coexist in equilibrium both P* and T are fixed, i.e., this particular equilibrium can exist at one and only one point in

P − T space.[1] Coexistence of more than three phases is prohibited. The Phase Rule does not specify the phases that are present. Quite clearly this question can be resolved only on the basis of whether particular phases are stable in regions where their chemical potentials have comparable values and, therefore, could even possibly have identical values at any specified P and T.

To examine such circumstances it will be instructive to enlarge on our capabilities to set down the chemical potential in terms of distribution numbers for any phase equilibrium constraint without following through the complete statistical argument. For this purpose we construct a condensed phase subject to particular constraints, which we illustrate with a simple example not obviously different from the gas–solid or gas–liquid equilibrium but leading to quite different results.

Consider the adsorption of a gas onto an inert solid surface.[2] There are again two phases, gas phase and adsorbed phase, in the equilibrium circumstance. In a closed, one-component system the Gibbs Phase Rule constraints require that there be one degree of freedom, i.e. if P is specified, then V (therefore phase density) is determined. We assume that there is a known density of adsorption sites on the solid and that the adsorption interaction is sufficiently strong to hold molecules immobile when adsorbed. In the simplest case the adsorbed molecule will have given up translational motion. In more complex cases some rotational motions may be eliminated. Internal vibration would seldom be perturbed but translational and rotational modes may be replaced by vibrations with respect to the surface. We consider that a partition function for adsorbed molecules can be constructed for various adsorption models. The adsorbed molecules exist in quantized energy states $\eta_1, \ldots, \eta_j$ with distribution numbers $N_{1,\mathscr{A}}, \ldots, N_{j,\mathscr{A}}$ and $\sum N_{j,\mathscr{A}} = N_{\mathscr{A}}$. For the gas phase, molecules occupy their normal energy states $\epsilon_1, \ldots, \epsilon_j$ with distribution numbers $\epsilon_1, \ldots, \epsilon_j$ and $\sum N_{j,g} = N_g$. We recognize, in general analogy to the gas-condensed phase equilibrium and Phase Rule considerations, that gaseous and adsorbed phases can coexist throughout T and P regions in which the two chemical potentials can be commonly valued.

Since the chemical potential of the gas phase increases with pressure

$$\mu = kT \ln \frac{z}{N} = -kT \ln \Lambda^{-3/2} \frac{kT}{P} z'$$

(using the ideal gas relationship as a reasonably secure guiding approximation), we require that the chemical potential of the adsorbed phase also increase with

[1] Proof of such invariance for T in any single component three-phase system was a promised feature (Chapter 2) in relating the PV vs β relationship to the Kelvin temperature scale.

[2] If the solid is unchanged as a result of the adsorption process, only the gas-adsorbate system enters into our considerations.

gas phase pressure to maintain equilibrium. Regardless of the exact form of the adsorbed phase partition function, we can assume that the (localized or partially localized) adsorbed phase chemical potential will increase with increase in adsorption since there will be a site permutation term in the combinatorial expression. We assume that there will be both occupied and unoccupied adsorption sites and that the number of occupied sites will increase with gas phase pressure. If we designate the total number of sites as M and if N_a of these are occupied by adsorbed molecules then $(M - N_a)$ are unoccupied and, if all of the original sites were identical, adsorbed molecules and unoccupied sites can be permuted to contribute to the overall combinatorial expression. The other contributions to the combinatorial expression are the common nonlocalized gas phase term and the localized adsorbed phase term giving

$$\Omega = \sum \frac{M}{(M - N_a)!N!} \frac{N_a!}{\prod N_{j,a}!} \frac{1}{\prod N_{j,g}!} \tag{9.18}$$

which should be maximized under the constraints

$$\sum N_{j,a} + \sum N_{j,g} = N$$
$$\sum N_{j,a}\eta_{j,a} + \sum N_{j,g}\epsilon_{j,g} = E$$

requiring only two undetermined multipliers, $-\alpha$ applied to the constraint on N and $-\beta$ applied to the constraint on E. It is not required, however, that we actually carry out the maximization process now that we recognize the concept of the equality of chemical potential for a system component. We can construct an expression for the Helmholtz free energy directly from the combinatorial term and proceed to the chemical potential,

$$A = E - TS = E - kT \ln \Omega$$
$$\mu_K = (\partial A/\partial N_K)_{V,T}. \tag{9.19}$$

From Eq. (9.18)

$$\ln \Omega = \ln M! - \ln(M - N_a)! - \sum N_{j,a} \ln N_{j,a} - \sum N_{j,g} \ln N_{j,g} + N. \tag{9.20}$$

The distribution numbers for particles among energy states are always given by an Eq. (9.4) form that reflects the number of particles to which a particular partition function is to be applied. Therefore with

$$N^*_{j,g} = \frac{N^*_g e^{-\epsilon_j/kT}}{z_g}, \qquad N^*_{j,a} = \frac{N^*_a e^{-\eta_j/kT}}{z_a}$$

we reexpress the $-\sum N_{j,g}^* \ln N_{j,g}^*$ and $\sum N_{j,a}^* \ln N_{j,a}^*$ as

$$
\begin{aligned}
-\sum N_{j,g}^* \ln \frac{N_g^* e^{-\epsilon_{j,g}/kT}}{z_g} &= -\sum\left[N_{j,g} \ln N_g^* + N_{j,g}^*\left(-\frac{\epsilon_j}{kT}\right) - N_{j,g} \ln z_g \right] \\
-\sum N_{j,a}^* \ln \frac{N_a^* e^{-\epsilon_{j,a}/kT}}{z_a} &= -\sum\left[N_{j,a}^* \ln N_a^* + N_{j,a}^*\left(-\frac{\eta_j}{kT}\right) - N_{j,a}^* \ln z_a \right].
\end{aligned} \tag{9.21}
$$

Combining Eqs. (9.19), (9.20), and (9.21) and collecting terms produces

$$
\ln \Omega = \frac{E}{kT} + \ln \frac{M!}{N_a!(M-N_a)!} + \ln \frac{z_g^{N_g}}{N_g!} + \ln z_a^{N_a}
$$

and

$$
A = -kT\left[\ln \frac{M!}{N_a!(M-N_a)!} z_a^{N_a} + \ln \frac{z_g^{N_g}}{N_g!} \right] = A_a + A_g. \tag{9.22}
$$

Comparison of this expression with the combinatorial term confirms experience with simpler systems. Individual terms in the combinatorial expression translate directly into the free energy expression

$$
\frac{N!}{\prod N_j!} \rightarrow A = -kT \ln z^N \tag{9.23}
$$

$$
\frac{1}{\prod N_{j,g}!} \rightarrow A_g = -kT \ln \frac{z_g^N}{N!} \tag{9.24}
$$

$$
\frac{M!}{(M-N_a)!N_a!} \frac{N_a!}{\prod N_{j,a}!} \rightarrow A_a = -kT \ln \frac{M!}{(M-N_a)!\,N_a!} z_a^{N_a}. \tag{9.25}
$$

Generally

$$
(\text{term!}) \frac{N!}{\prod N_j!} \rightarrow A = -kT \ln z^N (\text{term!}). \tag{9.26}
$$

The chemical potential follows from appropriate differentiation. Then

$$
\begin{aligned}
\mu_g &= -kT[\partial(\ln z_g^{N_a}/N_g!)/\partial N_g] = kT\left[\frac{\partial}{\partial N_g}(N_g \ln z_g - N_g \ln N_g + N_g) \right] \\
&= -kT(\ln z_g - \ln N_g)
\end{aligned} \tag{9.27}
$$

$$\begin{aligned}\mu_a &= -kT\{\partial[\ln M!/N_a!(M-N_a)!)z_a^{N_a}]/\partial N_a\} \\ &= -kT\{\partial[N_a \ln z_a - \ln M! - \ln(M-N_a)!]/\partial N_a\} \\ &= -kT\left(\ln\frac{M-N_a}{N_a} + \ln z_a\right).\end{aligned} \tag{9.28}$$

Equating the chemical potentials for the equilibrium circumstance produces

$$-kT(\ln z_a - N_g^*) = -kT\left(\ln\frac{M-N_a^*}{N_a^*} + \ln z_a\right) \tag{9.29}$$

$$N_g^* = \frac{N_a^*}{M-N_a^*}\frac{z_g}{z_a}$$

where N_g^* and N_a^* reflect the distribution of molecules between the gas and adsorbed phases $N_g^* + N_a^* = N$.[3]

Chemical equilibrium introduces no new statistical problems. We confine the present arguments to systems of gases for simplicity and to the proposition that particles of several species present in a thermodynamic system behave independently with respect to their molecular motions. Consider that two gases $\mathscr{A}$ and $\mathscr{B}$ can, under fortunate conditions (which may require the presence of a catalyst), produce a gas mixture in which $\mathscr{A}$, $\mathscr{B}$, and the molecule species $\mathscr{AB}$ coexist in an equilibrium circumstance in an isolated system. All prior statistical arguments would appear to apply, i.e., the equilibrium concentrations of $\mathscr{A}$, $\mathscr{B}$, and $\mathscr{AB}$ are those which maximize the combinatorial expression

$$\Omega = \sum \frac{1}{N_{\mathscr{A}}!N_{\mathscr{B}}!N_{\mathscr{AB}}!}\frac{N_{\mathscr{A}}!}{\prod n_{j,\mathscr{A}}!}\frac{N_{\mathscr{B}}!}{\prod n_{j,\mathscr{B}}!}\frac{N_{\mathscr{AB}}!}{\prod n_{j,\mathscr{AB}}!} \tag{9.30}$$

where the $n_{j,\mathscr{A}}, n_{j,\mathscr{B}}, \cdots$ are distribution numbers for occupancy of the available energy levels for $\mathscr{A}$ type molecules, etc. If the energy states $\epsilon_{j,\mathscr{A}}$ and $\epsilon_{j,\mathscr{AB}}$ are available to $\mathscr{A}$-type molecules and energy states $\epsilon_{j,\mathscr{B}}$ and $\epsilon_{j,\mathscr{AB}}$ are available to $\mathscr{B}$ type molecules, then the maximum term expression and the constraints are

$$\begin{aligned}t_{\max} &= \frac{1}{\prod n_{j,\mathscr{A}}^*}\frac{1}{\prod n_{j,\mathscr{B}}^*}\frac{1}{\prod n_{j,\mathscr{AB}}} \\ &\sum^j n_{j,\mathscr{A}}^* + \sum^j n_{j,\mathscr{AB}}^* = N_{\mathscr{A}} \\ &\sum^j n_{j,\mathscr{B}}^* + \sum^j n_{j,\mathscr{AB}}^* = N_{\mathscr{B}} \\ &\sum^j n_{j,\mathscr{A}}^*\epsilon_{j,\mathscr{A}} + \sum^j n_{j,\mathscr{B}}^*\epsilon_{j,\mathscr{B}} + \sum^j n_{j,\mathscr{AB}}^*\epsilon_{j,\mathscr{A},\mathscr{B}} - \sum n_{j,\mathscr{AB}}^* D_{0,\mathscr{AB}} = E\end{aligned} \tag{9.31}$$

[3] When the gas phase partition function is written as $\Lambda^{-3N/2}(NkT/P^*)z_g'$, then $P^* = [\Theta/(1-\Theta)](kTz_g'/z_a)$ where $\Theta(=N_a^*/M)$ is the fraction of adsorption sites covered. The adsorption equation $P = [\Theta/(1-\Theta)]f(T)$ is known as the Lagmuir adsorption equation and was originally obtained from kinetic argument.

where $N_{\mathscr{A}}$ and $N_{\mathscr{B}}$ are total number of $\mathscr{A}$- and $\mathscr{B}$-type atoms both combined and uncombined and $D_{0,\mathscr{A}\mathscr{B}}$ is the dissociation energy of $\mathscr{A}\mathscr{B}$ molecules to free atoms. The energy constraint arises from the consideration that $\mathscr{A}$, $\mathscr{B}$, and $\mathscr{A}\mathscr{B}$, molecules will not have the same energy zero. If $\mathscr{A}$ and $\mathscr{B}$ are monatomic species, their ground level energy states are $D_{0,\mathscr{A}\mathscr{B}}$ above the ground state of $\mathscr{A}\mathscr{B}$ molecules. If $\epsilon_{j,\mathscr{A}\mathscr{B}}$, the energy states of $\mathscr{A}\mathscr{B}$ molecules, are referenced to the ground states of the dissociated atoms $\mathscr{A}$,$\mathscr{B}$ then each energy state $\epsilon_{j,\mathscr{A}\mathscr{B}}$ must be corrected by the dissociation energy $D_{0,\mathscr{A}\mathscr{B}}$. This concept was illustrated in Figure 4.1.

To obtain t_{max} we follow the usual procedures with Lagrangian undetermined multipliers, which is set down with some detail in this instance in order to emphasize the role of chemical potential in chemical equilibrium. We construct

$$
\begin{aligned}
d\ln t = -\sum \ln N_{j,\mathscr{A}} dN_{j,\mathscr{A}} - \sum \ln N_{j,\mathscr{B}} dN_{j,\mathscr{B}} - \sum \ln N_{j,\mathscr{A}\mathscr{B}} dN_{j,\mathscr{A}\mathscr{B}} \\
-\alpha_{\mathscr{A}}[\sum dN_{j,\mathscr{A}} + \sum dN_{j,\mathscr{A}\mathscr{B}}] = 0, \qquad -\alpha_{\mathscr{B}}[\sum dN_{j,\mathscr{B}} + \sum dN_{j,\mathscr{A}\mathscr{B}}] = 0 \\
-\beta[\sum \epsilon_{j,\mathscr{A}d} N_{j,\mathscr{A}} + \sum \epsilon_{j,\mathscr{B}} dN_{j,\mathscr{B}} + \sum(\epsilon_{j,\mathscr{A}\mathscr{B}} - D_{0,\mathscr{A}\mathscr{B}}) dN_{j,\mathscr{A}\mathscr{B}}] = 0,
\end{aligned}
$$

we add the four expressions, set the sum equal to zero, and note that each separate term in each sum must be separately equal to zero for arbitrary variation $dN_{j,\mathscr{A}}$, etc. so that

$$
\begin{aligned}
-\alpha_{\mathscr{A}} - \beta\epsilon_{j,\mathscr{A}} - \ln N_{j,\mathscr{A}} &= 0 \\
-\alpha_{\mathscr{B}} - \beta\epsilon_{i,\mathscr{B}} - \ln N_{j,\mathscr{B}} &= 0 \\
(\alpha_{\mathscr{A}} + \alpha_{\mathscr{B}}) - \beta(\epsilon_{j,\mathscr{A}\mathscr{B}} - D_{0,\mathscr{A}\mathscr{B}}) - \ln N_{j,\mathscr{A}\mathscr{B}} &= 0
\end{aligned} \tag{9.32}
$$

From Eqs. (9.31) and (9.32) we obtain the distribution numbers that prevail at equilibrium

$$
\begin{aligned}
N^*_{j,\mathscr{A}} &= e^{-\alpha_{\mathscr{A}}} e^{-\beta\epsilon_{j,\mathscr{A}}} \\
N^*_{j,\mathscr{B}} &= e^{-\alpha_{\mathscr{B}}} e^{-\beta\epsilon_{j,\mathscr{B}}} \\
N_{j,\mathscr{A}\mathscr{B}} &= e^{-(\alpha_{\mathscr{A}} + \alpha_{\mathscr{B}})} e^{-\beta(\epsilon_j - D_{0,\mathscr{A}\mathscr{B}})}
\end{aligned}
$$

$$
N^*_{\mathscr{A}} = e^{-\alpha_{\mathscr{A}}} \sum^{j} e^{-\epsilon_{j,\mathscr{A}}/kT} \tag{9.33}
$$

$$
N^*_{\mathscr{B}} = e^{-\alpha_{\mathscr{B}}} \sum^{j} e^{-\epsilon_{j,\mathscr{B}}/kT}
$$

$$
N^*_{\mathscr{A}\mathscr{B}} = e^{-(\alpha_{\mathscr{A}} + \alpha_{\mathscr{B}})} e^{D_{0,\mathscr{A}\mathscr{B}}/kT} \sum^{j} e^{-\epsilon_{j,\mathscr{A}\mathscr{B}}/kT} \tag{9.34}
$$

where we have introduced $\beta = 1/kT$ as an accepted convention. Following the introduction of partition function notation for the sums over energy states we have

$$e^{-\alpha_{\mathscr{A}}} = \frac{N^*_{\mathscr{A}}}{z_{\mathscr{A}}}$$

$$e^{-\alpha_{\mathscr{B}}} = \frac{N^*_{\mathscr{B}}}{z_{\mathscr{B}}}$$

$$e^{-(\alpha_{\mathscr{A}} + \alpha_{\mathscr{B}})} = \frac{N^*_{\mathscr{A}\mathscr{B}}}{z_{\mathscr{A}\mathscr{B}} e^{D_0/kT}}$$

or

$$\frac{N^*_{\mathscr{A}}}{z_{\mathscr{A}}} \frac{N^*_{\mathscr{B}}}{z_{\mathscr{B}}} = \frac{N^*_{\mathscr{A}\mathscr{B}}}{z_{\mathscr{A}\mathscr{B}} e^{D_0/kT}}$$

then

$$\frac{N^*_{\mathscr{A}\mathscr{B}}}{N^*_{\mathscr{A}} N^*_{\mathscr{B}}} = \frac{z_{\mathscr{A}\mathscr{B}} e^{D_0/kT}}{z_{\mathscr{A}} z_{\mathscr{B}}}. \tag{9.35}$$

Equations (9.33), (9.34), and (9.31) suffice to completely fix the distribution numbers $N^*_{\mathscr{A}}, N^*_{\mathscr{B}}$, and $N^*_{\mathscr{A}\mathscr{B}}$ and therefore the exact composition of the equilibrium mixture. We note that each ideal gas partition function will contain a term V in the translational component and for $\rho = N/V$ we can write

$$\frac{\rho^*_{\mathscr{A}\mathscr{B}}}{\rho^*_{\mathscr{A}} \rho^*_{\mathscr{B}}} = \frac{z''_{\mathscr{A}\mathscr{B}} e^{D_{0,\mathscr{A}\mathscr{B}}}}{z''_{\mathscr{A}} z''_{\mathscr{B}}} \tag{9.36}$$

where the values ρ are now molecule densities at equilibrium and the terms z'' have no volume dependence. Then

$$\frac{\rho^*_{\mathscr{A}\mathscr{B}}}{\rho^*_{\mathscr{A}} \rho^*_{\mathscr{B}}} = K_{eq}(T) \tag{9.37}$$

and $K_{eq}(T)$ is a function of T only. We can refer to $K_{eq}(T)$ as the equilibrium constant.

Not all chemical equilibria will be so simple in statistical expression. Polyatomic molecules may be featured as some or all reactants and products and stoichiometric coefficients will not always be unity. In many real processes of interest the interaction system will not be isolated during the chemical process but may exchange energy with the surroundings and the energy exchange may be of quantitative concern. This will be an appropriate time to illustrate the utility of minimizing the system free energy at constant T and V to identify the equilibrium state.

Consider that the molecular species $\mathscr{A}_2\mathscr{B}$, $\mathscr{C}\mathscr{D}$, $\mathscr{A}\mathscr{D}$, and $\mathscr{C}_2\mathscr{B}$ can coexist in a system of specified T and V under such circumstances that the molecule ratios of the several species is a fixed quantity as observed above. Then any mixture of the molecule species will dissociate and recombine as required to produce the appropriate molecule ratios. The numbers of molecules which react under these conditions is fixed by the assumed stoichiometry

$$\mathscr{A}_2\mathscr{B} + 2\mathscr{C}\mathscr{D} \rightleftharpoons 2\mathscr{A}\mathscr{D} + \mathscr{B}\mathscr{C}_2.$$

We imagine the construction of such an equilibrium system through assembling some mixture of the four molecule species in a closed system of specified volume, which we will insist is large enough that the molecules behave as an ideal gas mixture. The closed, rigidly bounded system in now placed in diathermal contact with a large constant temperature bath and permitted to attain both chemical and thermal equilibrium. The Helmholz free energy of the particle system when it is in chemical equilibrium at specified T and V is

$$A = -kT \ln \mathscr{Z}_{T,V} = -kT\left[\ln \prod^{K}\left(\frac{z_K}{N_K}\right)^{N_K} + \sum N_K\right] \tag{9.38}$$

where the subscript K identifies the several species. The free energy minimum is given by

$$(dA)_{T,V} = -kT\sum^{K}(d\ln \mathscr{Z}_K)_{T,V} = -kT\sum\frac{\partial}{\partial N_K}(N_K \ln z_K - N_K \ln N_K + N_K)dN_K$$

$$= -kT\sum^{K}(\ln z_K - \ln N_K)dN_K = 0 \tag{9.39}$$

subject to the constraints imposed by stoichiometric considerations.

It will be required that we construct these constraints on the basis of the numbers of $\mathscr{A}, \mathscr{B}, \mathscr{C}$, and $\mathscr{D}$ atoms, which are present in the system since these identify the original mixture and are the constant feature of the process. The contraints are

$$2N_{\mathscr{A}_2\mathscr{B}} + N_{\mathscr{A}\mathscr{D}} = N_{\mathscr{A}}$$
$$N_{\mathscr{A}_2\mathscr{B}} + N_{\mathscr{B}\mathscr{C}_2} = N_{\mathscr{B}}$$
$$2N_{\mathscr{C}\mathscr{D}} + N_{\mathscr{B}\mathscr{C}_2} = N_{\mathscr{C}}$$
$$N_{\mathscr{C}\mathscr{D}} + N_{\mathscr{A}\mathscr{D}} = N_{\mathscr{D}}.$$

We then construct

$$-\alpha_{\mathscr{A}}[2dN_{\mathscr{A}_2\mathscr{B}} + dN_{\mathscr{A}\mathscr{D}}] = dN_{\mathscr{A}} = 0$$
$$-\alpha_{\mathscr{B}}[dN_{\mathscr{A}_2\mathscr{B}} + dN_{\mathscr{B}\mathscr{C}_2}] = dN_{\mathscr{B}} = 0$$
$$-\alpha_{\mathscr{C}}[dN_{\mathscr{C}\mathscr{D}} + dN_{\mathscr{B}\mathscr{C}_2}] = dN_{\mathscr{C}} = 0$$
$$-\alpha_{\mathscr{D}}[dN_{\mathscr{C}\mathscr{D}} + dN_{\mathscr{A}\mathscr{D}}] = dN_{\mathscr{D}} = 0,$$

recognize that each term in the partition function $\sum e^{-\epsilon_{j,\kappa}/kT}$ becomes $e^{-(\epsilon_{j,\kappa}-D_{0,\kappa})/kT}$ when the common base line of dissociated atoms is established, and obtain $\ln z_\kappa = \ln \sum e^{-\epsilon_{j,\kappa}/kT} e^{D_{0,\kappa}/kT}$ which we write as $\ln z_\kappa + (D_{0,\kappa}/kT)$. We then add the constraints and collect terms to write

$$
\begin{aligned}
&[-kT \ln z_{\mathcal{AB}}/N_{\mathcal{A}_2\mathcal{B}} - (2\alpha_{\mathcal{A}} + \alpha_{\mathcal{B}}) - D_{0,\mathcal{A}_2\mathcal{B}}]dN_{\mathcal{A}_2\mathcal{B}} + \\
&[-kT \ln z_{\mathcal{BC}_2}/N_{\mathcal{BC}_2} - (\alpha_{\mathcal{B}} + 2\alpha_{\mathcal{C}}) - D_{0,\mathcal{BC}_2}]dN_{\mathcal{BC}_2} + \\
&[-kT \ln z_{\mathcal{CD}}/N_{\mathcal{CD}} - (\alpha_{\mathcal{C}} + \alpha_{\mathcal{D}}) - D_{0,\mathcal{CD}}]dN_{\mathcal{CD}} + \\
&[-kT \ln z_{\mathcal{AD}}/N_{\mathcal{AD}} - (\alpha_{\mathcal{A}} + \alpha_{\mathcal{D}}) - D_{0,\mathcal{AD}}]dN_{\mathcal{AD}} = 0.
\end{aligned}
\tag{9.40}
$$

Variation of any species term, e.g., $dN_{\mathcal{A}_2\mathcal{B}}$, fixes the variation in the remainder of such terms through the process stoichiometry. Therefore Eq. (9.40) can be written

$$
\sum^{\kappa}\left(-kT \ln \frac{z_\kappa}{N_\kappa} - \sum \alpha_\kappa - D_{0,\kappa}\right) n\, dN_{\mathcal{A}_2\mathcal{B}} = 0 \tag{9.41}
$$

where in this case n is either $\pm 1, \pm 2$. If the variation is arbitrary, each term in the sum is zero by familiar argument so that

$$
\begin{aligned}
2\alpha_{\mathcal{A}} + \alpha_{\mathcal{B}} &= \alpha_{\mathcal{A}_2\mathcal{B}} = -kT \ln(z_{\mathcal{A}_2\mathcal{B}}/N_{\mathcal{A}_2\mathcal{B}}) - D_{0,\mathcal{A}_2\mathcal{B}} \\
2\alpha_{\mathcal{C}} + \alpha_{\mathcal{B}} &= \alpha_{\mathcal{BC}_2} = -kT \ln(z_{\mathcal{BC}_2}/N_{\mathcal{BC}_2}) - D_{0,\mathcal{BC}_2} \\
\alpha_{\mathcal{A}} + \alpha_{\mathcal{D}} &= \alpha_{\mathcal{AD}} = -kT \ln(z_{\mathcal{AD}}/N_{\mathcal{AD}}) - D_{0,\mathcal{AD}} \\
\alpha_{\mathcal{D}} + \alpha_{\mathcal{C}} &= \alpha_{\mathcal{CD}} = -kT \ln(z_{\mathcal{CD}}/N_{\mathcal{CD}}) - D_{0,\mathcal{CD}}
\end{aligned}
$$

from which

$$
2\alpha_{\mathcal{A}} = -kT \ln \frac{z_{\mathcal{A}_2\mathcal{B}}}{N_{\mathcal{A}_2\mathcal{B}}} - D_{0,\mathcal{A}_2\mathcal{B}} - \alpha_{\mathcal{B}} = 2\left(-kT \ln \frac{z_{\mathcal{AD}}}{N_{\mathcal{AD}}} - D_{0,\mathcal{AD}}\right) - \alpha_{\mathcal{D}}
$$

$$
2\alpha_{\mathcal{C}} = -kT \ln \frac{z_{\mathcal{BC}_2}}{N_{\mathcal{BC}_2}} - D_{0,\mathcal{BC}_2} - \alpha_{\mathcal{B}} = 2\left(-kT \ln \frac{z_{\mathcal{CD}}}{N_{\mathcal{CD}}} - D_{0,\mathcal{CD}}\right) - \alpha_{\mathcal{D}}
$$

and

$$
\begin{aligned}
\ln \frac{z_{\mathcal{A}_2\mathcal{B}}}{N_{\mathcal{A}_2\mathcal{B}}} - \ln \frac{z_{\mathcal{BC}_2}}{N_{\mathcal{BC}_2}} + \frac{D_{0,\mathcal{A}_2\mathcal{B}}}{kT} - \frac{D_{0,\mathcal{BC}_2}}{kT} &= \ln\left(\frac{z_{\mathcal{AD}}}{N_{\mathcal{AD}}}\right)^2 - \ln\left(\frac{z_{\mathcal{CD}}}{N_{\mathcal{CD}}}\right)^2 \\
&\quad + \frac{2D_{0,\mathcal{AD}}}{kT} - \frac{2D_{0,\mathcal{CD}}}{kT}.
\end{aligned}
$$

We collect these terms as

$$\frac{N^2_{\mathscr{A}\mathscr{D}} N_{\mathscr{B}\mathscr{C}_2}}{N_{\mathscr{A}_2\mathscr{B}} N^2_{\mathscr{C}\mathscr{D}}} = \frac{z^2_{\mathscr{A}\mathscr{D}} z_{\mathscr{B}\mathscr{C}_2}}{z_{\mathscr{A}_2\mathscr{B}} z^2_{\mathscr{C}\mathscr{D}}} e^{\Delta D_0/kT} \tag{9.42}$$

where

$$\Delta D_0 = (2D_{0,\mathscr{A}\mathscr{D}} + D_{0,\mathscr{B}\mathscr{C}_2}) - (D_{0,\mathscr{A}\mathscr{B}_2} + 2D_{0,\mathscr{C}\mathscr{D}})$$

i.e., the stoichiometric sum of dissociation energies of products less those of reactants. As before, the Eq. (9.12) ratio of the numbers of molecules of each species present at equilibrium is the equilibrium constant and is determined by the partition function ratio and the dissociation energy term. Each partition function is a function of T and V and the individual dissociation energies are constants. The molecule numbers are dependent on T and V but the molecule densities, which are proportional to pressure, are only dependent on temperature

$$\frac{(N_{\mathscr{A}\mathscr{D}}/V)^2 (N_{\mathscr{B}\mathscr{C}_2}/V)}{(N_{\mathscr{A}_2\mathscr{B}}/V)(N_{\mathscr{A}\mathscr{D}}/V)^2} = \frac{P^2_{\mathscr{A}\mathscr{D}} P_{\mathscr{B}\mathscr{C}_2}}{P_{\mathscr{A}_2\mathscr{B}} P^2_{\mathscr{C}\mathscr{D}}} = K_{eq}(T) \tag{9.43}$$

where each P_K is the partial pressure of that component in the equilibrium mixture. The example is sufficiently general to permit the equilibrium constant for any gas phase reaction to be set down by analogy.

Although we phrased the address to chemical equilibrium in terms of gas phase reactions, review of the arguments will make it clear that they are completely general. The equilibrium constant is just an expression in terms of partition functions and dissociation energies for the unique concentration expression that reflects the equality of chemical potential for each atomic species in its distribution among molecular species. Having established the equilibrium constant concept we can write for any general reaction having a balanced equation representation:

$$a\mathscr{A} + b\mathscr{B} + \cdots \rightleftharpoons c\mathscr{C} + d\mathscr{D} + \cdots$$

$$\frac{n^c_{\mathscr{C}} n^d_{\mathscr{D}}}{n^a_{\mathscr{A}} n^b_{\mathscr{B}}} = K'_{eq}(T) \tag{9.44}$$

where the n_K are any desirable concentration terms (e.g., moles liter^{-1}) and K'_{eq} is the appropriately modified partition function ratio such as Eq. (9.43). The equilibrium concentrations are usually available experimental quantities. This is fortunate since our computation of partitition functions for condensed phases, particularly in mixed condensed phases, is not reliable. There are also implications that are unique to mixed phase systems that we should now consider.

If the condensed phases are pure chemical species (i.e., not solutions), e.g.,

$$\mathscr{B}_{gas} + \mathscr{A}\mathscr{B}_{solid} \rightleftharpoons \mathscr{A}\mathscr{B}_{2\,solid}$$

the combinatorial term at equilibrium, utilizing notation introduced earlier in this chapter that reflects localization in the solids and non-localization in the gas, is

$$\Omega^* = t_{\max} = \frac{M_{\mathscr{A}\mathscr{B}}!}{\prod M^*_{j,\mathscr{A}\mathscr{B}}!} \frac{M_{\mathscr{A}\mathscr{B}_2}!}{\prod M^*_{j,\mathscr{A}\mathscr{B}_2}} \frac{1}{\prod N^*_{j,\mathscr{B}}}$$

where

$$\sum M^*_{j,\mathscr{A}\mathscr{B}} + 2\sum M^*_{j,\mathscr{A}\mathscr{B}_2} + \sum N^*_{j,\mathscr{B}} = N_{\mathscr{B}}$$

and

$$\sum M^*_{j,\mathscr{A}} + \sum M^*_{j,\mathscr{A}\mathscr{B}_2} = N_{\mathscr{A}}. \tag{9.45}$$

Equations (9.45) provide the constraints on interchange between species and require two multipliers, $-\alpha_{\mathscr{A}}$ and $-\alpha_{\mathscr{B}}$. The energy constraint is

$$\sum \eta_{j,\mathscr{A}\mathscr{B}} M^*_{j,\mathscr{A}\mathscr{B}} + \sum \eta_{j,\mathscr{A}\mathscr{B}_2} M^*_{j,\mathscr{A}\mathscr{B}_2} + \sum \epsilon_{j,\mathscr{B}} N^*_{j,\mathscr{B}} - D_{0,\mathscr{A}\mathscr{B}} \sum M_{j,\mathscr{A}\mathscr{B}} - D_{0,\mathscr{A}\mathscr{B}_2} \sum M_{j,\mathscr{A}\mathscr{B}_2} = E - E_0$$

and, as always, only one multiplier, $-\beta$, is required. The detail is familar and the results are

$$N^*_{j,\mathscr{B}} = e^{-\alpha_{\mathscr{B}}} e^{-\beta\epsilon_{j,\mathscr{B}}}$$
$$M^*_{j,\mathscr{A}\mathscr{B}} = e^{-(\alpha_{\mathscr{A}}+\alpha_{\mathscr{B}})} e^{-\beta\eta_{j,\mathscr{A}\mathscr{B}}} e^{\beta D_{0,\mathscr{A}\mathscr{B}}}$$
$$M^*_{j,\mathscr{A}\mathscr{B}_2} = e^{-(\alpha_{\mathscr{A}}+2\mathscr{B})} e^{-\beta\eta_{i,\mathscr{A}2\mathscr{B}}} e^{\beta D_{0,\mathscr{A}2\mathscr{B}}} \tag{9.46}$$

leading to, with $\beta = 1/kT$,

$$\alpha_{\mathscr{B}} = kT \ln(z_{\mathscr{B}}/N_{\mathscr{B}})$$
$$\alpha_{\mathscr{A}} + \alpha_{\mathscr{B}} = kT \ln z_{\mathscr{A}\mathscr{B}} + D_{0,\mathscr{A}\mathscr{B}}/kT$$
$$\alpha_{\mathscr{A}} + 2\alpha_{\mathscr{B}} = kT \ln z_{\mathscr{A}\mathscr{B}_2} + D_{0,\mathscr{A}\mathscr{B}_2}/kT$$

from which

$$\frac{1}{N^*_{\mathscr{B}}} = \frac{z_{\mathscr{A}\mathscr{B}_2}}{z_{\mathscr{A}\mathscr{B}} z_{\mathscr{B}}} e^{\Delta D_0/kT} \tag{9.47}$$

where $\Delta D_0 = (D_{0,\mathscr{A}\mathscr{B}_2} - D_{0,\mathscr{A}\mathscr{B}})$. The right-hand side of the Eq. (9.47) expression is the equilibrium constant and the partition function for all participating species appears in it. The left-hand side of Eq. (9.47) contains only the molecule numbers for the nonlocalized phase. The result is general. Molecule numbers for pure localized phases will not appear,

$$K_{eq} = \frac{1}{N_{\mathscr{B}}/V_{gas}} = \frac{1}{N_{\mathscr{B}}/V} = \frac{z_{\mathscr{A}\mathscr{B}_2}}{z_{\mathscr{A}\mathscr{B}} z'_{\mathscr{B}}} \Lambda_{\mathscr{B}}^{-3/2} e^{\Delta D_0/kT} \tag{9.48}$$

where $z'_{\mathscr{B}}$ relates to all energy states for component $\mathscr{B}$ other than translation.

If the localized phases are not pure substances, the results will not be the same. Suppose in the above example that $\mathscr{A}\mathscr{B}$ and $\mathscr{A}\mathscr{B}_2$ form a solid solution. The combinatorial term will be

$$\Omega = \frac{(M_{\mathscr{A}\mathscr{B}} + M_{\mathscr{A}\mathscr{B}_2})!}{M_{\mathscr{A}}! M_{\mathscr{B}}!} \frac{M_{\mathscr{A}\mathscr{B}}!}{\prod M_{j,\mathscr{A}\mathscr{B}}!} \frac{M_{\mathscr{A}\mathscr{B}_2}}{\prod M_{j,\mathscr{A}\mathscr{B}_2}!} \frac{1}{\prod N_{j,\mathscr{B}}!} \tag{9.49}$$

i.e., of a general form such as Eq. (9.8), which we constructed for phase equilibrium in which a mixed (solution) phase participated. The treatment will differ in the constraints that we apply since atoms $\mathscr{A}$ and $\mathscr{B}$ can be distributed among molecular species and phases rather than just between phases. The constraint expressions will not differ from those set down in Eq. (9.45) but, since the combinatorial term is different, the distribution number expressions will differ from those for the pure condensed phases at equilibrium. We can just write down the appropriate expressions from Eq. (9.49) and the constraints as

$$\begin{aligned}
N^*_{j,\mathscr{B}} &= e^{-\alpha} e^{-\beta\epsilon_{j,\mathscr{B}}} \\
M^*_{j,\mathscr{A}\mathscr{B}} &= (M^*_{\mathscr{A}\mathscr{B}} + M^*_{\mathscr{A}\mathscr{B}_2}) e^{-(\alpha_{\mathscr{A}} + \alpha_{\mathscr{B}}} e^{-\beta\eta_{j,\mathscr{A}\mathscr{B}}} e^{\beta D_{0,\mathscr{A}\mathscr{B}}} \\
M^*_{j,\mathscr{A}\mathscr{B}_2} &= (M^*_{\mathscr{A}\mathscr{B}} + M^*_{\mathscr{A}\mathscr{B}_2}) e^{-(\alpha_{\mathscr{A}} + 2\alpha_{\mathscr{B}})} e^{-\beta\eta_{j,\mathscr{A}\mathscr{B}_2}} e^{\beta D_{0,\mathscr{A}\mathscr{B}_2}}
\end{aligned} \tag{9.50}$$

leading to

$$\alpha_{\mathscr{B}} = kT \ln \frac{z_{\mathscr{B}}}{N^*_{\mathscr{B}}}$$

as before, but with

$$\begin{aligned}
\alpha_{\mathscr{A}} + \alpha_{\mathscr{B}} &= \ln \frac{M^*_{\mathscr{A}\mathscr{B}}}{M^*_{\mathscr{A}\mathscr{B}} + M^*_{\mathscr{A}\mathscr{B}_2}} + \ln z_{\mathscr{A}\mathscr{B}} + D_{0,\mathscr{A}\mathscr{B}}/kT \\
\alpha_{\mathscr{A}} + 2\alpha_{\mathscr{B}} &= \ln \frac{M^*_{\mathscr{A}\mathscr{B}_2}}{M^*_{\mathscr{A}\mathscr{B}} + M^*_{\mathscr{A}\mathscr{B}_2}} + \ln z_{\mathscr{A}\mathscr{B}_2} + D_{0,\mathscr{A}\mathscr{B}_2}/kT \\
\ln \frac{z_{\mathscr{B}}}{N_{\mathscr{B}}} &= \ln \frac{z_{\mathscr{A}\mathscr{B}_2}}{z_{\mathscr{A}\mathscr{B}}} + \frac{1}{kT}[D_{0,\mathscr{A}\mathscr{B}_2} - D_{0,\mathscr{A}\mathscr{B}}] + \ln \frac{M^*_{\mathscr{A}\mathscr{B}}}{M^*_{\mathscr{A}\mathscr{B}_2}} \\
\frac{M^*_{\mathscr{A}\mathscr{B}_2}}{M^*_{\mathscr{A}\mathscr{B}} N^*_{\mathscr{B}}} &= \frac{z_{\mathscr{A}\mathscr{B}_2}}{z_{\mathscr{A}\mathscr{B}} z_{\mathscr{B}}} e^{\Delta D_0/kT}.
\end{aligned} \tag{9.51}$$

In the solution system the condensed phase molecule numbers have not cancelled. Again, the result is general. Molecule distributions within a solution phase must appear in the expression and the equilibrium constant is given by the Eq. (9.51) form involving partition function ratios, although we do have some latitude, as always, in arriving at the most convenient form of expression.

We could, for example, have called the terms $M_{\mathscr{AB}}/(M_{\mathscr{AB}} + M_{\mathscr{AB}_2})$ and $M_{\mathscr{AB}_2}/(M_{\mathscr{AB}} + M_{\mathscr{AB}_2})$ molecule fractions $m_{\mathscr{AB}}$ and $m_{\mathscr{AB}_2}$ and written

$$\frac{m^*_{\mathscr{AB}_2}}{m^*_{\mathscr{AB}} N^*_{\mathscr{B}}/V} = \frac{z_{\mathscr{AB}_2}}{z_{\mathscr{AB}} \Lambda_{\mathscr{B}}^{-3/2} z^1_{\mathscr{B}}} e^{\Delta D_0/kT}. \tag{9.52}$$

We could have divided every molecule number term by $N_{\mathscr{A}}$ giving, with this particular stoichiometry

$$\frac{X^*_{\mathscr{AB}_2}}{X^*_{\mathscr{AB}} n^*_{\mathscr{B}}/V} = \frac{1}{N_1} \frac{z_{\mathscr{AB}_2}}{z_{\mathscr{AB}} \Lambda^{-3/2} z'_{\mathscr{B}}} e^{\Delta D_0/kT} = K_{eq}(T) \tag{9.53}$$

where $X^*_{\mathscr{AB}_2}$ and $X^*_{\mathscr{AB}}$ are mole fractions of species $\mathscr{A}$ and $\mathscr{B}$ in the condensed phase and $n_{\mathscr{B}}/V$ is the molar concentration in the gas phase. Conventional notation would identify the various numerical values of the equilibrium constant as K_C, K_P, K_X to indicate, respectively, the appropriate ratio of concentrations (N/V), pressures or mole fractions. The implications of adopting other choices is best examined in an extended framework, which we will address in the following chapters.

PROBLEMS

9.1. Calculate the vapor pressure of metallic lithium at 200 K given that Θ_D for the solid is 385.

9.2. When benzene is adsorbed on a solid surface the adsorption can be localized, in which case the normal translational and rotational states for the gas phase are converted to a single (planar) rotation of the molecule on the adsorption site. Internal motions would not be affected. Calculate a value for the equilibrium constant that would apply to the adsorbed phase gas phase equilibrium. The normal moments of inertia are $I_A = 2.93 \times 10^{-38}\,g\,cm^2$, $I_B = I_C = 1.46 \times 10^{38}\,g\,cm^2$, and the symmetry number for benzene is 12.

9.3. The iodine molecule dissociates at elevated temperatures. Calculate the equilibrium constant for the dissociation at 1200 K given the following spectroscopic information. The atomic ground state is 4-fold degenerate, the rotational constant for I_2 is $0.0373\,cm^{-1}$, $\tilde{\nu}$ is $214.36\,cm^{-1}$, and the dissociation energy is 1.5422 eV.

CHAPTER CONTENT SOURCES AND FURTHER READING

1. T. L. Hill, *Introduction to Statistical Thermodynamics*, Chapter 10. Addison-Wesley, Reading, MA, 1960.
2. O. K. Rice, *Statistical Mechanics, Thermodynamics and Kinetics*, Chapter 7. W. H. Freeman, San Francisco, 1967.
3. G. S. Rushbrooke, *Statistical Mechanics*, Chapters XI, XII, XIII, and XIV. Oxford, London, 1949.
4. C. Kittell and H. Kroemer, *Thermal Physics*, 2nd ed., Chapter 11. W. H. Freeman, San Francisco, 1980.

SYMBOLS: CHAPTER 9

$\epsilon_1, \epsilon_2 \cdots \epsilon_j$ = energy states for gas phase molecules

$\eta_1, \eta_2 \cdots \eta_j$ = energy states for condensed phase molecules

N_g, N_c = numbers of gas phase, condensed phase molecules

$\epsilon_{1,\mathscr{A}} \cdots \epsilon_{K,\mathscr{B}} \cdots \epsilon_{j,\mathscr{A}\mathscr{B}}$ = energy states for various species in a gas phase

$\eta_{1,\mathscr{A}} \cdots \eta_{K,\mathscr{B}} \cdots \eta_{j,\mathscr{A}\mathscr{B}}$ = energy states for various species in a condensed phase

$N_{1,\mathscr{A}} \cdots N_{K,\mathscr{B}} \cdots N_{j,\mathscr{A}\mathscr{B}}$ = number of molecules of various species in gas phase energy states

$M_{1,\mathscr{A}} \cdots M_{K,\mathscr{B}} \cdots M_{j,\mathscr{A}\mathscr{B}}$ = number of molecules in condensed phase energy states

$\kappa, \mathfrak{p}$ = number of components, number of phases in Gibbs Phase Rule expression

$K_{eq}(T)$ = the equilibrium constant

$\alpha_{\mathscr{A}}, \alpha_{\mathscr{B}}, \alpha_{\mathscr{A}\mathscr{B}}$ = undetermined multiplier applied to constraints on energy states available to various species

CARRYOVER SYMBOLS

Ω, α, β, Λ, P, V, k, T, z, $\mathscr{Z}$, μ, A, N, D_0, X_κ.

10

PHASE AND CHEMICAL EQUILIBRIUM II: MACROSCOPIC AND PROCESS CONSIDERATIONS, NONELECTROLYTE SYSTEMS

For phase equilibrium in a one-component system our statistical arguments have demonstrated that the chemical potential is equally valued in each phase. For gas-localized condensed phase equilibria

$$\mathrm{N}_g^* = \sum \mathrm{N}_{j,g}^* = e^{-\alpha} z_g \tag{9.4}$$

$$M_c^* = \sum \mathrm{M}_{j,c}^* = \sum M_j^* e^{-\alpha} z_c. \tag{9.5}$$

The gas phase and condensed phase expressions differ in form because of the localization consideration. Since the undetermined multiplier α is common to both expressions

$$\mathrm{N}_g^* = \frac{z_g}{z_c} \qquad \text{or} \qquad \frac{\mathrm{N}_g^*}{\mathrm{V}} = \frac{\Lambda^{-3/2} z_g'}{z_c} = \rho_g^*$$

where z_g' relates to all gas phase energy states other than translation and $\Lambda = h^2/2\pi m k\mathrm{T}$. The terms z_g, z_c, and Λ are all species specific and temperature dependent. Therefore, in any one component system in gas-condensed phase equilibrium the gas phase density ρ_g^* or gas phase pressure $\mathrm{P}^* = k\mathrm{T}(\mathrm{N}_g^*/\mathrm{V})$ is uniquely valued for a specified value of T. Also, $\mu_g = \mu_c$. Although we did not duplicate the arithmetic, the equilibrium involving two condensed phases does not impose new conditions with regard to the single undetermined multiplier for a one component system. We would obtain for localized condensed phases

ϑ and ζ,

$$M_{\vartheta}^{*} = \sum \mathrm{M}_{j\vartheta} e^{-\alpha} z_{\vartheta} \qquad \text{and} \qquad M_{\zeta}^{*} = \sum \mathrm{M}_{j,\zeta} e^{-\alpha} z_{\zeta}$$

and $-kT \ln \mathscr{Z}_{\vartheta} = -kT \ln \mathscr{Z}_{\zeta}$. There is no other unique constraint on the distribution between phases at specified T for two condensed phases in equilibrium.

The phase rule admits the coexistence of three phases in a one-component system. Again, without the arithmetic, when there is one value for the chemical potential and the chemical potentials are given by $-kT \ln z_c$ (condensed phase) or $-kT \ln (z_g/N_g)$ (gaseous phase) the phase requirements are that $F = 0$, i.e., there are no degrees of freedom and both P* and T are uniquely valued for a given molecule species.

On the basis of these statistical conclusions, it is required that we obtain those temperature and pressure values for which the chemical potentials are equally valued. The task is simple when we recognize that $G = N\mu$ and that $(\partial G/\partial T)_{P,N} = -S_{P,N}$ and $(\partial G/\partial P)_{T,N} = V_{T,N}$ [Eqs. (4.31) and (4.34)]. Taken with the concept of entropy as a measure of randomness, we can immediately observe that $\tilde{S}_g \gg \tilde{S}_\ell > \tilde{S}_s$.[1] Also we know that $\tilde{V}_g \gg \tilde{V}_\ell > \tilde{V}_s$, or rarely $\tilde{V}_s > \tilde{V}_\ell$, although in either case the difference between molar volumes of the condensed phases is small. The entropy difference between condensed phases will, however, be substantial.

On the basis of these observations we can construct a general argument for the chemical potential vs temperature relationship at some fixed, overall closed system (i.e., $N = N_g + \sum N_{c,i}$) pressure in a one-component system. Each chemical potential is represented by a line of slope $-S$ for that phase. The lines will intersect at the common value of T for which the chemical potentials are equal. Figure 10.1 reflects this circumstance.

For all $T > T_{\ell,g}$, the Gibbs free energy for the system is minimized when only the gas phase is present at pressure P_1, for which Figure 10.1 was constructed. At that pressure, when $T = T_{\ell,g}$ the chemical potentials of the gas and liquid phases are identical and the two phases coexist in equilibrium. Between $T_{\ell,g}$ and $T_{s,\ell}$ the liquid phase is the stable phase for which the free energy is minimized. At $T_{s,\ell}$, the liquid and solid phases coexist and below $T_{s,\ell}$ the solid is the stable phase.

For other pressures the μ vs T lines will be shifted in accordance with our observations regarding volume. This is demonstrated in Figure 10.2 constructed for $P = P_2 < P_1$ with dashed lines referencing the Figure 10.1 lines. Both $T_{s,\ell}$ and $T_{\ell,g}$ are shifted to new values with $T_{\ell,g}$ always being shifted to lower values as the pressure is decreased. As indicated in the enlargement of the $T_{s,\ell}$ region, $T_{s,\ell}$ may be shifted to higher or lower values as pressure is decreased. If $\tilde{V}_\ell > \tilde{V}_s$, $(\partial T_{s,\ell}/\partial P) > 0$; if $V_s > V_\ell$, $(\partial T_{s,\ell}/\partial P) < 0$. In either event $(\partial T_{\ell,g}/\partial P)$ is substantially greater than $(\partial T_{s,\ell}/\partial P)$ so that at successive still lower pressures the

[1] Recall that the tilde designates molar quantity in our notation.

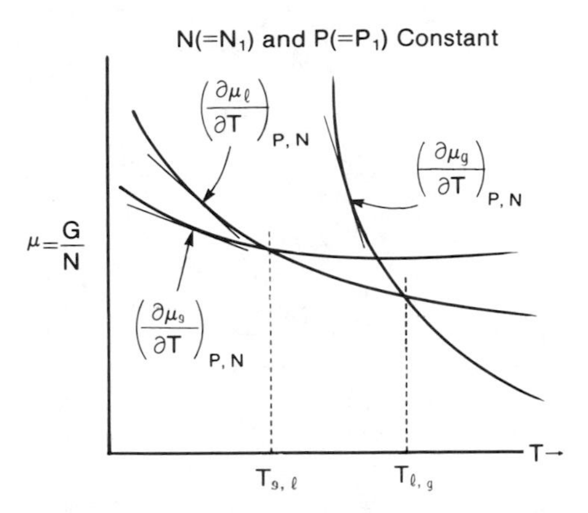

Figure 10.1. The variation in chemical potential with temperature for gas, solid, liquid phases in a closed isobaric system.

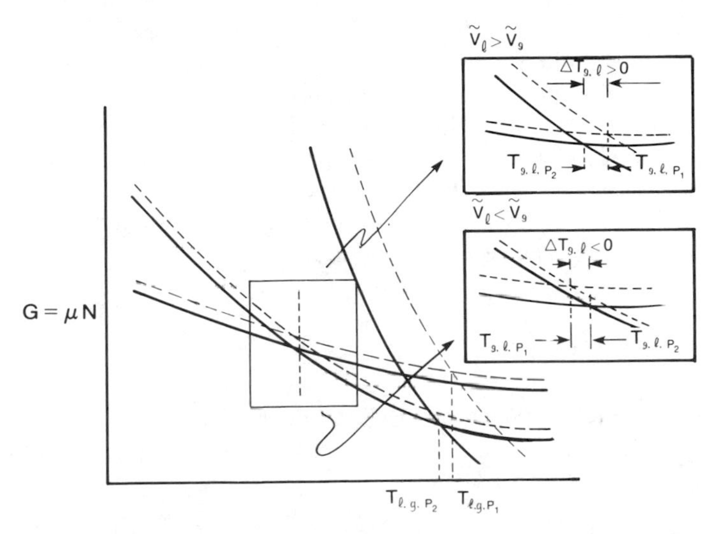

Figure 10.2. The change in chemical potential with pressure. Dashed lines reference Figure 10.1. Solid lines represent the change in chemical potential with temperature at a lower pressure.

$T_{\ell,g}$ point will move toward the $T_{\ell,s}$ point and there will exist some constant pressure plot for which the two separate two-phase equilibrium points will coincide reflecting the three-phase equilibrium state. The single pressure and temperature values describing this equilibrium state are known as triple point values. At still lower values of pressure the line representing the variation in gas phase chemical potential will move past the $T_{s,\ell}$ intersection point representing the two-phase gas–solid equilibrium relationships. In this region,

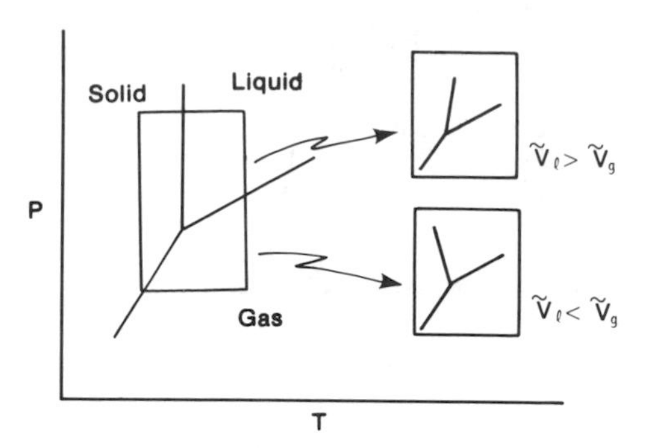

Figure 10.3. A one-component phase diagram identifying regions of phase stability and conditions governing two and three phase equilibrium.

where the gas phase pressure is less than the triple point pressure, the gas phase is stable at high temperature and the solid phase is stable at low temperature. The transition between phases is called sublimation. Figure 10.3 is a P, T plot reflecting the phase equilibria in an overall closed, one-component system. Two phases are in equilibrium at P, T values represented by the solid lines, three phases at the line intersections, and a single phase within the open regions.

Within any single phase region the state of the system is described by the appropriate species-specific partition function and the thermodynamic properties of that single phase system are given by the normal relationships at T and P (or V through the equation of state),

$$E = kT^2\left(\frac{\partial \ln \mathscr{Z}}{\partial T}\right)_{V,N}$$

$$H = E + PV \qquad (\cong E \text{ for condensed phases})$$

$$S = k \ln \mathscr{Z} + \frac{E}{T}$$

$$A = -kT \ln \mathscr{Z}$$

$$G = A + PV \qquad (\cong A \text{ for condensed phases}).$$

When two phases are present in an overall closed one-component system, the thermodynamic properties of the system are just the sum of those calculated on the basis of the two partition functions using the values of N that reflect the relative amount of each phase.

Since the two-phase region is a process path between two single phase regions there is an isothermal, isobaric discontinuity in thermodynamic properties associated with it. The discontinuity can be characterized as the difference between the calculated property values of the two phases under the transition conditions, e.g., for the gas–liquid transition

$$\Delta E_{vap} = (E_g - E_\ell) = kT^2\left[\left(\frac{\partial \ln \mathscr{Z}_g}{\partial T}\right)_{T,N} - \left(\frac{\partial \ln \mathscr{Z}_\ell}{\partial T}\right)_{T,N}\right]. \tag{10.1}$$

Normally it will be covenient to imagine the transfer of some fixed amount of material across the phase boundary. If we take this amount to be Avogadro's number of particles, then the transition values are molar quantities for more convenient comparison between species.

In addition to any statistical features of these process parameters imposed by the partition function expressions, all of those macroscopic features that we have identified will apply. For example, in the consideration of energy exchange processes we related an enthalpy change to the thermal quantity $Q_{P,N}$ that represented energy exchange between system and surroundings during an isobaric process. An energy discontinuity during a phase transition will require that energy be transferred into or out of the system at constant P and T and the enthalpy change for the process, if it is carried out at constant pressure, is

$$\int_{process} dH = \Delta H_{process} = \int dq_P = Q_P. \tag{10.2}$$

Since ΔH cannot be route dependent, Q_P cannot be route dependent, i.e., the process is reversible equivalent regardless of the manner by which it is accomplished. Then from $dS = dq^*/T$; $(\Delta S)_P = \Delta H/T$. So long as two phases (ϑ, ζ) coexist in the one component system $\mu_\vartheta = \mu_\zeta$ and $G_\vartheta/N_\vartheta = G_\zeta/N_\zeta$ and for any infinitesimal change in P, T or distribution between phases $dG_\vartheta = -dG_\zeta$. Then for an overall closed, two-phase, one-component system, e.g., gas–solid

$$-S_g dT + V_g dP + \mu_g dN_g = -S_s dT + V_s dP + \mu_s dN_s \tag{10.3}$$

with

$$\mu_s = \mu_g \qquad \text{and} \qquad dN_s = -dN_g$$

so

$$\frac{dP}{dT} = \frac{S_g - S_s}{V_g - V_s} \simeq \frac{S_g - S_s}{V_g} \cong \frac{(S_g - S_s)P}{NkT} \tag{10.4}$$

and

$$\left(\frac{d \ln P}{dT}\right)_{sub} = \frac{\Delta S_{sub}}{NkT} = \frac{\Delta H_{sub}}{NkT^2}. \tag{10.5}$$

The gas–liquid equilibrium line would be characterized by a similar expression,

$$\left(\frac{d \ln P}{dT}\right)_{vap} = \frac{S_g - S_\ell}{NkT} = \frac{\Delta H_{vap}}{NkT^2}. \tag{10.6}$$

The solid–liquid equilibrium line is characterized by a slightly different expression. Since the liquid and solid volumes differ only slightly

$$\frac{dP}{dT} = \frac{S_s - S_\ell}{V_s - V_\ell} = \left(\frac{\Delta S}{\Delta V}\right)_{\text{fus}} = \left(\frac{\Delta H}{T\Delta V}\right)_{\text{fus}}. \tag{10.7}$$

To the extent that the heats of vaporization and sublimation can be represented as constants, the ln P vs T^{-1} plots are linear

$$\frac{d \ln P}{d(1/T)} = -\frac{\Delta H}{N k}.$$

If the enthalpy, entropy, and volumes are molar values, $N k = N_A k = R$. Equations 10.5 and 10.7 are known as the Clausius-Clapeyron and Clapeyron equations respectively.

The (Figure 10.3) single component phase diagram relates to the Figure 6.5 PVT diagram as shown in Figure 10.4. The single component phase diagram (Figure 10.3) is the projection onto a PV surface of intersection lines related to a constant system volume plane that could lie anywhere in the two-phase region of interest. For the liquid–vapor equilibrium line, the reference plane should be chosen to pass through the critical point in order that the PV equilibrium line terminates there.

To complete arguments for the single component phase equilibrium, we note that it will always be possible to obtain a good value for the chemical potential of a gas phase from statistical calculation,

$$\mu(T) = \mu^\circ(T) + kT \ln P \qquad \text{(ideal gas)}$$

$$\mu(T) = \mu^\circ(T) + kT \ln f \qquad \text{(real gas).}$$

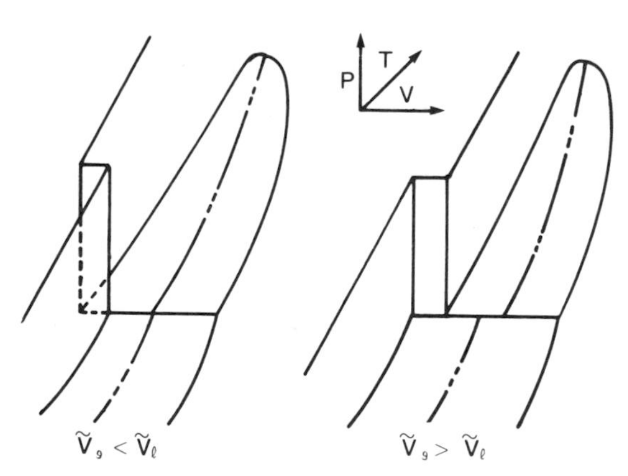

Figure 10.4. Illustration of intersecting PVT surfaces in a region including the point of three phase equilibrium.

These values will apply anywhere in the single phase region and along the two-phase equilibrium line where gas and condensed phases coexist. Since along these lines the gas phase and condensed phase chemical potentials are equal

$$\mu_c = \mu_g = \mu_g^\circ + kT \ln P^*(T) \tag{10.8}$$

where $P^*(T)$ is the vapor pressure of the condensed phase at T and μ_g° is the standard chemical potential of the gas phase at T as previously defined. At values of T substantially below the critical point we expect $P^*(T)$ to be low and the gas phase to behave ideally. At higher values of T fugacities would be required.

Since the condensed phase cannot exist at pressures less than $P^*(T)$, we will be interested in the condensed phase chemical potentials only at $P^*(T)$ and higher so that it will be operationally convenient to define the condensed phase standard chemical potentials in terms of Eq. (10.8). This will provide a more certain baseline in that we do not have large confidence in the generality of the models that we adopted for the condensed phase partition function calculations. We will therefore write

$$\mu_c^\circ(T) = \mu_g^\circ(T) + kT \ln P^*(T) \tag{10.9}$$

defining the condensed phase standard state. As a standard state it is no more or less precisely calculable than our original standard state defined by the temperature-dependent partition function components since we require the condensed phase partition function to calculate $P^*(T)$ but the term is readily measurable. In terms of the Eq. (10.9) defined standard state the molar chemical potential of a condensed phase anywhere in the condensed phase regions of stability is

$$N_A \mu_c = \tilde{\mu}_c(T, P) = \tilde{\mu}_c^\circ(T, P^*) + \int_{P=0}^{P=P^*} \tilde{V} dP \tag{10.10}$$

where $\tilde{V} = \tilde{V}^0(P^*, T) + \alpha\tilde{V}$, α being the isothermal compressibility.

To extend the gas/condensed phase argument to multicomponent systems we require only the phase distribution numbers from the statistical argument. For two components ($\mathscr{A}$ and $\mathscr{B}$) of any molecular complexity both present in ideal mixed phases,

$$N^*_{g,\mathscr{A}} = \frac{M^*_{c,\mathscr{A}}}{M^*_{c,\mathscr{A}} + M^*_{c,\mathscr{B}}} \frac{z_{g,\mathscr{A}}}{z_{c,\mathscr{A}}} \tag{10.11}$$

$$N^*_{g,\mathscr{B}} = \frac{M^*_{c,\mathscr{B}}}{M^*_{c,\mathscr{A}} + M^*_{c,\mathscr{B}}} \frac{z_{g,\mathscr{B}}}{z_{c,\mathscr{B}}}. \tag{10.12}$$

Replacing the molecule fraction terms by mole fraction and assuming ideal gas phase behavior for the two-component gaseous system we can write

$$\frac{N_{g,\mathscr{A}} \ell T}{V} = P_{\mathscr{A}} = \ell T \ln X_{\mathscr{A}} \Lambda_{\mathscr{A}}^{-3/2} \frac{z'_{g,\mathscr{A}}}{z_{c,\mathscr{A}}} \tag{10.13}$$

with all notation previously defined. An analogous result follows for component $\mathscr{B}$. Comparison of these expressions with Eq. (9.6) for the vapor pressure (P^*) of a single component two-phase system shows that

$$P_{\mathscr{A}} = X_{\mathscr{A}} P^*_{\mathscr{A}}, \qquad P_{\mathscr{B}} = X_{\mathscr{B}} P^*_{\mathscr{B}}.$$

This behavior, which is a direct result of the ideal equilibrium distribution of particles between phases, is known (on the basis of its empirical origin) as Raoult's Law. It can apply only under the same set of constraint conditions that apply to the distribution numbers, which are the same constraints we applied to the ideal mixing problem. These were that the normal gas phase and the normal liquid phase partition functions applied over the entire range of mixed compositions (otherwise we should have to construct a concentration-dependent variation in the partition functions). If the partition functions are unchanged there is no ΔE_{mix} and no ΔV_{mix}; $\Delta G_{mix} = T\Delta S_{mix} = \ell T(N_{\mathscr{A}} \ln X_{\mathscr{A}} + N_{\mathscr{B}} \ln X_{\mathscr{B}})$. We shall recognize Raoult's Law as an extension of the ideal mixing case to phase equilibrium. The vapor pressure composition diagram for a system that follows Raoult's Law is sketched in Figure 10.5 where the partial pressures of the two components are represented by dashed lines. In this example $P^*_{\mathscr{B}} > P^*_{\mathscr{A}}$. The composition axis represents the condensed phase composition. Quite clearly the vapor phase composition, which is determined by the partial pressures, will differ from the liquid phase composition. Consider the following arithmetic:

$$P = X_{\mathscr{A}} P^*_{\mathscr{A}} + X_{\mathscr{B}} P^*_{\mathscr{B}} \tag{10.14}$$

$$Y_{\mathscr{A}} = \frac{X_{\mathscr{A}} P^*_{\mathscr{A}}}{P} \quad \text{and} \quad Y_{\mathscr{B}} = \frac{X_{\mathscr{B}} P^*_{\mathscr{B}}}{P}$$

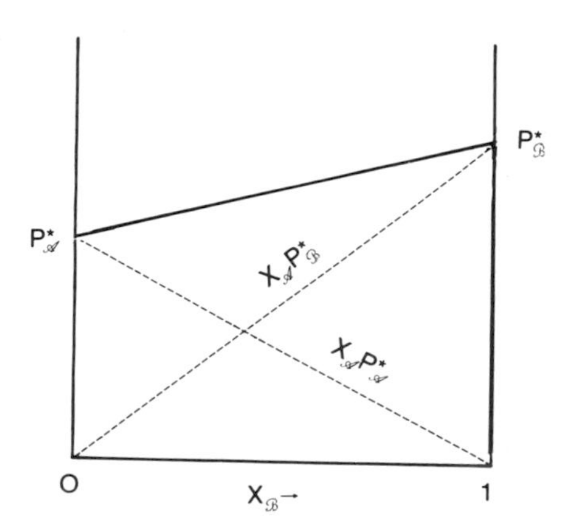

Figure 10.5. Illustration of Raoult's Law behavior in a two-component system.

where the Y are pressure fractions of the two components in the vapor phase and therefore (from ideal gas partial pressure considerations) mole fractions of the two components in the gas phase. If the overall mixture is $X_{\mathscr{A}} = X_{\mathscr{B}} = 0.5$, it is easy to see that

$$Y_{\mathscr{B}} = \frac{0.5P^*_{\mathscr{B}}}{P^*_{\mathscr{A}} + P^*_{\mathscr{B}}} > Y_{\mathscr{A}} = \frac{0.5P^*_{\mathscr{A}}}{P^*_{\mathscr{A}} + P^*_{\mathscr{B}}}. \tag{10.15}$$

A general conclusion follows from the three point argument $P = P^*_{\mathscr{A}}$ as $X_{\mathscr{A}} \to 1$, $P \to P^*_{\mathscr{B}}$ as $X_{\mathscr{B}} \to 1$, and $Y_{\mathscr{A}} < Y_{\mathscr{B}}$ for $X_{\mathscr{A}} = X_{\mathscr{B}} = 0.5$: the vapor phase is always richer than the liquid phase in the more volatile (higher vapor pressure) component. In the ideal (Raoult's Law) system the total pressure and vapor phase composition curves for the range in T common to liquid–vapor coexistence for the two components form an envelope bounded by the single component vapor–liquid equilibrium curves for the two components as shown in Figure 10.6. A conventional boiling point $(T \text{ vs } X)_P$ diagram results from the intersection of a constant pressure plane through the two-component phase diagram. Retaining our recognition of the PVT areas in the single component phase diagram in which single phases appear, Figure 10.5 and Figure 10.6 can be consolidated in two sketches (Figure 10.7) that represent $(P \text{ vs } X)_T$ and $(T \text{ vs } X)_P$. Figures 10.7a and 10.7b more clearly reflect the nature of the two-phase envelopes and the equilibrium phase compositions. The relative amounts of liquid and vapor phases for any overall composition $X'_{\mathscr{B}}$ lying within the two-phase region are related to the length of the line segments (Figure 10.7c). This follows readily from the argument that for n total moles distributed between the two phases

$$n_\ell + n_g = n$$

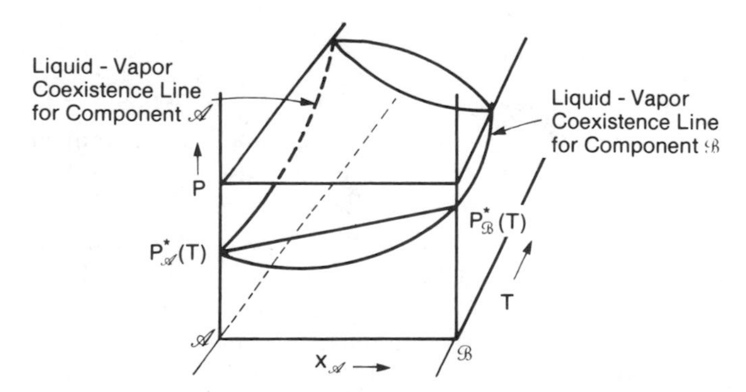

Figure 10.6. Illustration of the Raoult's Law vapor–liquid phase composition reguirements at constant pressure and at constant temperature.

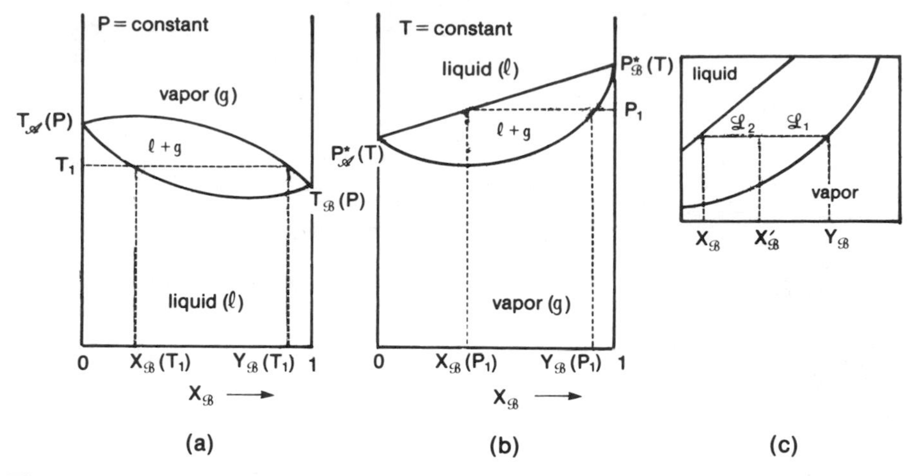

Figure 10.7. Conventional boiling point (a) and vapor pressure (b) diagrams for a two-component system illustrating equilibrium vapor phase and liquid phase composition and line length segments (c), which identify relative amounts of liquid and vapor phases for a given overall composition $X_{\mathscr{B}}$.

and $nX_{\mathscr{B}}$ of these are component $\mathscr{B}$ distributed as $n_\ell X_{\mathscr{B}}$ in the liquid and $n_g Y_{\mathscr{B}}$ in the vapor phases then, imagining that we condense the liquid–vapor system,

$$n_\ell X_{\mathscr{B}} + n_g Y_{\mathscr{B}} = nX'_{\mathscr{B}}$$

and

$$n_\ell X_{\mathscr{A}} + n_g Y_{\mathscr{A}} = n(1 - X'_{\mathscr{B}})$$

so that

$$-n_\ell(X_{\mathscr{B}} - X'_{\mathscr{B}}) = n_g(Y_{\mathscr{A}} - X'_{\mathscr{B}})$$

but $X_{\mathscr{B}} - X'_{\mathscr{B}}$ is line segment $-\mathscr{L}_2$ and $(Y_{\mathscr{A}} - X'_{\mathscr{B}})$ is line segment $\mathscr{L}_1$, so

$$\frac{n_\ell}{n_g} = \frac{\mathscr{L}_1}{\mathscr{L}_2}. \tag{10.16}$$

The method of intercepts (Figure 8.3 and related discussion) provides an excellent framework for address to condensed phase equilibria. We take P to be greater than P^*_Σ (the equilibrium vapor pressure of any mixed system of present interest) and P = 1 atm is sufficiently high to provide a wide range of mixed, condensed systems for our consideration. Recall that $(\partial G_\Sigma/\partial X_K)_{T,P,N_J} = \mu_K$ and that the general form of G_Σ will be that of a string suspended by two points ($\mu_{\ell,\mathscr{A}}$ and $\mu_{\ell,\mathscr{B}}$) for a two-component ($\mathscr{A}$ and $\mathscr{B}$) liquid solution. The precise G_Σ curve form will depend on the relative values $\mu_{\ell,\mathscr{A}}$ and $\mu_{\ell,\mathscr{B}}$ and deviations from ideality, but the general (ideal) form provides a complete argument which can be modified by intuitive adjustment. Reference to Figure 10.1

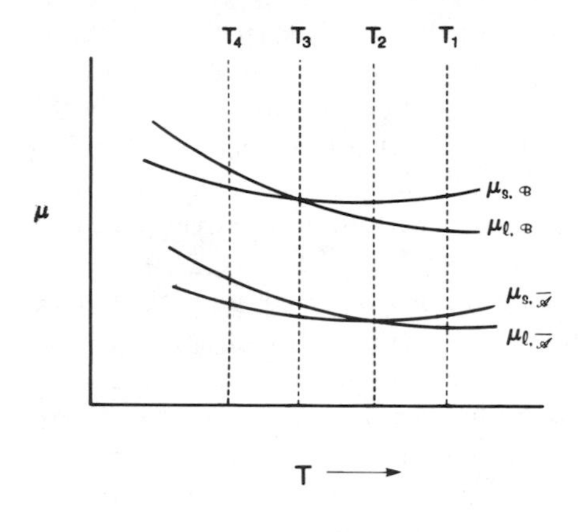

Figure 10.8. Solid and liquid chemical potentials for two pure species vs temperature referencing various temperature domains in which particular phase stabilities occur.

will also demonstrate that each condensed phase for each pure component will have a temperature range of thermodynamic stability in which the chemical potential for that phase is lowest valued for any phase option available to that species. Consider a temperature range $T_4 < T < T_1$ (Figure 10.8) that includes the solid–liquid coexistence point ($\mu_\ell = \mu_s$) for both $\mathscr{A}$ and $\mathscr{B}$. Only liquid systems will exist in the region $T_2 < T < T_1$. The solution phase chemical potentials for each species (we have assumed a near ideal form for G_Σ for illustration purposes) will be the lowest valued chemical potential for any phase available to that species over the entire range of mixed systems. This will be true over the entire temperature range $T_2 < T < T_1$ (Figure 10.9a). In the

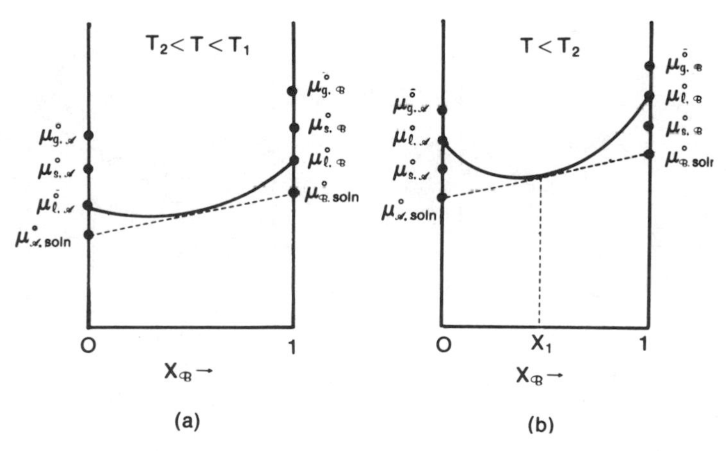

Figure 10.9. Illustration of the chemical potential as reflected by the intercept method for individual components in a two-component solution system in (a) a temperature domain in which the solution is the thermodynamically stable phase over all compositions and (b) in a temperature domain where single (solution) phase stability is limited to a particular range of compositions.

temperature range $T < T_3$ the solid phases are the thermodynamically stable phases for the pure components ($\mu_s < \mu_\ell < \mu_g$) but at any particular temperature (say T_4) solutions can exist in a particular range of compositions fixed by the particular temperature (around X_1 in Figure 10.9b) for which the chemical potential $\mu_{\mathscr{A},\text{soln}} < \mu_{s,\mathscr{A}}$ and $\mu_{\mathscr{B},\text{soln}} < \mu_{s,\mathscr{B}}$. The solutions are stable liquid phases although the pure component liquid is not. Clearly this is not true over the entire composition range. As we move from X_1 to lower $X_{\mathscr{B}}$ values, $\mu_{\mathscr{A},\text{soln}}$ increases while $\mu_{\mathscr{B},\text{soln}}$ decreases. There is a particular composition (X_2 Figure 10.10a) for which $\mu_{\mathscr{A},\text{soln}} = \mu_{\mathscr{A},s}$. The solution of this composition will coexist with solid $\mathscr{A}$ and solutions in which $X_{\mathscr{B}} < X_2$ will not exist. Further decreases in temperature that would move $\mu_{s,\mathscr{A}}$ and $\mu_{\ell,\mathscr{A}}$ to higher values (with an always increasing difference) have the effect of moving the solution concentration that will coexist with solid $\mathscr{A}$ to higher $X_{\mathscr{B}}$ values. The same arguments apply to solutions in which $X_1 < X_{\mathscr{B}}$ and the temperature is some particular value, say $T_5 < T_4$. There is also some particular temperature T_e in the general range below T_3 where, for a particular composition χ_e (Figure 10.10b), $\mu_{s,\mathscr{A}} = \mu_{\mathscr{A},\text{soln}}$ and $\mu_{s,\mathscr{B}} = \mu_{\mathscr{B},\text{soln}}$ so that both solid phases ($\mathscr{A}$ and $\mathscr{B}$) can coexist with the solution. The temperature T_e and composition χ_e are referred to as eutectic values. At the eutectic temperature for all overall concentrations $X_{\mathscr{B}}$ there is a three phase equilibrium involving solid $\mathscr{A}$ solid $\mathscr{B}$ and a solution of composition χ_e. This is an invariant point in a two-component system. The temperature cannot be changed through heating or cooling until at least one of the three phases vanishes. On cooling the solution phase vanishes. On heating one of the solid phases vanishes (for $X_{\mathscr{B}} > \chi_e$ solid $\mathscr{A}$ vanishes.)

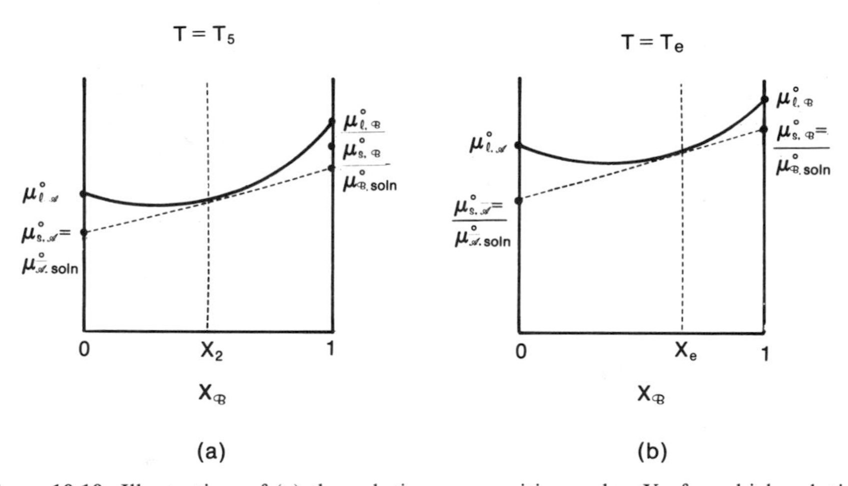

Figure 10.10. Illustration of (a) the solution composition value X_2 for which solution and the pure solid phase for one component in a two-component system will coexist and (b) the solution composition value X_e at T_e for which three phase coexistence will occur.

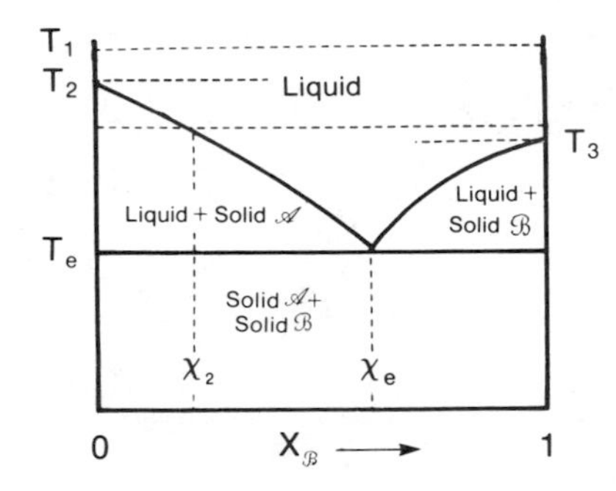

Figure 10.11. A two-component phase diagram identifying temperature/composition domains in which single phases and two phase equilibria exist.

The above arguments extended to various values of T and also applied to component $\mathscr{B}$, will produce the Figure 10.11 two-component solid–liquid phase equilibrium diagram. The Figure 10.11 phase diagram is typical of systems that are completely miscible throughout the liquid region but are completely immiscible throughout the solid region. Limited solubility in the liquid phase is commonly encountered and slight solubility in the solid phase is often encountered.

Limited miscibility is a reflection of extreme deviation from the ideal mixing case. We have observed that although volume changes on mixing are not arithmetically very important, they are a measure of nonideality in that the associated changes in the potential wells that define interaction energies can be large. We approach our understanding of limited miscibility through the energy of mixing. We have obtained an expression for the excess (nonzero) energy of mixing in a two-component (species 1 and 2) regular solution

$$^{E}(\Delta E_{mix}) = c\omega_{12}(N_1 + N_2)X_1X_2 \tag{8.95}$$

which is based on the exchange energy (exchanging 12 interactions for 11 or 22 interactions). In a general nonideal solution we will replace $c\omega_{12}$ with the term $\Delta\phi$ denoting a more complex mixing process[2] and construct the partition function for mixing as

$$\Delta\mathscr{Z}_{mix} = z_{1,12}^{N_1} z_{2,12}^{N_2} \frac{(N_1 + N_2)!}{N_1!N_2!} e^{-\Delta\phi X_1 X_2/kT} - z_1^{N_1} z_2^{N_2} \tag{10.17}$$

defining

$$^{E}(\Delta\mathscr{Z}_{mix}) = e^{\Delta\phi X_1 X_2/kT}. \tag{10.18}$$

[2] In this more general case there is an assumed temperature dependence through volume and $^{E}(\Delta S_{mix}) \neq 0$; $\Delta\phi$ represents a convenient notation for a general discussion of nonideality.

Then

$$
\begin{aligned}
{}^{E}(\Delta G_{\text{mix}}) &= \mathscr{k}T \ln {}^{E}(\Delta \mathscr{Z}_{\text{mix}}) \\
&= \Delta\phi X_1 X_2
\end{aligned}
\tag{10.19}
$$

and, in analogy to Eqs. (8.96) and (8.97),

$$
\ln \gamma_1 = \frac{\Delta\phi X_2^2}{\mathscr{k}T}
\tag{10.20}
$$

$$
\ln \gamma_2 = \frac{\Delta\phi X_1^2}{\mathscr{k}T}.
\tag{10.21}
$$

The vapor phase partial pressure of each component in equilibrium with the mixed condensed phase is a reflection of the chemical potential and therefore of the activity coefficient of that component in the mixture. Taking Raoult's Law as reflecting ideality, we construct

$$
N_K \mu_{K,g} = N_K \mu^{\circ}_{K,g} + N_K \mathscr{k}T \ln P_K = G_{K,g}
\tag{10.22}
$$

$$
N_K(\mu_{K,g})_R = \mu_K \mu^{\circ}_{K,g} + N_K \mathscr{k}T \ln P^*_K X_K = (G_{K,g})_R
\tag{10.23}
$$

where P_K is the partial pressure of component K in the gas phase in equilibrium with a nonideal solution and $P^*_K X_K$ is the partial pressure of component K over an ideal solution. Then $G_K - (G_K)_R$ is the excess free energy in the gas phase, which derives from nonideality in the liquid mixture so that, on the basis of our definition of liquid phase activity coefficients,

$$
\ln \frac{P_K}{P^*_K X_K} = {}^{E}(\Delta G_{\text{mix}}) = \mathscr{k}T \ln \gamma_K.
\tag{10.24}
$$

In our two-component mixture from Eqs. (10.20) and (10.21)

$$
\frac{P_1}{P^*_1 X_1} = e^{\Delta\phi X_2^2/\mathscr{k}T} \quad \text{and} \quad \frac{P_2}{P^*_2 X_2} = e^{\Delta\phi X_1^2/\mathscr{k}T}.
\tag{10.25}
$$

Since $\Delta\phi$ can be positive or negative, the deviations from Raoult's Law can be positive or negative and the deviations will be in the same direction for both components. Also $P_1 \rightarrow P^*_1 X_1$ as $X_2 \rightarrow 0$ $(X_1 \rightarrow 1)$ and $P_2 \rightarrow P^*_2 X_2$ as $X_2 \rightarrow 1$. The form of positive and negative deviations is indicated in Figure 10.12 where dashed lines represent Raoult's Law behavior.

Where experimental data are available and the partial pressures are known, a Figure 10.12-type plot will resolve the exponential term in Eqs. (10.24) and (10.25) as activity coefficients γ_1 and γ_2 from line segment lengths (Figure 10.12a)

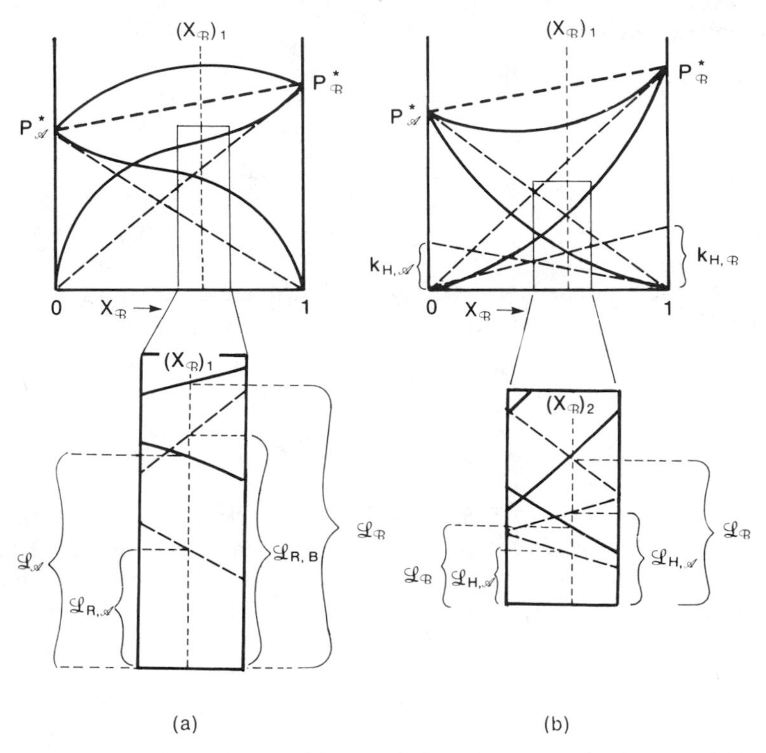

Figure 10.12. Illustration of (a) positive deviations from Raoult's Law and identifying (exploded diagram) line length segments that relate the deviations to activity coefficients and (b) negative deviations from Raoult's Law also relating partial pressures to Henry's Law and identifying line length segments that relate deviations from Henry's Law to activity coefficients.

as they relate to line segments $\mathscr{L}_{\mathscr{A}}$ and $\mathscr{L}_{\mathscr{B}}$

$$\mathrm{P}_{\mathscr{A}} = \gamma_{\mathrm{R},\mathscr{A}} \mathrm{X}_{\mathscr{A}} \mathrm{P}^*_{\mathscr{A}} \qquad \mathscr{L}_{\mathscr{A}}/\mathscr{L}_{\mathrm{R},\mathscr{A}} = \gamma_{\mathrm{R},\mathscr{A}}$$

$$\mathrm{P}_{\mathscr{B}} = \gamma_{\mathrm{R},\mathscr{B}} \mathrm{X}_{\mathscr{B}} \mathrm{P}^*_{\mathscr{B}} \qquad \mathscr{L}_{\mathscr{B}}/\mathscr{L}_{\mathrm{R},\mathscr{B}} = \gamma_{\mathrm{R},\mathscr{B}}.$$

Since partial pressure variation with respect to composition at constant temperature is equivalent in magnitude to chemical potential variation with composition and since we require (Gibbs–Duhem relationship) $\sum \mathrm{N}_{\mathrm{K}} d\mu_{\mathrm{K}} = 0$, it follows that in any region where the partial pressure varies linearly with composition for one component it must also vary linearly for the other where only two components are present. Since Raoult's Law is always observed for the nearly pure major component (solvent), in that region the minor component (solute) must exhibit linear deviations from Raoult's Law when $\Delta\phi$ is nonzero. This behavior is the basis for a second activity coefficient definition normally applied only to the solute and known as Henry's Law. We construct a framework for this concept by writing

$$P_K = k_H X_K \tag{10.26}$$

where

$$k_H = \left(\frac{dP_i}{dX_i}\right)_{X_i \to 0} \tag{10.27}$$

representing the straight line slope for the linear region as the Henry's Law constant. Then activity coefficients on this basis are defined by

$$P_{\mathscr{A}} = \gamma_{H,\mathscr{A}} k_{H,\mathscr{A}} X_{\mathscr{A}} \qquad \mathscr{L}_{\mathscr{A}}/\mathscr{L}_{H,\mathscr{A}} = \gamma_{H,\mathscr{A}} \tag{10.28}$$

$$P_{\mathscr{B}} = \gamma_{H,\mathscr{B}} k_{H,\mathscr{B}} X_{\mathscr{B}} \qquad \mathscr{L}_{\mathscr{B}}/\mathscr{L}_{H,\mathscr{B}} = \gamma_{H,\mathscr{B}} \tag{10.29}$$

and, again, can be resolved from experimental partial pressures in the geometrical framework illustrated in Figure 10.12b, which represents a negative deviation from Raoult's Law. Either activity convention can, of course, be applied to positive or to negative deviations from ideality.

A particular consideration arises when there are large positive deviations from ideal mixing. To examine this circumstance we return to a complete expression for the free energy of mixing such as we would obtain from

$$\Delta G_{mix}(T) = N\ell T \ln(X_1 \ln X_1 + X_2 \ln X_2) + NX_1X_2\Delta\phi. \tag{10.30}$$

The first term in Eq. (10.30) will always be negative. When $\Delta\phi > 0$, since the second term is maximum valued at $X_1 = X_2 = 0.5$, there will be a range of compositions over which ΔG_{mix} becomes positive and in this range solutions are not thermodynamically stable. In terms of the chemical potential Figure 10.13 illustrates the region of immiscibility and the existence of two phases in which the chemical potential of each component is single valued. The two phases coexist as a minimum free energy system for any composition $\chi < X_2 < \chi'$. Since the first term in Eq. (10.30) increases with temperature we anticipate that the

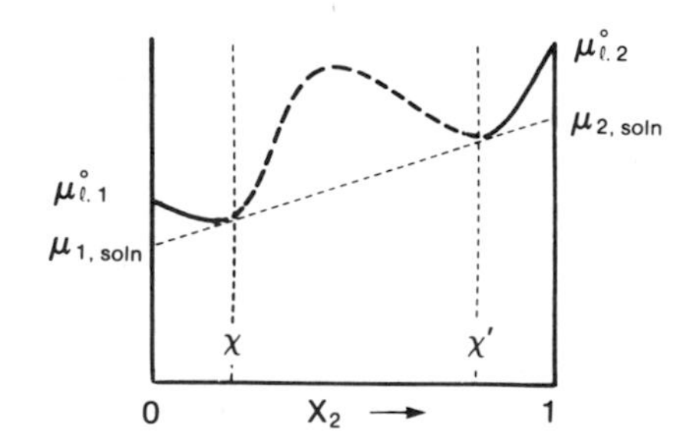

Figure 10.13. Illustration of the chemical potential vs composition relationship leading to immiscibility in a two-component system.

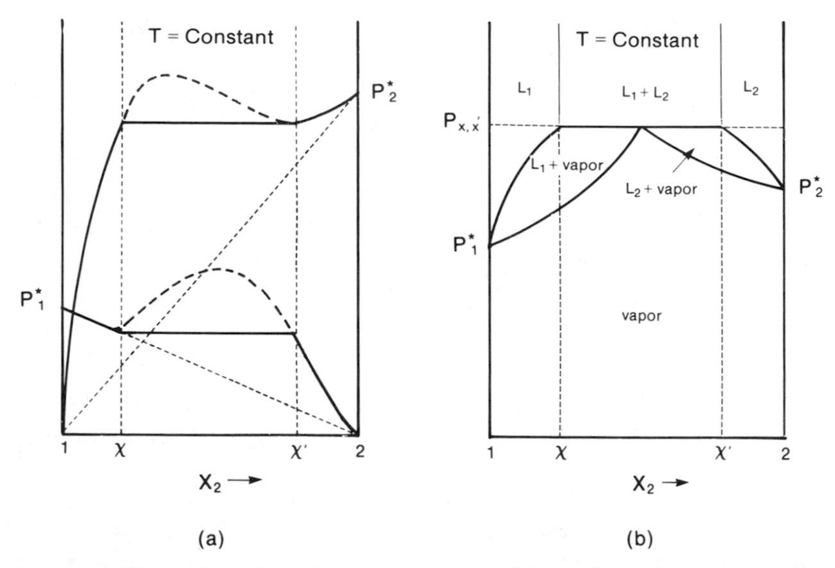

Figure 10.14. Illustration of (a) the vapor pressure diagram in a two-component system exhibiting a region of immiscibility and (b) the liquid/vapor composition envelope for the same system.

region of immiscibility will shrink with increasing temperature. Any temperature dependence associated with $\Delta\phi$ would be superimposed in this consideration.

The partial pressures of each component in the vapor phase will be characteristic of either of the two phases since chemical potentials of each component are characteristic of either phase. The partial pressures of each component in the two condensed phase, gas phase system can be represented in Figure 10.14a curve forms which satisfy the requirement that the partial pressure of a component be the same for each phase. The system pressure is, as always, the sum of partial pressures and is constant, as is each partial pressure over the range X to X′ as illustrated in Figure 10.14b. In each of the single phase regions all prior argument relating to the vapor phase composition apply. At any pressure greater than $P_{XX'}$ a single liquid phase is present for compositions $X_2 < \chi$ or $X_2 > \chi'$ and two liquid phases are present for $\chi < X_2 < \chi'$. The three-dimensional phase diagram can be visualized from Figure 10.15b by extending a temperature axis into the page. A constant pressure plane passed through this construct produces a Figure 10.15-like phase diagram, which could be referred to as a boiling point diagram (normal boiling point diagram if $P = 1$ atm). In this construct we do not recognize a variation in the phase compositions χ and χ' with decreasing T.

Although many mixed solid systems exhibit zero or very low miscibility, those that do form solid solutions (alloys) are of great interest. The form of the solid solution phase diagram will not differ from that for liquid mixtures at pressures greater than the mixed system vapor pressure. A simple phase diagram

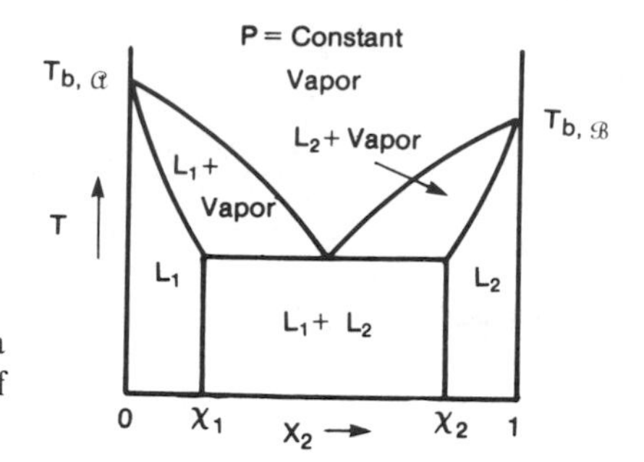

Figure 10.15. A boiling point diagram for a two-component system exhibiting a region of immiscibility.

for solid–liquid equilibrium where the solid phases are partially miscible would assume the Figure 10.16 form where $(\mathscr{A}\mathscr{B})_1$ and $(\mathscr{A}\mathscr{B})_2$ represent various solid solutions of components $\mathscr{A}$ and $\mathscr{B}$.

Often, starting with two components, one or more compounds may form. This behavior should be regarded as the ultimate result of extreme negative deviation from ideal mixing. Extended discussion is not required in terms of phase behavior in that we can regard the system phase diagram as two separate systems as indicated in Figure 10.17a for $\mathscr{A}$, $\mathscr{B}$, and $\mathscr{A}\mathscr{B}$ components in the solid–liquid region of stability. We have assumed complete liquid miscibility and no miscibility for the solid phases. Any additional complexity encountered up to this point could be readily introduced. A completely new complexity can, however, be encountered if the compound formed is not stable over the entire range of thermal concern. Such behavior is illustrated in Figure 10.17b where it is assumed that the compound $\mathscr{A}\mathscr{B}$ decomposes (at $T = T_D$) before reaching its melting point. We again encounter an invariance point at $T = T_D$ where the three phases are solution of composition χ_D, solid $\mathscr{A}\mathscr{B}$, and solid $\mathscr{B}$. The same arguments apply to this invariant point and to the eutectic point.

In many solution systems our interest will be confined to a very narrow composition range, e.g., dilute solutions. Here we are often concerned only with the properties of the solvent. The presence of a (dilute) second component in a

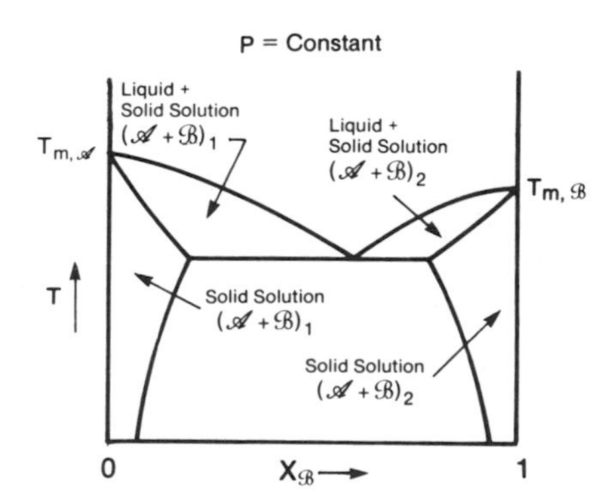

Figure 10.16. A condensed phase diagram in which there is limited miscibility in the the regions os solid phase stability.

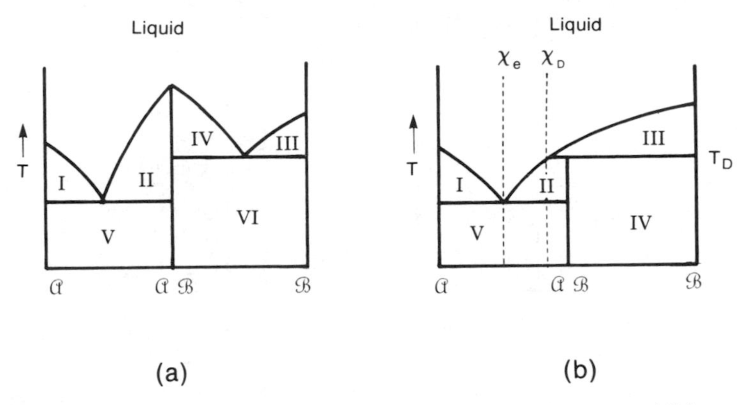

Figure 10.17. A condensed phase diagram where (a) the mixed system exhibits compound formation and (b) the compound decomposes at a temperature below its melting point. Areas designated I, II, III, and IV are regions in which liquid and vapor coexist. Areas V and VI are solid phase mixtures.

liquid phase will always lower the vapor pressure contribution of the major component. For the pure solvent,

$$\mu_\ell(T) = \mu_g(T) = \mu_g^\circ(T) + kT \ln P^* = \mu_\ell^\circ(T) \tag{10.31}$$

while for the solvent (component 1) in a solution system[3]

$$\mu_1(T) = \mu_\ell^\circ(T) + kT \ln X_1 \tag{10.32}$$

so $\mu_\ell(T) < \mu_\ell^\circ(T)$ for all T. If the solute (component 2) has no measureable vapor pressure, then the only gas phase component is solvent. At any specified pressure the gas phase chemical potential is determined. The solid phase chemical potential is unchanged if the solute is insoluble in the solid phase of the solvent (which is commonly true). Then, in terms of our single component chemical potential vs temperature plot, Figure 10.18 results when μ_ℓ for the solvent is displaced to lower values by the solute presence. The solid–displaced liquid lines intersect at a lower value of T and the displaced liquid–gas lines intersect at a higher value of T. When the pressure is 1 atmosphere, the pure solvent intersection points are the normal boiling point and the normal freezing points and ΔT_f and ΔT_b are the changes in those values. With $\mu_\ell = \mu_g$ for all solid–liquid equilibria, any change in molar Gibbs free energy will follow from the expression

$$N_A[\mu_{\ell,1} = \mu_{g,1}^\circ + kT \ln P_1^* X_1] \tag{10.33}$$

$$N_A[d\mu_{\ell,1} - d\mu_{\ell,1}^\circ = kT\, d \ln X_1] \tag{10.34}$$

[3] For general dilute solution discussion we will find the component 1 (solvent), component 2 (solute) notation more convenient than the species specific designations $\mathcal{A}$, $\mathcal{B}$, ...

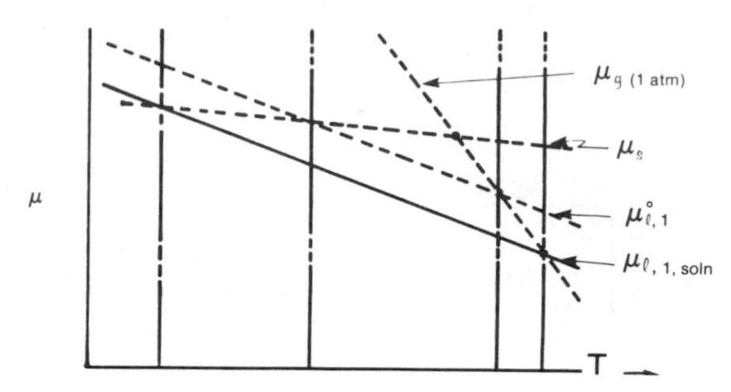

Figure 10.18. Illustration of the displacement in boiling point and freezing point of a solvent when its chemical potential is lowered by the presence of a solute.

substituting $G = N\mu$ with $d\ln X_1 = (1/X_1)(-dX_2) \rightarrow (-dX_2)$ when X_2 is small

$$d(\Delta G_{vap}/T) = -R\,dX_2$$

$$\frac{1}{R}\frac{d(\Delta G_{vap}/T)}{dT} = -\frac{dX_2}{dT}$$

and, utilizing the Gibbs–Helmholz equaton (4.36),

$$\frac{dX_2}{dT} = -\frac{\tilde{H}^0_{g,1} - \bar{H}_{\ell,1}}{RT^2}. \tag{10.35}$$

Integrating Eq. (10.35) over the range $X_2 = 0$, $T = T^0_b$ (the normal pure component boiling point) to X_2, $T = T_b$ (the normal boiling point of the solution) and setting $\tilde{H}^0_{g,1} - \bar{H}_{\ell,1} = \Delta H_{vap}$

$$X_2 = -\frac{\Delta H_{vap}}{R}\left(\frac{1}{T_b} - \frac{1}{T^0_b}\right) = \frac{\Delta H_{vap}}{R}\left(\frac{T^0_b - T_b}{T_b T^0_b}\right). \tag{10.36}$$

In terms of the boiling point elevation, $\Delta T_b = T_b - T^0_b$ (a positive number), and setting $T_b T^0_b = (T^0_b)^2$ (a very close approximation since ΔT_b will not be large) we rewrite Eq. (10.36) in the form

$$\Delta T_b = \frac{R(T^0_b)^2}{\Delta H_{vap}} X_2. \tag{10.37}$$

The boiling point elevation of the solvent[4] has a solvent specific (through T_b

[4] Properties of the solvent that reduce to a simple dependence on solute concentration boiling point elevation, freezing point depression, together with vapor pressure lowering and osmotic pressure are known as colligative properties. Boiling point elevation and freezing point depression are always referenced to normal (1 atm) values.

and ΔH_{vap}) dependence on the solute concentration. A parallel argument starting with the expression

$$N_A[\mu_{\ell,1} = \mu_{s,1} = \mu^\circ_{\ell,1} + kT \ln X_1] \tag{10.38}$$

will produce an expression for the lowering of the solvent freezing point $\Delta T_f = T_f^0 - T_f$ (a positive number)

$$\Delta T_f = \frac{R(T_f^0)^2}{\Delta H_{fus}} X_2 \tag{10.39}$$

where $\bar{H}_{\ell,1} - \tilde{H}^0_{s,1}$ is the enthalpy change on fusion.

In the arguments developed above we have treated the solvent as an ideal solution component taking the standard state as that of the pure substance [Eqs. (10.32)(10.33)]. If the solution is not ideal, it is still most appropriate to retain that standard state for the solvent expressing the chemical potential as

$$\mu_{1,soln} = \mu^\circ_{1,\ell} + kT \ln a_1 = \mu^\circ_{1,\ell} + kT \ln \gamma_1 X_1 \tag{10.40}$$

where a_1 is solvent activity (the solution equivalent of fugacity) with general definition $a_K = \gamma_K X_K$. The activity coefficient will approach one as the mole fraction approaches one. For the solute the most appropriate standard state will be the dilute solution behaving in accordance with Henry's Law

$$\mu_{2,soln} = (\mu^\circ_{2,soln})_X + kT \ln a_2 = \mu^\circ_{2,soln} + kT \ln \gamma_{2,X} X_2. \tag{10.41}$$

For the solute the activity coefficient will approach one as the mole fraction approaches zero since it is in the limit of infinite dilution where the solution will always obey Henry's Law. It will always be more convenient when nonvolatile solutes are a solution component to express their concentrations in molal or molar terms in which case the analogous expressions for chemical potential would be

$$\mu_{2,soln} = (\mu^\circ_{2,soln})_m + kT \ln \gamma_{2,m} m \tag{10.42a}$$

or

$$\mu_{2,soln} = (\mu^\circ_{2,soln})_c + kT \ln \gamma_{2,c} c. \tag{10.43a}$$

Clearly the standard states and the related activity coefficients will be different than those defined by Eq. (10.41). The complete expressions for concentration relationships are, for the molal unit

$$X_2 = \frac{mM_1}{1000 + mM_1} \tag{10.44}$$

where M_1 is the solvent mole weight, and for the molar unit

$$\mathrm{X}_2 = \frac{cM_1}{c(M_1 - M_2) + 1000\rho} \tag{10.45}$$

where M_2 is the solute mole weight and ρ the solution density. For quite dilute solutions

$$\mathrm{X}_2 = \frac{\mathfrak{m}M_1}{1000} = \frac{cM_1}{1000\rho^0} \tag{10.46}$$

where ρ^0 is the solvent density. For dilute aqueous solutions there is negligible distinction between molarity and molality. For such solutions we can construct relationships between the standard chemical potentials by making the substitution of Eq. (10.44) into Eq. (10.41)

$$\mu_{2,\mathrm{soln}} = \mu^{\circ}_{2,\mathrm{soln}} + kT \ln \gamma_{2,\mathrm{m}} + kT \ln \mathfrak{m} + kT \ln \frac{M_1}{1000} \tag{10.42b}$$

or by the substitution of Eq. (10.45) into Eq. (10.41)

$$\mu_{2,\mathrm{soln}} = \mu^{\circ}_{2,\mathrm{soln}} + kT \ln \gamma_{2,c} + kT \ln c + kT \ln \frac{M_1}{1000\rho^0}. \tag{10.43b}$$

Operational expressions for the dilute solutions include the terms $kT \ln(M_1/1000)$ or $kT \ln(M_1/1000\rho^0)$ into the standard chemical expression. The different activity coefficients are related as

$$\gamma_{\mathrm{X}} = \gamma_{\mathrm{m}}(1 + .001M_1) = \gamma_c\left(\frac{\rho}{\rho^0} + \frac{0.001(M_1 - M_2)}{\rho^0}\right) \tag{10.47}$$

and are seen to be identical at low concentration levels. The term γ_{X} is often referred to as the rational activity coefficient.

In the dilute solution range where ideal behavior for the solvent produced Eqs. (10.37) and (10.39) we can utilize the Eq. (10.46) limiting forms to obtain

$$\Delta \mathrm{T_b} = \left(\frac{\mathrm{R}(\mathrm{T_b^0})^2 M_1}{1000\Delta \mathrm{H_{vap}}}\right)\mathfrak{m} = K_{\mathrm{b}}\mathfrak{m} \tag{10.48}$$

and

$$\Delta \mathrm{T_f} = \left(\frac{\mathrm{R}(\mathrm{T_f^0})^2 M_1}{1000\Delta \mathrm{H_{fus}}}\right)\mathfrak{m} = K_{\mathrm{f}}\mathfrak{m} \tag{10.49}$$

TABLE 10.1. Cryoscopic and Ebullioscopic Constants for Some Solvents

Solvent	K_f	K_b
Water	1.86	0.52
Acetic acid	3.57	3.07
Carbon tetrachloride		5.02
Dioxane	4.71	
Ethyl alcohol		1.19
Benzene	5.10	2.60
Cyclohexane		2.79
Phenol	7.40	3.56
Nitrobenzene	7.00	5.24
Acetic acid	3.57	3.07

From a collection of data in *The Handbook of Chemistry and Physics*, 67th ed., Chemical Rubber Press.

where K_b and K_f are known, respectively, as the ebullioscopic constant and the cryoscopic constant. Representative values for several common solvents are listed in Table 10.1. In a more general treatment, if the solution is not dilute and does not behave ideally, the Eq. (10.35) expression would be phrased in terms of the solvent activity $a_1(=\gamma X_1)$, the approximation $T_f T_f^0 = (T_f^0)^2$ will not be sufficient, and it will not be appropriate to regard ΔH_{vap} as a constant over the range of integration. The general treatment will provide the basis for obtaining activity coefficients and the most precise data will follow from freezing point depression measurements.

We rephrase Eq. (10.38) as

$$N_1[\mu_{s,1} = \mu_{\ell,1} = \mu_{\ell,1}^{\circ} + kT \ln a_1] \tag{10.50}$$

from which

$$N_A\left[\frac{\partial \mu_{s,1}/T}{\partial T} - \frac{\partial \mu_{\ell,1}^{\circ}/T}{\partial T}\right] = -\frac{\tilde{H}_s - \tilde{H}_\ell^0}{RT^2} = \frac{\partial \ln a_1}{\partial T}. \tag{10.51}$$

Integration of Eq. (10.51), recognizing that

$$\left(\frac{\partial \Delta H_{fus}}{\partial T}\right) = C_{P,\ell} - C_{P,s} = \Delta C_P$$

and

$$\Delta H_{fus}(T) - \Delta H_{fus}(T_f^0) = \Delta C_P(T - T_f^0) = \Delta C_P \theta \tag{10.52a}$$

where the freezing point depression is represented by θ. Substitution into Eq. (10.51) with $T_f^0 - \theta = T_f$; $dT = -d\theta$ gives

$$-d\ln a_1 = \frac{-\Delta C_P \theta}{R(T_f^0 - \theta)^2} d\theta. \qquad (10.52b)$$

Equation (10.52) can be simplified for integration by expressing $1/(T_f^0 - \theta)^2$ as a power series which takes the form

$$\frac{1}{(T_f^0 - \theta)^2} = \frac{1}{(T_f^0)^2}\left(1 + \frac{2\theta}{T_f^0} + \frac{3\theta^2}{(T_f^0)^2} + \cdots\right).$$

The revision of Eq. (10.52), neglecting the term in θ^2, and integrating gives

$$-\int_{a_1=1}^{a_1} d\ln a_1 = \frac{1}{R(T_f^0)^2}\int_0^{\theta}\left(1 + \frac{2\theta}{T_f^0} + \cdots\right)(\Delta H_{\text{fus},T_f^0} - \theta\Delta C_P)d\theta$$

$$-\ln a_1 = \frac{\Delta H_{\text{fus},T_f^0}\theta}{RT_f^0} + \frac{\theta^2}{R(T_f^0)^2}\left(\frac{\Delta H_{\text{fus},T_f^0}}{T_f^0} - \frac{\Delta C_P}{2}\right). \qquad (10.53)$$

Inserting appropriate constants for water we obtain

$$\ln a_1 = -9.702 \times 10^{-3}\theta - 5.2 \times 10^{-6}\theta^2. \qquad (10.54)$$

The activity of the solute follows from application of the Gibbs–Duhem equation (8.64)

$$d\ln a_2 = -\frac{N_1}{N_2} d\ln a_1 = -\frac{n_1}{n_2} d\ln a_1 = -\frac{1000}{mM_1} d\ln a_1. \qquad (10.55)$$

On the basis of Eq. (10.55) we can reconstruct and simplify Eq. (10.53) by inserting K_f as defined by Eq. (10.49) for some of the collection of constants to obtain

$$d\ln a_2 = \frac{1}{K_f}\frac{d\theta}{dm} + \frac{b\theta d\theta}{m} + c\frac{\theta^2 d\theta}{m}. \qquad (10.56)$$

Normally it will be possible to drop the term in θ^2. The integration of Eq. (10.56) is approached by defining a function j

$$j = 1 - \frac{\theta}{K_f m} \qquad (10.57)$$

which, when differentiated, produces

$$d\mathscr{j} = \frac{\theta}{K_f m^2} dm - \frac{1}{K_f m} d\theta = -(\mathscr{j} - 1)\frac{dm}{m} - \frac{d\theta}{K_f m} \tag{10.58}$$

so that

$$\frac{d\theta}{K_f m} = (1 - \mathscr{j})\frac{dm}{m} - \mathscr{j}. \tag{10.59}$$

Combining Eqs. (10.56) and (10.59) we have, for $\gamma_2 = a_2/m$,

$$d \ln a_2 = (1 - \mathscr{j}) d \ln m - d\mathscr{j} + b\frac{\theta d\theta}{m}$$

$$d \ln \gamma_2 = d \ln a_2 - d \ln m$$

$$= -\mathscr{j} d \ln m - d\mathscr{j} + b\frac{\theta d\theta}{m}. \tag{10.60}$$

In accordance with our standard state convention, $\gamma = 1$ for $m = 0$ so integration between limits m and 0 gives

$$\ln \gamma_2 = -\int_0^m \mathscr{j} d \ln m - b \int_0^m \frac{\theta d\theta}{m} - \mathscr{j}. \tag{10.61}$$

Both of the integrals can be resolved from freezing point data, the first by plotting the experimental values of $\mathscr{j}$ vs $\ln m$ (or $\mathscr{j}/m$ vs m) and the second by plotting θ/m vs θ and evaluating the area under each curve. Parallel arguments can be constructed for obtaining solute and solvent activities, and therefore activity coefficients from boiling point elevations. In either case the activities relate to the temperature of the measurement, either a narrow range just below the normal freezing point or a narrow range just above the boiling point. We are normally interested in 298 K values since this is our conventional standard state for most calculations and we should adjust activity coefficients to that temperature.

The variation in activity of a solution component with temperature is addressed as follows. The molar chemical potential of the solvent is

$$N_A[\mu_{1,soln} = \mu_1^\circ + kT \ln a_1]$$

and the variation of solvent activity is

$$\frac{\partial \ln a_1}{\partial T} = \frac{N_A}{R}\left[\left(\frac{\partial \mu_{1,soln}/T}{\partial T}\right)_P - \left(\frac{\partial \mu_1^\circ/T}{\partial T}\right)_P\right] = -\frac{\bar{H}_1 - \tilde{H}_1^0}{RT^2} = -\frac{\bar{L}_1}{RT^2} \tag{10.63}$$

where $\bar{H}$, is the partial molar enthalpy of solvent in the solution, $\tilde{H}_1^0$ is the molar enthalpy of the solvent in the standard state (since the standard state is pure solvent), and $\bar{L}_1$ is a commonly used notation referred to as the partial molar heat content. Then

$$\frac{\ln a_1''}{a_1'} = \frac{1}{R}\int_{T'}^{T''} \frac{\bar{L}_1}{T^2} dT. \tag{10.64}$$

The partial molar heat content will not be constant over any wide range in temperature but, in analogy to Eq. (10.52a) we can write

$$\bar{L}_1 = \bar{L}_1'' + (\bar{C}_{P,1} - \tilde{C}_{P,1}^0)(T - T'') \tag{10.65}$$

where $\bar{C}_{P,1}$ is the solvent partial molar heat capacity and $\tilde{C}_{P,1}^0$ the solvent molar heat capacity. The difference term is relatively constant. Making this substitution in Eq. (10.64) and performing the integration produces

$$\ln\frac{a''}{a'} = -\frac{\bar{L}_1''(T'' - T')}{T'T''} + \frac{\Delta\bar{C}_{P,1}}{R}\left(\frac{T'' - T'}{T'} - \ln\frac{T''}{T'}\right). \tag{10.66}$$

This relationship will permit the solvent activity as determined at the solution freezing point T′ (Eq. (10.54)] to be adjusted to any temperature T″, e.g., 25°C. The Gibbs–Duhem relationship for obtaining the solute activity coefficient at the T″ temperature requires only small address to accommodate an additional term since, from Eq. (10.55), we obtain

$$d\ln a_2 = -\frac{1000}{mM_1}(d\ln a_1 - dx)$$

where x is $\ln a''/a'$. The activity coefficient follows from

$$d\ln\gamma'' = d\ln\gamma' - \frac{1000}{mM_1}dx$$

$$\ln\gamma'' = \ln\gamma' - \frac{1000}{M_1}\int_0^m \frac{1}{m}dx \tag{10.67}$$

and would be obtained by graphical integration of the $1/m$ vs X curve.

To this point we have confined our attention to the extension of chemical potential concepts, in particular, the equality of a species chemical potential in all (equilibrium) phases in which it is present. As we saw in Chapter 9, chemical equilibrium does not differ in its bounding constraints from phase equilibrium except that we focus on the equality of chemical potential of all atomic species in all of their molecular associations. Operationally, rather different machinery is required and this we should develop to complete the arguments for this chapter. We will construct this machinery on the basis of our ability to calculate thermodynamic functions for gaseous systems at low to moderate pressure and to extend the calculations with experimental data to higher pressure gaseous systems and to liquid and solid phases. When coupled with experimental phase transition and heat capacity data, which will be more precise than many of our condensed phase calculations, all relevant thermodynamic information is available to us. It is convenient to tabulate values for common substances and it will be necessary to construct a framework that minimizes the tabulation effort. For this purpose we consider a general gas phase reaction of formation for a molecule species from its constituent atoms. It will be necessary to specify material quantities, temperature, and pressure, and provide for variations in each from that point.

Consider the process:

$$\mathscr{A}_{(g)} + \mathscr{B}_{(g)} \rightarrow \mathscr{A}\mathscr{B}_{(g)}.$$

We can specify any convenient temperature e.g., 25°C (= 298 K). In the reaction system during any real process the partial pressures of reactants and products will vary in the course of the process. We imagine the Figure 10.19 artificial system. The membrane passes $\mathscr{A}\mathscr{B}$ molecules quantitatively as they are formed but not $\mathscr{A}$ or $\mathscr{B}$ atoms, the pistons ($\mathscr{P}$) can move in such a manner as to maintain $P_{\mathscr{A}} = P_{\mathscr{B}} = 1$ atm during the process and also to maintain $P_{\mathscr{A}\mathscr{B}}$ at 1 atm as $\mathscr{A}\mathscr{B}$ is formed. The remaining boundaries are diathermal. We assume that 1 mol of $\mathscr{A}\mathscr{B}$ molecules is formed. The process is nonreversible; there will be changes in the state functions G, H, S, etc., but these changes depend only on the initial state ($\mathscr{A}$, $\mathscr{B}$ atoms) and the final state ($\mathscr{A}\mathscr{B}$ molecules). Since both molecules and atoms are involved we shall have to recognize the different energy

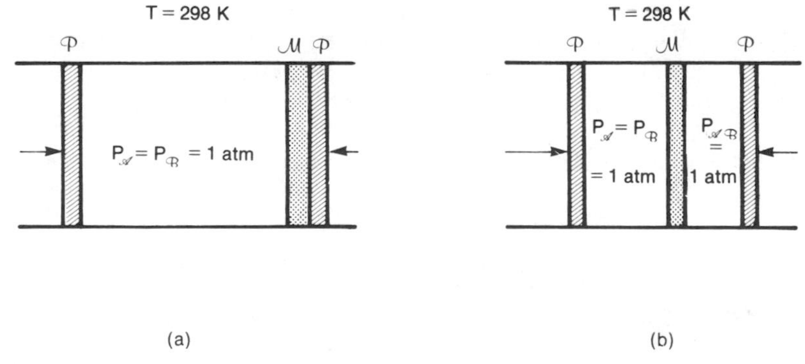

Figure 10.19. An imagined process by which reactants at 1 atm partial pressures are converted to a product at 1 atm pressure.

zeros. For each species

$$G = H - TS = E - E_0 + PV - TS$$

$$G - E_0 = -\mathscr{k}T \ln \mathscr{Z} + PV \qquad \text{for atoms}$$

$$= -\mathscr{k}T \ln \mathscr{Z} + PV - D_0 \qquad \text{for molecules}$$

where the electronic and nuclear ground states for atoms are the energy zero. For any process of concern the E_0 terms cancel and

$$\Delta G = \sum (N\mu)_{\text{products}} - \sum (N\mu)_{\text{reactants}}. \tag{10.68}$$

For our process

$$\Delta G = N_A \mu^\circ_{\mathscr{A}\mathscr{B}}(298) - N_A[\mu^\circ_{\mathscr{A}}(298) + \mu^\circ_{\mathscr{B}}(298)] - D_{0,\mathscr{A}\mathscr{B}} \tag{10.69}$$

where the $\mu^\circ(298)$ are standard chemical potentials at 298 K and 1 atm pressure.

It is common practice to designate this expression as $\Delta G^\circ_{f,\mathscr{A}\mathscr{B}}$, identifying the ΔG so defined as the standard free energy of formation. Since the μ° terms are only functions of (T), the specification of (298 K) completely defines $\Delta G^\circ_{f,\mathscr{A}\mathscr{B}}$. For any real process in which the partial pressures are not 1 atm the ΔG process term is

$$\Delta G = N_A \mu^\circ_{\mathscr{A}\mathscr{B}}(298) - N_A[\mu^\circ_{\mathscr{A}}(298) + \mu^\circ_{\mathscr{B}}(298)]$$

$$+ N_A \mathscr{k}T \ln \frac{\prod^{\kappa}(P_\kappa)_{\text{products}}}{\prod^{\kappa}(P_\kappa)_{\text{reactants}}} - D_{0,\mathscr{A}\mathscr{B}} \tag{10.70}$$

which, of course, reduces to ΔG°_f when all partial pressures are 1 atm. In the notation which we have developed, for the general case

$$\Delta G = \Delta G^\circ_f + RT \ln \frac{\prod^{\kappa}(P_\kappa)_{\text{products}}}{\prod^{\kappa}(P_\kappa)_{\text{reactants}}}. \tag{10.71}$$

In a more complex process the standard free energy terms for atoms cancel, e.g.,

$$\mathscr{A}\mathscr{B}_2 + 2\mathscr{C}\mathscr{D} \rightarrow 2\mathscr{B}\mathscr{C} + \mathscr{A}\mathscr{D}_2 \text{ at 298 K and all } P_\kappa = 1 \text{ atm}$$

and produces a ΔG for the process

$$\begin{aligned}
\Delta G = \{ & N_A[\mu^\circ_{\mathscr{A}\mathscr{D}_2} + \mu^\circ_{\mathscr{B}\mathscr{C}}] - N_A[\mu^\circ_{\mathscr{A}\mathscr{B}_2} + 2\mu^\circ_{\mathscr{C}\mathscr{D}_2}] \\
& - (D_{0,\mathscr{A}\mathscr{C}_2} + 2D_{0,\mathscr{C}\mathscr{D}_2}) + (D_{0,\mathscr{B}\mathscr{C}} + D_{0,\mathscr{A}\mathscr{D}_2})\}_{298} \\
= & [\Delta G^\circ_{f,\mathscr{A}\mathscr{D}_2} + 2\Delta G^\circ_{f,\mathscr{B}\mathscr{C}}]_{298} - [\Delta G^\circ_{f,\mathscr{A}\mathscr{B}_2} + 2\Delta G^\circ_{f,\mathscr{C}\mathscr{D}}]_{298} \\
& - N_A[\mu^\circ_{\mathscr{A}} + 2\mu^\circ_{\mathscr{D}} + 2\mu^\circ_{\mathscr{B}} + 2\mu^\circ_{\mathscr{C}}]_{298} + N_A[\mu^\circ_{\mathscr{A}} + 2\mu^\circ_{\mathscr{B}} + 2\mu^\circ_{\mathscr{C}} + 2\mu^\circ_{\mathscr{D}}]_{298} \\
= & \sum \Delta G^\circ_f(298) \text{ products} - \sum \Delta G^\circ_f(298) \text{ reactants}.
\end{aligned} \tag{10.72}$$

Clearly, the same argument can be made for the change in enthalpy for the process since

$$H = E - E_0 + PV$$

$$H - E_0 = kT^2\left(\frac{\partial \ln \mathscr{Z}}{\partial T}\right)_{S,V} + PV \qquad \text{for atoms}$$

$$= kT^2\left(\frac{\partial \ln \mathscr{Z}}{\partial T}\right)_{S,V} + PV - D_0 \qquad \text{for molecules}$$

and, defining a $\Delta H_f^\circ(298)$ in the same manner as the $\Delta G_f^\circ(298)$, leads to

$$\Delta H_{process}(298) = \sum \Delta H_f^\circ(298) \text{ products} - \sum \Delta H_f^\circ(298) \text{ reactants.} \qquad (10.73)$$

Operationally, the arithmetic embodied in Eqs. (10.72) and (10.73) leads to the observation that standard free energies of formation $\Delta G_f^\circ(298)$ and standard enthalpies of formation $\Delta H_f^\circ(298)$ for molecular species only should be included in process sums. Elemental species have no formation associated free energies or enthalpies and, where they appear in process equations, contribute zero to the Eq. (10.72) or (10.73) sums. Also, the arithmetic leads to the cancellation of all E_0 and D_0 quantities. Extended tables of ΔG_f° and ΔH_f° values are available and selected values are given in Table 10.2. These values have, in many cases, been obtained from direct calorimetric measurements of ΔH_f° coupled with calculated entropy changes for the process to produce

$$\Delta G_f^\circ = \Delta H_f^\circ - T\Delta S_{process}^\circ.$$

Since we have observed in connection with our statistical arguments relating to pure and perfectly ordered solids that the $T = 0\,K$ entropy value is zero and that calculated entropy values referred to the perfectly ordered solid at $T = 0\,K$ are absolute values, the $\Delta S_{process}^\circ$ value is

$$\sum_{\text{Products}} S^\circ(T) - \sum_{\text{Reactants}} S^\circ(T) = \Delta S(T)_{process}.$$

The basis for calculation of absolute entropies from experimental data is

$$\int_0^T dS = S(T) - S(0) = S(T) = \int_0^T \frac{C_P}{T} dT \qquad (7.62)$$

appropriately amended for the inclusion of entropy changes associated with phase transitions. These are, from Eq. (10.2),

$$\Delta H_{process} = Q_{p,process} = T\Delta S_{process}$$

since Q_P is reversible equivalent regardless of the process detail.

TABLE 10.2. Standard Gibbs Free Energies and Enthalpies of Formation together with Third Law Entropies of Formation at 298 K for Selected Substances

Substance (phase)	ΔG_f° (kJ mol^{-1})	ΔH_f° (kJ mol^{-1})	ΔS_f° (J mol^{-1})
AgCl (solid)	−109.72	−127.04	96.11
Al_2O_3 (α, solid)	−1576.4	−1669.8	50.99
Br_2 (liquid)	0	0	152.30
C (graphite)	0	0	5.69
C (diamond)	2.87	1.90	2.44
CO_2 (gas)	−394.4	−393.5	213.6
$CaCO_3$ (solid)	−1128.8	−1206.9	92.9
Cl_2 (gas)	0	0	223.0
HCl (gas)	−95.27	−92.3	187.0
H_2 (gas)	0	0	131.0
H_2O (gas)	−228.6	−241.8	188.7
H_2O (liquid)	−237.2	−285.8	69.9
Mg (solid)	0	0	32.51
MgO (solid)	−569.6	−601.8	26.78
$MgCO_3$ (solid)	−1029.3	−1112.9	65.69
N_2 (gas)	0	0	191.5
NH_3 (gas)	−16.6	−46.3	193.0
O (gas)	230.10	294.4	160.95
O_2 (gas)	0	0	205.0
S (solid, rhombic)	0	0	31.88
S (solid, monoclinic)	0.10	0.30	32.55
SO_2 (gas)	−300.4	−296.1	248.52
SO_3 (gas)	−370.4	−395.2	256.22
CH_4 (gas)	−50.8	−74.85	186.19
C_2H_6 (gas)	−32.89	−84.7	229.49
CH_3OH (liquid)	−166.31	−238.7	126.78
C_2H_5OH (liquid)	−174.18	−276.98	161.04
C_6H_6 (liquid)	124.52	49.04	124.52

A large body of thermodynamic properties for elements and their compounds is collected in *The Handbook of Physics and Chemistry*, 67th ed., Chemical Rubber Press. Also see end of chapter data sources.

The enthalpy change associated with any constant temperature process can, in principle, be calculated from the Eq. 10.1 analog

$$(\Delta H - \Delta E_0)_{\text{process}} = (\Delta E + \Delta PV)_{\text{process}}$$

$$\Delta H_{\text{process}} = \Delta\left[\mathscr{k}T^2\left(\frac{\partial \ln \mathscr{Z}}{\partial T}\right)_{V,N} + \mathscr{k}TV\left(\frac{\partial \ln \mathscr{Z}}{\partial V}\right)_{T,N}\right]. \quad (10.74)$$

If the process is a phase transition the partition function derivatives will be those characteristic of the two phases. If $Q_{p,process}$ is a measured quantity then $\Delta S_{process}$ is $Q_{p,process}/T$. From third law considerations and the Debye low temperature extrapolation

$$S^\circ(T_1) = \int_0^{T_1} AT^2\,dT$$

where A is the collection of constants appearing in the Debye T^3 Law (Eq. 7.47).

$$S^\circ(T) = \int_0^{T_1} AT^2\,dT + \int_{T_1}^{T_2} \frac{C_P}{T}\,dT + \sum \frac{\Delta H_{TR}}{T_{TR}} + \sum \int_{T_\chi}^{T_{\chi'}} \frac{C_{P,\chi}}{T}\,dT \qquad (10.75)$$

where the integrations over 0 to T_1 and T_1 to T_2 represent the absolute entropy from 0 K to T_2, which is the temperature at which the first phase transition involving the solid occurs. This may be a solid-to-solid transition or the solid-to-liquid transition. The $\sum^{\kappa}(\Delta H_{Tr}/T_{Tr})_{\kappa}$ term represents the entropy change associated with all phase transitions (κ) in the interval T to T_2 and the notation $T_\chi T_{\chi'} C_{P,\chi}$ is intended to convey that each integral involves a phase χ and T_χ to $T_{\chi'}$ is the interval in T over which that phase is the thermodynamically stable phase under 1 atmosphere pressure. Equation (10.75) can, of course, be extended into the gas phase region of stability at $T > T_{\ell,g}$ and $P \geqslant 1$ atmosphere with heat capacity measurements or statistical calculation. Selected values of the absolute (or third law) entropies are included in Table 10.2.

In the application of ΔG_f°, ΔH_f° and third law entropies to process calculations we should note that the Table 10.2 data have incorporated certain conventions and/or simplifications. When elements can exist in more than one form, the form that has the lowest free energy at 298 K and 1 atmosphere pressure (the thermodynamically stable form) is written into the equation of formation. As an example, we recognize that certain elemental forms exist as diatomic gases at 298 K and 1 atmosphere pressure (N_2, H_2, O_2) and that, while the element carbon can exist in a crystal structure known as graphite and another known as diamond, graphite is the thermodynamically stable form under these conditions. The thermodynamically stable phase for CO_2 at 298 K and 1 atmosphere pressure is gaseous. Then for $\Delta G_f^\circ(298)$ and $\Delta H_f^\circ(298)$ for CO_2 gas, the equation for defining the formation term is conventionally taken to be

$$O_{2(g)} + C_{graphite} \rightarrow CO_{2(g)}.$$

This will not be a matter of concern since

$$\Delta G^\circ(298) = N_A\mu^*_{CO_2}(298) - N_A\mu^*_C(298) - N_A\mu^*_{O_2}(298) + D_{0,O_2} - D_{0,CO_2} \qquad (10.76)$$

and with all oxygen containing elements defined in the same manner, it is the

$\mu^*_{O_2}(298)$ and D_{0,O_2} terms, which will always have a stoichiometric counterpart leading to cancellation. So long as carbon is always graphite in the formation term the same conclusion follows. The third law entropies given in Table 10.4 include gaseous substances. In these cases the entropies were obtained from statistical calculations [Eq. (3.15)] and do not reflect nonideality. Correction can be accomplished through the use of any reliable equation of state.

Equation (10.71) demonstrates that we are not limited to 1 atmosphere pressure in the calculation of ΔG for a process. Neither are we limited to 298 K in that the μ° terms can be calculated for any temperature. If all substances of interest are gaseous our statistical calculations for any prescribed value of T will provide $H - E_0$ and $G - E_0$, usually, however, written as $H - H_0$, and $G - H_0$.[5] To minimize the effort of tabulation, the function $[G^\circ(T) - H_0^\circ]/T = \Phi(T)$ is normally calculated and tabulated. If we also calculate and tabulate the functions $[H^\circ(T) - H_0^\circ\zeta/T]$ and $\Delta H_f^\circ(0)$, it is possible to confine the tabulation to rather wide intervals of T and interpolate within such intervals since neither $[G^\circ(T) - H_0^\circ]/T$ nor $[H^\circ(T) - H_0^\circ]/T$ changes greatly with T. In terms of the free energy and enthalpy functions we can obtain $\Delta G_f^\circ(T)/T$ and $\Delta H_f^\circ(T)/T$ for any substance using the following prescription:

$$\frac{\Delta G_f^\circ(T)}{T} = \frac{1}{T}\Big\{\Delta H_f^\circ(298) - [H^\circ(298 - H_0^\circ] + \sum_{\text{elements}} [H^\circ(298) - H_0^\circ]$$

$$+ [G^\circ(T) - H_0^\circ] - \sum_{\text{elements}} [G^\circ(T) - H_0^\circ] \qquad (10.77)$$

$$= \frac{\Delta H_f^\circ(0)}{T} - \frac{[G(T) - H_0^\circ]}{T} - \sum_{\text{elements}} \frac{[G^\circ(T) - H_0^\circ]}{T} \qquad (10.78)$$

$$\frac{\Delta H_f^\circ(T)}{T} = \frac{\Delta H_f^\circ(0)}{T} - \frac{[H^\circ(T) - H_0^\circ]}{T} - \sum_{\text{elements}} \frac{[H^\circ(T) - H_0^\circ]}{T}. \qquad (10.79)$$

Tables 10.3 and 10.4 list representative selections of molar values. Values of $\Delta H_f^\circ(T)$ and $\Delta G_f^\circ(T)$ for the reactants and products for a process occurring at T combine according to the conventions outlined for Eqs. (10.72) and (10.73) to yield $\Delta H^\circ(T)$ and $\Delta G^\circ(T)$ for the process.

Values of $\Delta G_f(T)$ and $\Delta H_f(T)$ are particularly useful in defining the equilibrium constant and its variation with temperature for those processes which do not go to completion. As we have noted in Chapter 7, using the simple example

$$\mathscr{A} + \mathscr{B} \rightleftharpoons \mathscr{A}\mathscr{B}$$

the equilibrium constant can be defined by

$$K_{eq}(T) = \frac{N_{\mathscr{A}\mathscr{B}}/V}{N_{\mathscr{A}}N_{\mathscr{B}}/V^2} = \frac{z'_{\mathscr{A}}e^{-D_{0,\mathscr{A}\mathscr{B}}}}{z'_{\mathscr{A}}z'_{\mathscr{B}}} \qquad (10.80)$$

[5]Since S = 0 for all pure substances there is no distinction between A_0, G_0, H_0, and E_0.

TABLE 10.3. The Free Energy Function $(G_f^\circ - H_0^\circ)/T$ at Various Temperatures for Selected Substances

Substance	Temperature (K)						
	298.16	400	500	600	800	1000	1500
methane	36.46	38.86	40.75	42.39	45.21	47.65	52.84
ethane	45.27	48.24	50.77	53.08	57.29	61.11	69.46
ethylene	43.98	46.61	48.74	50.70	54.19	57.29	63.94
propane	52.73	56.48	59.81	62.93	68.74	74.10	85.86
propylene	52.95	56.39	59.32	62.05	67.04	71.57	81.43
n-butane	58.54	63.51	67.91	72.01	79.63	86.60	101.95
butene (l)	59.25	63.64	67.52	71.14	77.82	83.93	97.27
Isobutane	56.08	60.72	64.95	68.95	76.45	83.38	98.64
Isobutene	56.47	60.90	64.77	68.42	75.15	81.29	94.66
n-pentane	64.52	70.57	75.94	80.96	90.31	98.87	117.72
n-hexane	70.62	77.75	84.11	90.06	101.14	111.31	133.64
acetylene	39.976	42.451	44.508	46.313	49.400	52.005	57.231
graphite	0.5172	0.824	1.146	1.477	2.138	2.771	4.181
H_2	24.423	26.422	27.947	29.203	31.186	32.738	35.590
$H_2O(g)$	37.165	39.505	41.293	42.766	45.128	47.010	50.598
$CO(g)$	40.350	42.393	43.947	45.222	47.254	48.860	51.864
$CO_2(g)$	43.555	45.828	47.667	49.238	51.895	54.109	58.481
$N_2(g)$	38.817	40.861	42.415	43.688	45.711	47.306	50.284
$O_2(g)$	42.061	44.112	45.675	46.968	49.044	50.697	53.808

Taken from V. Fried, H. F. Hameka, and U. Blukis, *Physical Chemistry*. Macmillan, New York, 1977.

TABLE 10.4. The Enthalpy Function at Various Temperatures and the Standard Heat of Formation for Selected Substances

Substance	$(\Delta H^\circ_f)_0$ (kcal mole^{-1})	$(H^\circ_T - H^\circ_0)/T$ (cal K^{-1} mole^{-1}) Temperature (K)						
		298.16	400	500	600	800	1000	1500
methane	−15.987	8.039	8.307	8.73	9.249	10.401	11.56	14.09
ethane	−16.517	9.578	10.74	12.02	13.36	15.95	18.28	23.00
ethylene	14.522	8.47	9.28	10.23	11.22	13.10	14.76	18.07
propane	−19.482	11.78	13.89	16.08	18.22	22.20	25.67	32.43
propylene	8.468	10.86	12.48	14.15	15.82	18.94	21.69	27.05
n-butane	−23.67	15.58	18.35	21.19	23.96	29.08	33.54	42.18
butene(l)	4.96	13.79	16.21	18.68	21.08	25.46	29.25	36.56
Isobutane	−25.30	14.34	17.41	20.50	23.45	28.76	33.31	42.03
Isobutene	−0.98	13.69	16.30	18.83	21.25	25.62	29.37	36.67
n-pentane	−27.23	18.88	22.38	25.94	29.38	35.71	41.19	51.75
n-hexane	−30.91	22.21	26.45	30.72	34.82	42.35	48.85	61.34
acetylene	54.329	8.021	8.853	9.582	10.212	11.249	12.090	13.694
graphite	0	0.8437	1.257	1.642	1.997	2.602	3.075	3.876
$H_2(g)$	0	6.7877	6.8275	6.8590	6.8825	6.9218	6.9658	7.1295
$H_2O(g)$	−57.107	7.941	7.985	8.051	8.137	8.362	8.608	9.232
$CO(g)$	−27.2019	6.951	6.959	6.980	7.016	7.125	7.257	7.572
$CO_2(g)$	−93.9686	7.506	7.987	8.446	8.871	9.612	10.222	11.336
$N_2(g)$	0	6.9502	6.9559	6.9701	6.9967	7.0857	7.2025	7.5024
$O_2(g)$	0	6.942	6.981	7.048	7.132	7.320	7.497	7.851

Taken from V. Fried, H. F. Hameka, and U. Bluķis, *Physical Chemistry*. Macmillan, New York, 1977.

where the z' are the temperature dependent elements of the partition function. These are just those elements which we have incorporated into the $\mu^\circ(T)$ term together with the 1 atmosphere conversion factor so that

$$\begin{aligned} kT \ln K_{eq}(T) &= kT \ln z'_{\mathscr{AB}} - kT \ln z'_{\mathscr{A}} - kT \ln z'_{\mathscr{B}} - D_{0,\mathscr{AB}} \\ &= \Delta\mu^\circ(T) - D_{0,\mathscr{AB}} \end{aligned} \tag{10.81}$$

and

$$N_A kT \ln K_{eq}(T) = -\Delta G_f^\circ(T). \tag{10.82}$$

This will, of course, be applicable to processes of higher complexity with $-\Delta G^\circ(T)$ constructed on the basis of all molecular forms present.

Since the temperature dependence of the equilibrium constant follows that of $\Delta G^\circ/T$, it will always be most convenient to take advantage of the Gibbs-Helmholz result to obtain

$$\left(\frac{\partial \Delta G^\circ/T}{\partial T}\right)_{P,N} = -\frac{\Delta H^\circ}{T^2} = -N_A k\left(\frac{\partial \ln K_{eq}}{\partial T}\right)_{P,N}. \tag{10.83}$$

The equilibrium constant will be known at 298 K from tabulated values of $\Delta G^\circ(298)$. Integration of Eq. (10.83), assuming that $\Delta H^\circ(298)$ is constant, produces

$$\ln K_{eq}(T) = \ln K_{eq}(298) + \frac{\Delta H^\circ(298)}{R}\left[\frac{1}{T} - \frac{1}{298}\right]. \tag{10.84}$$

More exact calculations follow from the use of free energy and enthalpy functions.

In the event that enthalpy functions are not available, other procedures can sometimes be followed. If heat capacities for all reactants and products are known and $\Delta H_{process}$ is known at one temperature $[\Delta H^\circ(T)]$,

$$\text{Reactants}(T_1) \rightarrow \text{Products}(T_1) \qquad \Delta H^\circ(T_1)$$

can be imagined to be one step in a cyclic process

$$\begin{array}{ccccc} \text{Reactants}(T_1) & \xleftarrow{(4)} & \text{Products}(T_1) & \qquad & -\Delta H^\circ(T_1) \\ {\scriptstyle(1)}\downarrow & & {\scriptstyle(3)}\uparrow & & \\ \text{Reactants}(T_2) & \xrightarrow{(2)} & \text{Products}(T_2). & & \Delta H^\circ(T_2) \end{array}$$

Enthalpy changes for the cyclic process must sum to zero, since enthalpy is a state function, so

$$\Delta H_{cycle} = \Delta H_1^\circ + \Delta H_2^\circ + \Delta H_3^\circ + \Delta H_4^\circ \tag{10.85}$$

$$\Delta H_1^\circ = \int_{T_1}^{T_2} \sum_{\text{Reactants}} C_P dT, \qquad \Delta H_2^\circ = \Delta H_{T_2}^\circ,$$

$$\Delta H_3^\circ = \int_{T_2}^{T_1} \sum_{\text{Products}} C_P dT, \qquad \Delta H_4^\circ = -\Delta H_{T_1}^\circ \tag{10.86}$$

and

$$\Delta H^\circ(T_2) = \Delta H^\circ(T_1) + \int_{T_1}^{T_2} \left(\sum_{\text{Products}} C_P - \sum_{\text{Reactants}} C_P \right) dT. \tag{10.87}$$

Experimental heat capacities are widely available, usually expressed in a three constant form as we noted in Chapter 5,

$$\tilde{C}_P = a + bT + cT^{-2} \tag{5.87}$$

or we can obtain heat capacities for gases and solids by direct calculation using the methods of Chapter 5. Representative heat capacity data for common gases were listed in Table 5.5. Selected values for several liquids and solids are listed in Table 10.5.

TABLE 10.5. Temperature Dependence of Heat Capacities for Selected Solids and Liquids $\tilde{C}_P = a + bT + cT^{-2}$

Substance, phase	a (J K^{-1} mole^{-1})	$b(\times 10^{-3})$ (J K^{-1} mole^{-1})	$c(\times 10^{5})$ (J K^{2} mole^{-1})
H_2O, liquid	46.86	30.0	
C_6H_{14}, liquid	195.0		
CH_3OH, liquid	81.6		
C_2H_5OH, liquid	113.0		
CCl_4, liquid	133.0		
$CHCl_3$, liquid	116.0		
Al, solid	20.67	12.38	
Al, liquid	29.29		
C, solid (graphite)	16.86	4.77	−8.54
Cu, solid	22.64	6.28	
Mg, solid	22.30	10.25	−0.431
Mg, liquid	33.47		
Al_2O_3, solid	109.3	18.36	−30.41
NaCl, solid	45.94	16.32	

Taken from a larger collection in *The Handbook of Chemistry and Physics*, 67th ed., Chemical Rubber Press.

The relationships developed in this chapter are not confined to nonelectrolyte systems, but the electrical features of ionic species in solution impose additional complications. The framework for addressing the electrical features is the subject of the next chapter.

PROBLEMS

10.1. The change in any thermodynamic function for any change in thermodynamic state can, in principle, be calculated from the difference between statistical expressions for the substance in the two states. Calculate the molar enthalpy change for sublimation of metallic lithium at 200 K using the data given in problem 9.1. What is the change in entropy for the process involving 1 mole of lithium?

10.2. Where experimental data relating to changes in state are available we would utilize the data in the condensed difference expressions developed in this chapter, e.g., the integrated from of Eq. (10.5), to provide ΔH_{tran}. Using the following vapor pressure data for H_2O,

T(K)	293	303	313	323	333	343	353	363
P*(atm)	0.02308	0.04183	0.07278	0.1217	0.1965	0.03075	0.4672	0.6918

obtain ΔH_{vap} for H_2O at 30, 50, 70, 80°C and extropolate the data to 100°C to obtain ΔH_{vap} at 100°C.

10.3. The heat of fusion of ice is 6.01 kJ mole^{-1} and the density of ice is 0.92 g ml^{-1}. The density of liquid water can be taken as 1.00 and the melting point under one atmosphere pressure as 273.2 K. Assuming that there are no phase changes in the ice structure below 100 atmospheres, and that ΔH_{fus} and ρ_{ice} do not change significantly with pressure, what is the melting point of ice under 100 atmospheres pressure?

10.4. Raoult's Law applies to ideal mixed systems. If we assume that hexane (1) and heptane (2) form an ideal mixture, what is the vapor pressure above an equimolar mixture of the two liquids at 25°C? The vapor pressures are (1) 148 mm Hg and (2) 51.3 mm Hg. If a sample taken from the vapor phase at 25°C were condensed, what would its composition be?

10.5. Mixtures of acetone and chloroform exhibit negative deviations from Raoult's Law. At 25°C the following partial pressure data (mm Hg) are observed:

X_{CHCl_3}	0	0.2	0.4	0.6	0.8	1.0
P_{CHCl_3}	0	30.2	69.7	132	203	283
$P_{acetone}$	348	259	168	100	42	0

Calculate activity coefficients on the Raoult's Law basis for each component at each concentration. Plot the data and obtain activity coefficients at 0.2 mol fraction for each component as solute on the Henry's Law basis.

10.6. From Figure 10.17, imagine that a single phase liquid mixture at (roughly) 0.2 mole fraction of component $\mathscr{B}$ is cooled slowly. Identify the phases and phase compositions that would be encountered.

10.7. A solution of 9.0 g of naphthalene in 1000 g of benzene freezes at 5.07°C. The normal freezing point of benzene is 5.42°C. Calculate the heat of fusion of benzene.

10.8. The freezing point depression of aqueous solutions of urea is measured to be

molality	0.03241	0.646	1.521	3.360
ΔT_f(°C)	0.5953	1.170	2.673	5.490

The heat capacity of liquid water at 0°C is 1.0074 cal/g °C and for ice is 0.5085 cal g^{-1} °C^{-1}. These data can be utilized in the Eqs. (10.53) and (10.61) framework to obtain activity coefficients. Obtain activity coefficients for urea and for water in 0.05, 1.0, and 2.0 molal solutions of urea.

10.9. In a two-component (1, 2) system the following enthalpy changes per mole of mixture were observed during mixing studies:

X_2	0.1	0.2	0.4	0.5	0.6	0.8	0.9
ΔH_{mix}(cal)	−290	−560	−1110	−1180	−1180	−720	−400

From the problem 10.8 result, determine $\bar{H}_1 - \tilde{H}_1^\circ$ and $\bar{H}_2 - \tilde{H}_2^\circ$ for each component at 0.2 and 0.6 mol fraction of component 2.

10.10. If we have interest in the reaction at 298 K

$$N_2 + 3H_2 \rightleftharpoons NH_3$$

we could calculate the value of the equilibrium constant from our statistical argument but that calculation would be tedious and in this case not required because $\Delta G_f(298)$ data are readily available. With reference to Table 10.2, calculate the value of the equilibrium constant for the reaction at 298 K.

10.11. If we are interested in the reaction at a much higher temperature we could, again, calculate any thermodynamic function including the equilibrium constant from statistical argument. However, if free energy functions have been calculated and are available we could use those data with a considerable savings in effort. Many tabulations are available but not always in the same form as our Tables 10.2, 10.3, and 10.4. Given the following data for NH_3, calculate the equilibrium constant at 500 K.

	$[G^\circ(500\,K) - H^\circ(298)]$	$[H^\circ(298) - H_0^\circ]$	ΔH_0°
NH_3	176.9	9.92	−39.2
H_2	116.9	6.196	0
N_2	177.5	8.67	0

10.12. For moderate temperature differences over which can imagine that ΔH° for the process will remain constant we can use Eq. (10.84) to obtain the change in the equilibrium constant. Assuming that this will be true, calculate the change in K_{eq} for the problem 10.10 reaction if it were carried out at 350 K rather than 298 K.

10.13. We can always utilize heat capacity data in the Eq. (10.87) form to obtain $\Delta H_{process}$ values at temperatures different from 298 K. Using data in Table 5.5, calculate the change in enthalpy for the problem 10.1 process if it were carried out at 800 K.

CHAPTER CONTENT SOURCES AND FURTHER READING

1. H. Eyring, ed., *Annual Review of Physical Chemistry*, Volumes 23 (1972), 25 (1974) and 27 (1976). Annual Reviews, Palo Alto, CA.
2. I. Klotz, *Chemical Thermodynamics*, Chapters 13, 14, 17, 18, and 19. W. A. Benjamin, New York, 1964.
3. J. M. Prasnitz, R. N. Lichtenthaler, and E. G. Alverado, *Molecular Thermodynamics of Fluid-Phase Equilibria*, 2nd ed., Chapters 6, 7, 8, and 9. Prentice-Hall, Englewood Cliffs, NJ, 1986.
4. R. S. Berry, S. A. Rice, and J. Ross, *Physical Chemistry*, Chapter 25. Wiley, New York, 1980.
5. E. A. Moelwyn-Hughes, *Physical Chemistry*, 2nd ed., Chapter XVII. Pergamon Press, London, 1961.
6. C. Kittel and H. Kroemer, *Thermal Physics*, 2nd ed., Chapter 11. W. H. Freeman, San Francisco, 1980.
7. W. E. Acree, Jr., *Thermodynamic Properties of Nonelectrolyte Solutions*, Academic Press, San Francisco, 1984.
8. G. N. Lewis and M. Randall, *Thermodynamics*, 2nd ed., Chapters 17–21, 16, 34, revised by K. S. Pitzer and L. Brewer. McGraw-Hill, New York, 1961.

DATA SOURCES

1. E. W. Washburn, ed., *International Critical Tables*. McGraw-Hill, New York, 1926–1933.
2. *JANAF Thermochemical Tables*, NSRDS-NBS, Catalog No. C 13.48.37. U.S. Government Printing Office, Washington, D.C., 1948.

3. F. D. Rossini, D. D. Wagman, W. H. Evans, S. Lervine, and I. Jaffe, *Selected Values of Chemical Thermodynamic Properties*, NBS Circular 500. U.S. Government Printing Office, Washington D.C., 1952.
4. R. R. Hultgren, L. Orr, P. D. Anderson, and K. K. Kelley, *Selected Values for the Thermodynamic Properties of Metals and Alloys.* Wiley, New York, 1963.
5. D. D. Wagman, *Selected Values of Chemical Thermodynamic Properties, Tables for Elements* (Vol. I, II, III), NBS TN 270-4, TN 270-5, TN 270-6. U.S. Government Printing Office, Washington, D.C., 1968–1971.
6. E. S. Domalski, *Selected Values of Heats of Combustion and Heats of Formation of Organic Compounds, J. Phys. Chem. Ref. Data*, **1**, 221 (1972).
7. F. R. Bichowsky and F. D. Rossini, *Thermochemistry of the Chemical Substances.* Reinhold, New York, 1936.

SYMBOLS: CHAPTER 10

New Symbols

ϑ, ζ, χ = general designation for various phases

$V_{\mathscr{g}}, V_{\ell}, V_{\mathscr{s}}$ = a thermodynamic property for gaseous, liquid, solid phase

$T_{\mathscr{s},\ell}, T_{\ell,\mathscr{g}}$ = value of a property where two phases coexist in equilibrium

$\Delta E_{vap} \cdots \Delta H_{sub} \cdots \Delta S_{fus}$ = change in a thermodynamic property on vaporization, sublimation, fusion (melting)

$P^*(T), P^*_{\mathscr{A}}, P^*_{\mathscr{B}}$ = vapour pressure of a single component system at T, of various species

$P_K, P_{\mathscr{A}}, P_{\mathscr{B}}$ = partial pressure of various species in a gas phase mixture, general, specific

$X_K, X_{\mathscr{A}}, X_{\mathscr{B}}$ = liquid phase mole fraction in a mixture for a general component, for specific components

$Y_{\mathscr{A}}, Y_{\mathscr{B}}$ = vapor phase mole fraction in a mixture

$n_{\mathscr{A}}, n_{\mathscr{B}}, n_1, n_2$ = number of moles, of various components ($\mathscr{A}, \mathscr{B}$ or 1, 2)

$\mu_{\mathscr{s},\mathscr{A}}, \mu_{\mathscr{g},1}, \mu_{\ell,\mathscr{B}}$ = chemical potential of a species in a phase

$\mu_{\mathscr{A},soln}$ = chemical potential of a solution component

$\gamma_R, \gamma_H, \gamma_{R,1} \cdots \gamma_{H,2}$ = activity coefficient on Raoult's Law basis, on Henry's Law basis, for a particular species

k_H = Henry's Law constant

X_e, T_e = values at a eutectic point in a phase diagram

X_D, T_D = values at compound decomposition point in a phase diagram

T_b^0, T_f^0, T_b, T_f = normal boiling, freezing points of solvent, boiling freezing points of solvent in a mixture

$\Delta T_b, \Delta T_f$ = change in boiling or freezing point of solvent from pure state to solution

M, M_{K}, M = molecular weight, of a component, of the solvent
$c, \mathfrak{m}$ = concentration in molarity, molality
$\gamma_{\mathrm{X}}, \gamma_{c}, \gamma_{\mathfrak{m}}, \gamma_{M}$ = activity coefficient for chemical potential expressions in various concentration units
$K_{\mathrm{b}}, K_{\mathrm{f}}$ = ebullioscopic, cryoscopic constants
$\mathrm{a}_{\mathrm{K}}, \mathrm{a}_{1}, \mathrm{a}_{2} \cdots$ = activity of a general solution component, of specific solution components
b, c = a collection of constants in an expression relating freezing point depression to solute activity (one time use)
$\mathscr{j}$ = a function constructed to simplify the treatment of freezing point data (one time use)
$\bar{\mathrm{L}}_{1}, \bar{\mathrm{L}}_{2} \cdots$ = partial molar heat content for a component in solution
$\Delta \mathrm{G}_{\mathrm{f}}, \Delta \mathrm{G}_{\mathrm{f}, \mathscr{A}\mathscr{B}}, \Delta \mathrm{G}^{\circ}_{\mathrm{f}, \mathscr{A}\mathscr{B}}$ = the free energy of formation, formation of the $\mathscr{A}\mathscr{B}$ species from constituent atoms and under standard conditions
$\Delta \mathrm{H}_{\mathrm{f}}, \Delta \mathrm{H}_{\mathrm{f} \mathscr{A}\mathscr{B}}, \Delta \mathrm{H}^{\circ}_{\mathrm{f}, \mathscr{A}\mathscr{B}}$ = the enthalpy of formation, of the $\mathscr{A}\mathscr{B}$ species from constituent atoms and under standard conditions
A = collection of constants in the Debye T^3 Law

CARRYOVER SYMBOLS

z_{g}, z_{c}, $\mathscr{Z}$, P, T, V, N, $\mathrm{N_A}$, μ, μ°, G, $\tilde{\mathrm{G}}$, S, $\tilde{\mathrm{S}}$, $\Delta \mathrm{G}_{\mathrm{mix}}$, $\Delta \mathrm{S}_{\mathrm{mix}}$, $(\Delta \mathrm{E}_{\mathrm{mix}})$, ω_{12}, ϕ, $\mathrm{D}_{0, \mathscr{A}\mathscr{B}}$, χ, K_{eq}

11

PHASE AND CHEMICAL EQUILIBRIUM III: ELECTROLYTE SYSTEMS

Structural studies show that many substances exist in their solid state as a network of positive and negative ions. These solids when placed in certain liquids, particularly those that consist of highly polar molecules such as water, dissolve to a small or substantial extent to form solutions in which the ions are dispersed and mobile as indicated by the ability of the solution to conduct electric current. Such substances are referred to as electrolytes. Colligative and electrical properties of the solutions indicate that the dispersal of the individual ions is effectively complete at least in dilute solutions.

There is a further class of substances that is molecular in its solid, liquid, or gaseous state but that does not remain so when it dissolves in particular liquids to form solutions. To a small or substantial extent these substances dissociate to provide ions. Both ions and molecules contribute to the colligative properties and ions contribute to the electrical properties of the solution. In some cases these molecular substances appear to dissociate completely to exhibit solution properties similar to those characteristic of solutions involving ionic solids. Since in solution partially dissociated substances behave as electrolytes but make a smaller ionic contribution to the electrical properties of the solution they are referred to as weak electrolytes. Ionic solids and completely dissociated molecular substances are then, in solution, strong electrolytes.

We would anticipate that the solubility and extent of dissociation of weak electrolytes and the solubility of strong electrolytes depend on attractive interactions involving solvent molecules. The ionization of molecular substances, whether or not they qualify as strong electrolytes, involves the dissociation of chemical bonds while the solubility of ionic solids and the dispersal of the ions require that Coulombic forces holding the individual ions in their rigid crystalline network be overcome. Although the electrical features may be the same in that the electrolyte solution contains solvent molecules and ions, in the weak electrolyte case there are also undissociated molecules and on the basis of our prior arguments we would understand that chemical equilibrium requirements must be satisfied at all levels of solution concentration. Such equilibrium requirements are relevant only in the solubility limit (saturation) for ionic solids where both solid and saturated solution phases are present. Arguments relating to the equality of chemical potentials will apply to all equilibria.

We anticipate that the strong solute–solvent interactions that are required to produce or stabilize electrolyte solutions are features that will lead to nonideal behavior even in dilute solutions. Electrostatic interactions between two ions, between an ion and a dipole and between two dipoles, lead to interaction energies which vary as r^{-1}, r^{-2}, and r^{-3}, respectively. Interaction forces vary as r^{-2}, r^{-3}, and r^{-4}, respectively.[1] Even when ions are dispersed in solution we would anticipate that the relatively long-range ion–ion interactions will be large. Approximations (nearest neighbor, etc.) that we have developed for nonelectrolyte solutions have no applicability and ion–dipole interactions may impose new boundaries on the concept of the ideal solution. Solvent character will clearly be very important since strong interactions would generally derive from or be controlled by specific electrical features of the solvent such as dipole moment and dielectric constant. There are other complications, however, as the data in Table 11.1 illustrate. Neither dipole moment nor dielectric constant

TABLE 11.1. Dielectric Constant (D) and Dipole Moment μ for Common Solvents

Substance	D	μ Debye Unit[a]
Water	78.5	1.87
Ammonia	16.9	1.3
Sulfur dioxide	17.6	1.6
Chloroform	4.8	1.01
Methyl alcohol	32.6	1.70
Ethyl alcohol	24.3	1.69
Formamide	109.	3.73
Tetrahydrofuran	3.0	1.63
Pyridine	12.3	2.19
Phenol	9.8	1.45
Analine	6.8	1.53
Nitro Benzene	34.8	4.22

[a] Debye = 10^{-18} esu cm or 3.36×10^{-30} Cm.

[1] See Appendix D on Intermolecular Forces.

uniquely identifies the most effective solvent (water) in terms of solubility of ionic solids or dissociation of molecular solutes although water does have high (but not uniquely high) values of both dielectric constant and dipole moment. Formamide excels in both measures but is not a particularly good solvent for ionic species.

The solvent role in producing ionic species by dissociation derives from chemical interaction and will be as varied as the spectrum of solvent–solute structures and bonding features. Weak electrolyte concern in aqueous systems is concentrated on dissociations that, in one form or another, involve hydrogen ion (so called weak acid–weak base equilibria) and solvent water. The chemistry is rather clear on a molecular basis given the ability of water molecules to accommodate hydrogen ion through hydrogen bonding with small disruption to general structure features. Some other substances in the Table 11.1 listing are hydrogen bonded or accommodate hydrogen ion but have limited suitability as ionizing solvents. There would also be chemistries as unique to some other solvent systems as hydrogen ion is to aqueous solutions.

In sum it appears that the variety of features involved in electrolyte solution systems severely limits any completely general approach. In this chapter we will rely largely on procedures through which our general thermodynamic understanding can provide a somewhat detailed appreciation of the microscopic features. Attention will be confined to water as a solvent in choosing examples but the thermodynamic relationships apply generally and would provide the same basis for microscopic understanding of any solvent system were the same data available. Since we choose to confine arguments largely to water as a solvent, it will be useful to consider what has been learned about the liquid water structure to better appreciate what can be addressed by intuitively meaningful calculation.

In Chapter 7 we noted that there is short-range order in all liquids. There is substantial order in water. The most successful models for liquid water are those that assume a cluster structure involving a variable number of hydrogen bonds. These approaches with the number of adjustable parameters they contain are only indicative of the structural complexity of polar hydrogen bonded liquids. No single model can be taken as providing a definitive structure. Indeed it appears that there is no such structure in water. Computer simulation[2] in which a representative partition function expression for energy is minimized for a variety of structures will produce instantaneous configurations such as those shown in Figure 11.1. Many such configurations are energetically indistinguishable.

We cannot have a close picture of water structure beyond the general conclusion that structured regions having coordination number of about 4 (almost independent of temperature) extending over about an 8 Å distance from

[2]In the Monte Carlo simulation technique starting with a random configuration, molecules are moved about one at a time under rules that ensure that acceptable configurations are associated with probability proportional to $\exp(-u/RT)$ where u is the potential energy of the system.

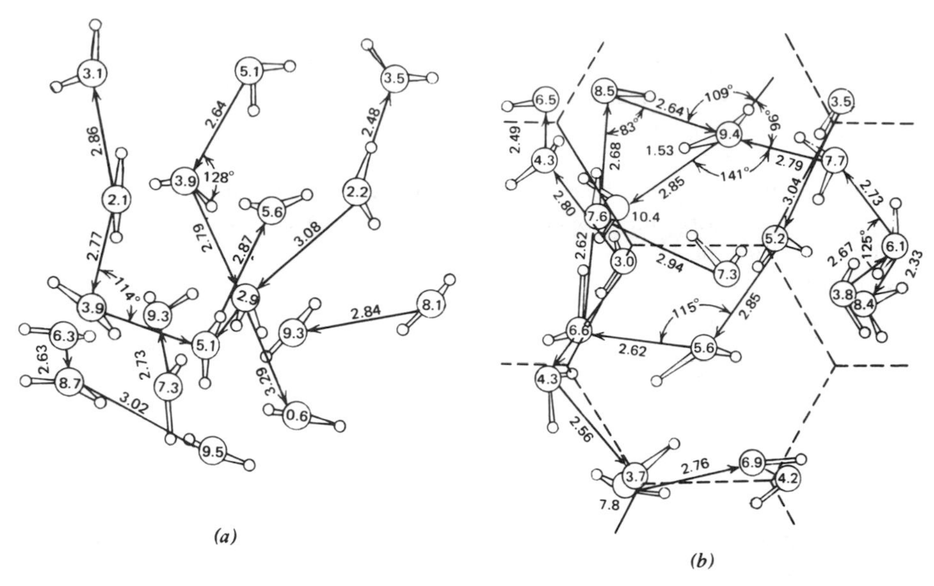

Figure 11.1. Water structures obtained by computer simulation. Taken from R. S. Berry, S. A. Rice, and J. Ross, *Physical Chemistry*, Wiley, New York, 1980. Reprinted with permission by John Wiley and Sons Inc.

a given molecule exist. This is consistent with loose tetrahedral coordination of water molecules through a network of hydrogen bonds in which oxygen centers are 2.7 to 3 Å apart. There is a spectrum of hydrogen bond energies and many structures that represent, in sum, permissible deformations. Such structured regions may consist of a few tens of molecules and appear to have lifetimes of (0) 10^{-11} sec (10^2 vibrations). Since hydrogen bond energies are only (0) 5 K cal $mole^{-1}$, local fluctuations in energy density provide for a constantly changing pattern of order. At any instant most of the molecules in the system are involved in a structure.

When any substance is dissolved in water the structural features of the water should be altered through mechanisms involving interactions in the vicinity of the solute ion or molecule. Computer simulation studies for inert gas molecules indicate that very substantial changes in order exist in the first hydration sphere even in these weakly interacting cases. We would anticipate much larger effects with ionic species in solution and for large solute molecules or anions that necessarily must impose large deformations in the solvent order. Extreme examples of the latter case are the clathrate structures in which a caged guest molecule (e.g., Xe or CH_4 among others) stabilize an expanded ice-like structure. If the solute is polar the complications should extend beyond cavity size but dipole interactions do not always have unique effects; $CHCl_3$ like CCl_4 forms clathrate compounds and the melting points are not very different. Extreme examples of ionic species having local influence on the solvent structure are found in the case of many transition metal ions that coordinate water molecules through interactions that are effectively chemical bonds.

Since it is quite difficult to construct physically meaningful expressions for the solvent alone, the incorporation of solute effects on solvent structure clearly cannot be addressed in our framework. Any solute behavior model that we can directly address will be confined to very dilute systems where the solvent can be treated as a continuum but, even in dilute solution, the long-range ion–ion interactions and complex ion–solute interactions preclude any simple modelistic approach. For example, we can construct a simple address to the enthalpy of ion solvation as follows. Consider that the enthalpy of ion hydration is the enthalpy change for the process

$$M^{+}(\text{gas}) + X^{-}(\text{gas}) \rightarrow M^{+}(\text{solv}) + X^{-}(\text{solv}).$$

From electrostatics we take the charge on a spherical conductor to be concentrated on the surface and the potential at the surface of a sphere in vacuum to be q/b where q is charge and b is the radius. We imagine a process in which the charge is increased from 0 to q so that the potential at any point in the process is $q\xi b$ where ξ is a charging parameter. The work required to increase the charge by $qd\xi$ is $q^2\xi d\xi/b$ so the work associated with the charging process is

$$w_{e,0} = \int_0^1 \frac{q^2}{b}\xi d\xi = \frac{1}{2}\frac{q^2}{b}. \tag{11.1}$$

If the sphere had been charged in a medium of dielectric constant D, the work would be

$$w_{e,D} = \frac{1}{2}\frac{q^2}{bD} \tag{11.2}$$

so that the transfer of a charged sphere from vacuum to a medium having dielectric constant D is

$$w_{e,\text{solv}} = -\frac{q^2}{2b}\left[1 - \frac{1}{D}\right] = \Delta\mu_{\text{solv}}. \tag{11.3}$$

We have set the electrical work associated with solvation equal to the change in chemical potential using the argument that change in Gibbs free energy is related to work other than mechanical work. The free energy change associated with the transfer of an ion collection from vacuum to a solution, which we will take to be infinitely dilute to minimize ion–ion interactions, is

$$\left(\sum^{i} (\Delta G_i)_{\text{solv}}\right)_{m\to 0} = \left(\sum^{i} N_i(\Delta\mu_i)_{\text{solv}}\right)_{m\to 0} = \left(\sum^{i} \frac{N_i q_i}{b_i}\left[1 - \frac{1}{D}\right]\right)_{m\to 0} \tag{11.4}$$

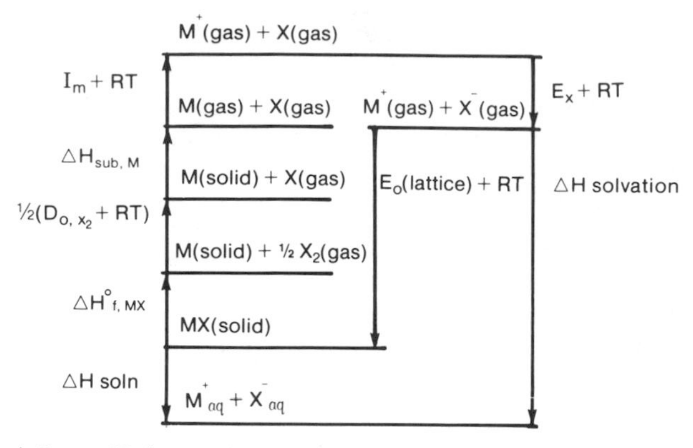

Figure 11.2. A Born–Haber cycle including enthalpies of solution and ion solvation.

and, via the Gibbs–Helmholz equation, the associated thermal quantity is

$$\left[\sum^{i}\left(\frac{\partial(\Delta G_i)_{solv}/T}{\partial T}\right)_P\right]_{m\to 0} = -\left(\sum^{i}\frac{(\Delta H_i)_{solv}}{T^2}\right)_{m\to 0} \tag{11.5}$$

where $\mathscr{q}_i$ is the individual ion charge $= \pm z_i \mathscr{q}_e$ with $\mathscr{q}_e$ being the magnitude of the unit (electron) charge, and b_i is the individual ion radius. In a general expression we would recognize that N molecules of a dissolved electrolyte $\mathscr{A}_{v_-}\mathscr{B}_{v_+}$ contain $(v_+ + v_-)N_i/N_A$ moles of solute ions and that the molar enthalpy of solvation is

$$(\Delta\tilde{H}_{solv})_{m\to 0} = \sum^{i}\frac{(v_+ + v_-)^2 z_i^2 \mathscr{q}_e^2}{b_i}\frac{N}{N_A}\left[1 + T^2\left(\frac{2\ln D}{\partial T}\right)_P\right]. \tag{11.6}$$

The enthalpy of the gas phase ion species to which the enthalpy of solvation relates is associated with the enthalpy of formation of the MX solid phase and to the conventional enthalpy of solution through a series of process steps (known as the Born–Haber cycle) as illustrated in Figure 11.2. The lattice energy is the major term in E_0 for the solid [Eq. (7.22)] and can be calculated quite precisely for ionic solids of known crystal structure.[3] All other terms in the cycle except the electron affinity term, E_x, are known with moderately high accuracy and E_x can be obtained from the sum over the cycle. Our simple calculation applied for example to NaF provides a calculated value of $-1214\,\text{kJ mole}^{-1}$ for the enthalpy of solvation while a best cyclic value is $-809\,\text{kJ mole}^{-1}$. Other cases compare even less favorably. The calculation is not without merit, however, in that we will continue to use and expand the notation.

Clearly, in the argument leading to Eq. (11.6) many considerations are

[3] See reference 3 (Berry et al., 1980), pp. 460 ff, 569 ff or reference 4 (Moelwyn-Hughes, 1961), p. 554 ff.

ignored. It is implicitly assumed that the energy levels of the gaseous ion were not altered by the change in environment. Also it is assumed that the solvent is a homogeneous dielectric substance and no account has been taken of the interaction between ions. Short of quantum calculations, which would require more awareness of the solvent structure than we can hope to have, address to the possibility of change in energy levels does not promise a useful improvement. In our framework we will have to be content with first-order characterization of the solvent by its dielectric constant only and characterization of the ion by its charge. We can then address the long-range ion–ion interactions assuming that the specific solvent interactions will be reflected as nonideal behavior features of electrolyte solutions even in those dilution domains where we would anticipate ideal behavior for nonelectrolyte solutions. The ideal format will be the expression for chemical potential. For any solution species i

$$\mu_i = \mu_i^\circ + kT \ln X_i + kT \ln \gamma_i \tag{11.7}$$

where γ_i is the activity coefficient ($\gamma_i X_i = a_i$). Following the argument that, if it were not for the ion charges the dissolved species would behave ideally at low concentrations, the electrical features of the ionic solution will reflect the term $kT \ln \gamma_i$. The problem is that of constructing a realistic treatment for the electrical work associated with charging an individual ion in the solution system. The Debye–Hückel theory, following and extending arguments introduced by Milner, is an approach to this problem.

The fundamental assumption is that the ion–ion interactions lead to each ion being surrounded by an ion atmosphere of opposite sign. The ion atmosphere is constantly fluctuating as a result of thermal motion but the probability of encountering an ion of opposite charge in the volume element surrounding a given ion is larger than the probability of encountering an ion of the same charge. Electrical neutrality requires that the net charge of the ion atmosphere will always be equal but opposite to that of the central (reference) ion. It is further assumed that the solvent can be treated as a uniform dielectric and that conventional electrostatic argument can be applied.

A conservative force is defined as the negative gradient of the potential energy.[4] The x-component of the force acting at a point where the potential energy is $\mathscr{V}(x, y, z)$ is

$$f_x = -\frac{\partial}{\partial x}[\mathscr{V}(x, y, z)].$$

The force components f_y, f_z are similarly expressed and the force vector at that point is

$$\mathbf{f} = -\nabla \mathscr{V}(x, y, z)$$

[4]See Appendix A.

where ∇ is the Laplacian operator representing $(\partial/\partial x)(\partial/\partial y)(\partial/\partial z)$. If the force arises from the interaction of two electric charges $\mathcal{q}_1$ and $\mathcal{q}_2$ separated by a distance r the potential energy is $\mathcal{V}(r) = \mathcal{q}_1\mathcal{q}_2/r$ and the force vector is

$$\mathbf{f} = -\mathcal{q}_1\mathcal{q}_2\nabla\left(\frac{1}{r}\right) = \mathcal{q}_1\mathcal{q}_2\frac{\mathbf{r}}{r^3}$$

with $r = [(x - x')^2 + (y - y')^2 + (z - z')^2]^{1/2}$ and $\mathbf{r}$ pointing in the direction of the line joining the two charges placed at x, y, z and x', y', z'.

The electric potential (φ) at any point is defined as the potential energy that a unit positive charge would have if placed at that point in the presence of other charges $\mathcal{q}_i$ located at distances r_i from that point

$$\mathcal{V}(r) = \varphi = \sum_i \frac{\mathcal{q}_i}{r_i}. \tag{11.8}$$

The field is defined as the vector sum of all forces arising from the charges other than the reference charge

$$\mathbf{F} = \sum -\mathcal{q}_i\nabla\left(\frac{1}{r}\right) = -\nabla\cdot\varphi. \tag{11.9}$$

If there is a sufficient density of charges making up the field the sum of charges can be represented by an integral. Where ρ represents charge denstity

$$\mathcal{q} = \iiint \rho\, dx\, dy\, dz.$$

Taken over the whole of a charge balanced system consisting of many charges the integral would be zero. The field from Eq. (11.9) is

$$\mathbf{F} = \iiint \rho'\nabla\left(\frac{1}{r}\right) dx'dy'dz'$$

and

$$\nabla\cdot\mathbf{F} = \nabla^2\varphi = -\iiint \rho'\nabla^2\left(\frac{1}{r}\right) dx'dy'dz'. \tag{11.10}$$

If the reference charge position at x, y, z is at the origin of a spherically symmetrical charge distribution the value of the integral is $4\pi\rho$ and

$$\nabla^2\varphi = -4\pi\rho. \tag{11.11}$$

This is the Poisson equation that, as outlined, applies to a continuum of charge in vacuo. If the charge is distributed throughout a medium of dielectric constant D, then

$$\nabla^2\varphi = -\frac{4\pi\rho}{D}. \tag{11.12}$$

As applied to an electrolyte system we would expect the charge distribution to be uniform throughout the system if there were no interactions between ions, $\rho = \sum q_K/\mathrm{V}$ where V is the solution volume. With interactions the charge density is unbalanced in the vicinity of any selected ion in the system. In terms of ion pair interactions we assume that the charge distribution can be represented through a Boltzmann factor $\exp(-\mathrm{E}_{\mathscr{A}\mathscr{B}}/k\mathrm{T})$ where $\mathrm{E}_{\mathscr{A}\mathscr{B}}$ is the interaction energy between two ions separated by distance r. The charge density at distance r from ion $\mathscr{A}$ due to the presence of all other ions $\mathscr{B}$ then is

$$\rho_r = \sum_{\mathscr{B}} \frac{q_{\mathscr{B}}}{\mathrm{V}} e^{-q_{\mathscr{B}}\varphi_r/k\mathrm{T}}$$

for pair wise interaction energies $\mathrm{E}_{\mathscr{A}\mathscr{B}} = q_{\mathscr{B}}\varphi_r$. The interaction energy is the work required to bring ion $\mathscr{A}$ from infinity to its central position in the presence of the ion atmosphere. Applying the ρ_r distribution function in the Poisson equation framework, we have

$$\nabla^2\varphi = -\frac{4\pi}{D\mathrm{V}}\sum^{\mathrm{i}} \mathrm{N_i} q_{\mathrm{i}} e^{-q_{\mathrm{i}}\varphi_r/k\mathrm{T}} \tag{11.13}$$

where $\mathrm{N_i}$ is the number of ions having charge q_{i}. If the potential energy ($q_1\varphi_r$) due to ion interactions is small compared to $k\mathrm{T}$ then the exponential term becomes $1 - q_{\mathrm{i}}\varphi_r/k\mathrm{T}$ and

$$\nabla^2\varphi = -\frac{4\pi}{D\mathrm{V}}\sum^{\mathrm{i}} \mathrm{N_i} q_{\mathrm{i}}(1 - q_{\mathrm{i}}\varphi_r/k\mathrm{T}) \tag{11.14}$$

but $\sum \mathrm{N_i} q_{\mathrm{i}} = 0$ so that

$$\nabla^2\varphi = \frac{4\pi\varphi_r}{D\mathrm{V}k\mathrm{T}}\sum^{\mathrm{i}} \mathrm{N_i} q_{\mathrm{i}}^2 = \kappa^2\varphi_r \tag{11.15}$$

with constants collected as

$$\kappa^2 = \frac{4\pi}{\mathrm{DV}k\mathrm{T}}\sum \mathrm{N_i} q_{\mathrm{i}}^2. \tag{11.16}$$

A solution for Eq. (11.15) is

$$\varphi_r = \frac{\mathrm{A}}{r} e^{-\kappa r} + \frac{\mathrm{B}}{r} e^{\kappa r} \tag{11.17}$$

with the second term vanishing (B = 0) to meet the condition that the potential be finite as r goes to infinity. It must also be assumed that there is some distance (b) of closest approach between the reference ion ($\mathscr{A}$) and any other ion ($\mathscr{B}$). Within the spherical volume associated with ion $\mathscr{A}$ the charge is $q_{\mathscr{A}}$ and the integrated charge in the remainder of the system is $-q_{\mathscr{A}}$ so

$$\int_b^\infty \rho_r 4\pi r^2 dr = -q_{\mathscr{A}}.$$

From Eqs. (11.12) and (11.15), $\rho_r = \kappa^2 D\varphi_r/4\pi$ and

$$\rho_r = \frac{A}{r} e^{-\kappa r}\left(\frac{\kappa^2 D}{4\pi}\right) \tag{11.18a}$$

so

$$q_{\mathscr{A}} = D\kappa^2 \mathrm{A} \int_b^\infty r e^{-\kappa r} dr. \tag{11.18b}$$

Integration by parts gives

$$\mathrm{A} = \frac{q_{\mathscr{A}} e^{\kappa b}}{D(\kappa b + 1)} \tag{11.19}$$

so that

$$\varphi_r = \frac{q_{\mathscr{A}} e^{\kappa b} e^{-\kappa r}}{Dr(\kappa b + 1)} \tag{11.20}$$

and the potential at the surface of the sphere ($r = b$) is

$$\varphi_{\mathrm{b}} = \frac{q_{\mathscr{A}}}{Db(\kappa b + 1)}. \tag{11.21}$$

This potential derives from the charge on the central ion and the charge in the remaining ion atmosphere. We are required to subtract the contribution of the central ion since it is only the potential of the ion atmosphere that contributes to the work term that represents charging the central ion and establishing the interaction energy term. We subtract $q_{\mathscr{A}}/Db$ from Eq. (11.21) to obtain

$$\varphi_{\mathscr{A}} = \frac{q_{\mathscr{A}}}{Db(1+\kappa b)} - \frac{q_{\mathscr{A}}}{Db} = -\frac{q_{\mathscr{A}}\kappa}{D(1+\kappa b)} \tag{11.22}$$

which is the potential at the surface of the charge free cavity of radius b in the presence of the ion atmosphere.

The free energy change associated with the process of charging the central ion in the presence of the ion atmosphere is the change in chemical potential associated with the electrical work process of increasing the central ion charge from zero to q_{A} with the charge at any point in the process being $\xi q_{\mathscr{A}}$

$$\mu_{\mathscr{A},\mathrm{el}} = \int_0^1 \varphi_{\mathscr{A}} \xi q_{\mathscr{A}} d\xi = -\frac{q_{\mathscr{A}}^2 \kappa (1+\kappa b)^{-1}}{D} \int_0^1 \xi d\xi = -\frac{q_{\mathscr{A}}^2 \kappa}{2D(1+\kappa b)}. \tag{11.23}$$

This is the component of the chemical potential which relates to ion interactions in the solution system. If it were not for such ion interactions the chemical potential of the ion in solution would be just that of an ideal mixture component

$$\mu_{\mathrm{i,ideal}} = \mu_{\mathrm{i}}^{\circ} + \mathscr{k}\mathrm{T}\ln \mathrm{X}_{\mathrm{i}}. \tag{11.24}$$

Ion interactions will produce the additional contribution from Eq. (1.23)

$$\mu_{\mathrm{i}} = \mu_{\mathrm{i}}^{\circ} + \mathscr{k}\mathrm{T}\ln \mathrm{X}_{i} + \mu_{\mathrm{i,el}}. \tag{11.25}$$

Since we would normally represent the nonideal behavior in terms of an activity coefficient

$$\mu_{\mathrm{i}} = \mu_{\mathrm{i}}^{\circ} + \mathscr{k}\mathrm{T}\ln a_{\mathrm{i}} = \mu_{\mathrm{i}}^{\circ} + \mathscr{k}\mathrm{T}\ln \mathrm{m}_{\mathrm{i}} + \mathscr{k}\mathrm{T}\ln \gamma_{\mathrm{i}}. \tag{11.26}$$

Equation (11.23) will produce the activity coefficient following some address to compatible units.

We note that μ_{el}, through κ, contains the term $\sum \mathrm{N}_{\mathrm{i}} q_{\mathrm{i}}^2/2\mathrm{V}$ representing a summation over all charges in the system. This term we replace with $(\mathrm{N}_0/1000)(\frac{1}{2}\sum \mathrm{c}_{\mathrm{i}} z_{\mathrm{i}}^2)$ where c_{i} is molar concentration $[\mathrm{N}_{\mathrm{i}}/\mathrm{N}_0\mathrm{V}(\text{liters})]$ and z_{i} the individual ion charges. We refer to $\frac{1}{2}\sum \mathrm{c}_{\mathrm{i}} z_{\mathrm{i}}^2$ as the ionic strength (**I**) of the solution.[5] The molar concentration unit is not compatible with the Eq. (11.25) expression in mole fraction where we have called the γ_{i} rational activity coefficients. However, we have noted [Eqs. (10.44) through (10.47)] that there is little difference between the various activity coefficients $(\gamma_x, \gamma_{\mathrm{m}}, \gamma_{\mathrm{i}})$ in dilute solutions and that the expressions for chemical potential for the dissolved species

[5] The ionic strength as defined by $\frac{1}{2}\sum \mathrm{c}_{\mathrm{i}} z_{\mathrm{i}}^2$ has historical significance as a measure of effective charge concentration. Note that for 1 molar NaCl, I = 1 molar $\mathrm{LaCl_3}$, I = 6.

require only minor redefinition of the standard chemical potential. For reference

$$\mu_{2,\mathrm{soln,m}} = \mu^{\circ}_{2,\mathrm{soln,m}} + kT \ln \gamma_{2,\mathrm{m}} + kT \ln \mathrm{m} \tag{10.44}$$

$$\mu_{2,\mathrm{soln,c}} = \mu^{\circ}_{2,\mathrm{soln,c}} + kT \ln \gamma_{2,\mathrm{c}} + kT \ln \mathrm{c}$$

where

$$\mu^{\circ}_{2,\mathrm{soln,m}} = \mu^{\circ}_{2,\mathrm{soln},x} + kT \ln \frac{M_1}{1000}. \tag{10.45}$$

and

$$\mu^{\circ}_{2,\mathrm{soln,c}} = \mu^{\circ}_{2,\mathrm{soln},x} + kT \ln \frac{M_1}{1000\rho_1}.$$

As indicated by these expressions, it will always be most reasonable to define $\mu^{\circ}_{\mathrm{soln}}$ as the chemical potential of the solution behaving ideally ($\gamma = 1$); in these cases at 1 molar or 1 molal concentration so that $\mu^{\circ}_{\mathrm{soln}}$ refers to a fictitious state. This will not be of concern when changes in chemical potential are the feature of interest.

Solute concentration terms normally refer to the charge neutral ion collection, i.e., we cannot add only positive or only negative ions to a solution but Eq. (11.22) produces a chemical potential for single ion species. In these terms

$$\mu_{\mathrm{solute}} = \mu_2 = \nu_+ \mu_+ + \nu_- \mu_-$$

where ν_+ and ν_- relate to the number ratio of positive and negative ions represented in the charge neutral collection. For a single dissolved component (2) in a solution system where we use molal concentration units

$$\mu_2 = \mu^{\circ}_{2(\mathrm{soln})} + \nu_+ kT \ln \gamma_+ \mathrm{m}_+ + \nu_- kT \ln \gamma_- \mathrm{m}_-$$

$$= \mu^{\circ}_{2(\mathrm{soln})} + kT \ln \gamma_+^{\nu_+} \gamma_-^{\nu_-} + kT \ln \mathrm{m}_+^{\nu_+} \mathrm{m}_-^{\nu_-}$$

$$= \mu^{\circ}_{2(\mathrm{soln})} + kT \ln \gamma_{\pm}^{\nu_+ + \nu_-} + kT \ln \mathrm{m}_{\pm}^{\nu_+ + \nu_-} \tag{11.27}$$

calling $\gamma_{\pm}$ the mean activity coefficient and $\mathrm{m}_{\pm}$ the mean molality, each defined in terms of Eq. (11.27)

$$\gamma_{\pm} = (\gamma_+^{\nu_+} \gamma_-^{\nu_-})^{1/(\nu_+ + \nu_-)}$$

$$\mathrm{m}_{\pm} = (\mathrm{m}_+^{\nu_+} \mathrm{m}_-^{\nu_-})^{1/(\nu_+ + \nu_-)}. \tag{11.28}$$

In terms of individual ion charges Eq. (11.23) would be

$$\mu_{\mathrm{i,el}} = -\frac{q_{\mathrm{i}}^2 \kappa}{2D(1 + \kappa b)} = -\frac{z_+^2 q_{\mathrm{e}}^2 \kappa}{2D(1 + \kappa d)} \quad \text{or} \quad -\frac{z_-^2 q_{\mathrm{e}}^2 \kappa}{2D(1 + \kappa b)} \tag{11.29}$$

and $\gamma_\pm$ would be, for

$$\mu_{+,\mathrm{el}} = \nu_+ \mathscr{k}\mathrm{T}\ln\gamma_+ = -(z_-\mathscr{k}\mathrm{T}\ln\gamma_+), \qquad \mu_{-,\mathrm{el}} = \nu_-\mathscr{k}\mathrm{T}\ln\gamma_- = z_+\mathscr{k}\mathrm{T}\ln\gamma_-$$

$$\ln\gamma_\pm = -\left(\frac{z_-}{z_+ - z_-}\right)\ln\gamma_+ + \left(\frac{z_+}{z_+ - z_-}\right)\ln\gamma_-$$

$$= \frac{z_-(z_+)^2 - z_+(z_-)^2}{(z_+ - z_-)}\left[\frac{q_e^2\kappa}{2D\mathscr{k}\mathrm{T}(1+\kappa b)}\right]$$

$$\ln\gamma_\pm = -z_+(-z_-)\left[\frac{q_e^2\kappa}{2D\mathscr{k}\mathrm{T}(1+\kappa b)}\right]. \tag{11.30}$$

In operational units with b in Angstrom units[6]

$$\ln\gamma_\pm = -1.172|z_+z_-|\frac{\mathrm{I}^{1/2}}{[1+(b/3)\mathrm{I}^{1/2}]} \tag{11.31a}$$

or, in high dilution, limiting form where our approximations would be most valid

$$\ln\gamma_\pm = -1.172|z_+z_-|\mathrm{I}^{1/2}. \tag{11.31b}$$

Equation (11.31b), known as the Debye–Hückel limiting law, has been studied in great depth both theoretically and experimentally. Typical behavior of electrolyte systems is illustrated in Figure 11.3. The form of activity coefficient variation with concentration is that obtained experimentally by techniques that are discussed later in this chapter. Generally the (linear) limiting law is not a good description of system behavior above about 10^{-3} molal.

Actually we should not expect the expression for the potential at the central ion due to the ion atmosphere [Eq. (11.19)] to be valid at any but the lowest level of concentration. Since the limiting form expression $\varphi_{\mathscr{A}} = -q_{\mathscr{A}}\kappa/D$ is just that for the potential arising from an equal but opposite charge placed at a distance $1/\kappa^{-1}$ from the central ion, we can regard the ion atmosphere as the equivalent charge distributed over a thin spherical shell at a distance $1/\kappa$ from the central ion; κ^{-1} is, then, an effective radius of the ion atmosphere and from Eq. (11.16) and $\mathrm{I} = \frac{1}{2}\sum c_i z_i^2$ so that, for an aqueous 1:1 electrolyte solution at 298 K

$$\kappa = \left(\frac{4\pi}{VD\mathscr{k}\mathrm{T}}\sum \mathrm{N}_i q_i^2\right)^{1/2} = \left(\frac{8\pi c q_e^2 \mathrm{I}}{D\mathscr{k}\mathrm{T}\,1000}\right)^{1/2} = \frac{50.29}{(D\mathrm{T})^{1/2}}\mathrm{I}^{1/2}\,\text{Å}^{-1}$$

so that $1/\kappa \cong 100$ Å for $c \cong 10^{-3}$ molar or about 30 Å for $c \cong 10^{-2}$ molar. At concentration levels much above these values we should not expect that the

[6] Most experimental data is in the form $\log\gamma_\pm$ for which the constant is 0.509

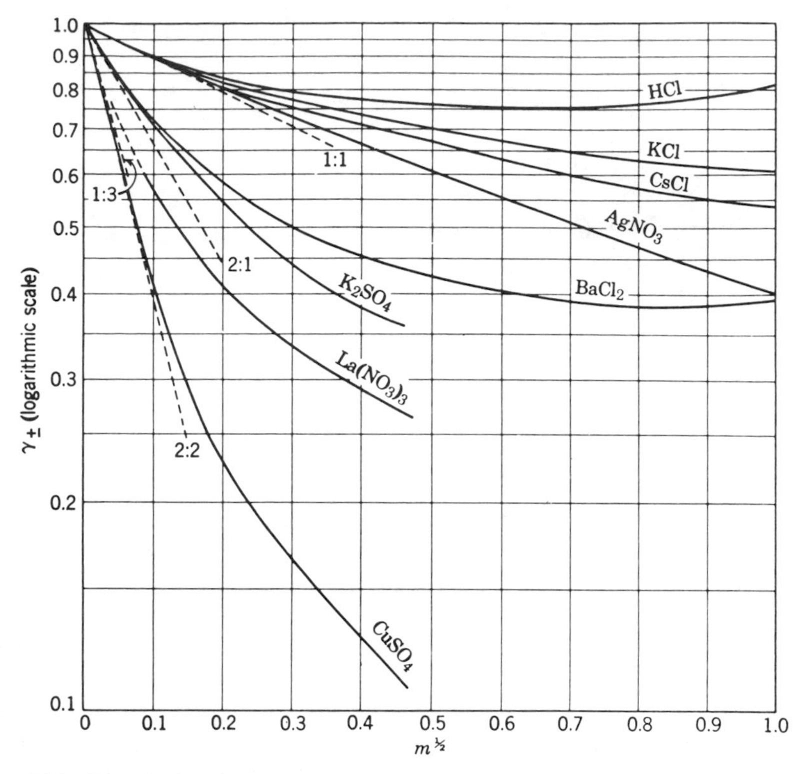

Figure 11.3. Experimental activity coefficients for various electrolytes compared with the Debye–Hückle limiting law (dashed lines). Taken from M. A. Paul, *Principles of Thermodynamics*, McGraw-Hill, New York, 1951. Reproduced with permission of McGraw Hill Inc.

concept of a homogeneous dielectric and diffuse ion atmosphere would be valid. We should not, however, abandon the model without making reasonable attempts to remove some approximations that were introduced for arithmetic convenience.

The linear form that we obtained as the limiting law derived from the neglect of them term κb in comparison with unity. Clearly this is less reasonable as concentration increases. We note however that Eq. (11.31) in its full form, or more precisely in the form

$$\ln \gamma_{\pm} = -\frac{\mathrm{A}(z_{+}(-z_{-})\varphi_e^2 \mathrm{I}^{1/2}}{1 + b\mathrm{B}\mathrm{I}^{1/2}} \tag{11.32}$$

where constants are collected as A and B, requires that $\ln \gamma_{\pm}$ must approach a limiting form

$$\ln \gamma_{\pm} = -\frac{A(z_{+})(-z_{-})q_e^2}{bB}$$

at high concentrations assuming that the distance of closest approach (b) is constant. Experimental values of $\gamma_{\pm}$ (or ($\ln \gamma_{\pm}$) actually go through a minimum at high electrolyte concentrations. Additional terms involving κ^2 and higher orders would result from a more reasonable expansion of the exponential in Eq. (11.13). Although such terms do not in themselves improve the fit greatly, the inclusion of a term involving ionic strength (proportional to κ^2) does provide a framework for quite good empirical extension.

$$\ln \gamma_{\pm} = -\frac{A(z_{+})(-z_{-})q_e^2 I^{1/2}}{1 + bBI^{1/2}} + CI. \tag{11.33}$$

Using experimental values of $\ln \gamma_{\pm}$ we can write for Eqn. (11.32)

$$\frac{-A(z_{+})(-z_{-})q_e^2 I^{1/2}}{\log \gamma_{\pm}} = 1 + bBI^{1/2}$$

so that, at low concentrations, a plot of the left-hand side of the expression vs $I^{1/2}$ produces the term b/A from the straight line slope; the constant C is resolved by conventional curve fit techniques.

Other improvements to the model are obvious but also difficult to address in anything other than an empirical framework. For example, we would imagine that even at moderate concentrations ion pairing could be significant. In a solution where the concentration of the ith ion species is n_i molecules cm^{-3} the number of ions of type i in a spherical shell of volume $4\pi r^2 dr$ at a distance r from a central ion j is

$$dN_i = n_i 4\pi r^2 e^{-z_i z_j q_e^2/Dr\ell T} dr.$$

The radial distribution of ions is dN_i/dr. When z_i and z_j are of opposite sign dN_i/dr is positive and will pass through a minimum as r increases. The distance r_m corresponding to the minimum is

$$r_m = \frac{z_i z_j q_e^2}{2D\ell T}.$$

We assume that at this distance of separation where the electrostatic interaction energy is equal to $2\ell T$, an ion pair will be formed from oppositely charged ions. The ion pair will have at least a reduced charge and, if neutral, will be removed from the Debye distribution altogether. The equilibrium is represented as

$$\mathscr{A}^{z+} + \mathscr{B}^{z-} = \mathscr{A}\mathscr{B}^{(z_+ + z_-)}$$

and the equilibrium constant as

$$K_{\mathscr{B}_j} = 4\Pi \int_b^{r_m} e^{-u_{\mathscr{A}\mathscr{B}}/kT} r^2 dr \qquad (11.34)$$

for a hard sphere distance of closest approach b if we assume that all internal degrees of motion are unchanged in the ion pair as compared to the original ions and $u_{\mathscr{A}\mathscr{B}}$ is the ion pair energy of formation. Setting $u = -(z_+z_-)q_e^2/DrkT$ and $dr = -[(z_+z_-)q^2/y^2DkT]dy$

$$K_{\mathscr{B}_j} = 4\pi\left[\frac{(z_+z_-)q_e^2}{DkT}\right]^3 \int_2^{y_b} e^y y^{-4} dy = 4\pi\left[\frac{(z_+z_-)q_e^2}{DkT}\right]^3 Q(y_b). \qquad (11.35)$$

Since y is a function of the degree of association, the evaluation of the integral can be accomplished only in the framework of successive approximation. Various tabulations for $Q(y_b)$ are available.[7] Typical values are $y_b = 3$, $Q = 0.33$; $y_b = b$, $Q = 1.04$. Following the argument that the intent of the ion pair treatment is to extend the Debye–Hückel theory we can utilize that result for any charged particle (single ion or charged ion pair) so

$$\frac{N_{\mathscr{A}\mathscr{B}}}{N_{\mathscr{A}}N_{\mathscr{B}}}\frac{\gamma_{\mathscr{A}}\gamma_{\mathscr{B}}}{\gamma_{\mathscr{A}\mathscr{B}}} = K_{\mathscr{B}_i}$$

and from Eq. (11.39)

$$\frac{N_{\mathscr{A}\mathscr{B}}}{N_{\mathscr{A}}N_{\mathscr{B}}} = K_{\mathscr{B}_j}\frac{\gamma_{\mathscr{A}\mathscr{B}}}{\gamma_{\mathscr{A}}\gamma_{\mathscr{B}}} = e^{(z_+z_-)\kappa'(1+\kappa r)} \qquad (11.36)$$

with

$$(\kappa')^2 = \frac{4\pi q_e^2}{2kT}[N_{\mathscr{A}}z_+^2 + N_{\mathscr{B}}z_-^2 + N_{\mathscr{A}\mathscr{B}}(z_+ + z_-)^2].$$

The $N_{\mathscr{A}}$, $N_{\mathscr{B}}$, $N_{\mathscr{A}\mathscr{B}}$ must satisfy the material balance

$$N'_{\mathscr{A}} = \nu_+ N = N_{\mathscr{A}} + N_{\mathscr{A}\mathscr{B}} \qquad (11.37)$$

$$N'_{\mathscr{B}} = \nu_- N = N_{\mathscr{B}} + N_{\mathscr{A}\mathscr{B}} \qquad (11.38)$$

[7] See reference 6 (Davidson, 1962), p. 508ff for original reference and further detail.

where N is the formal molecular concentration of $\mathscr{A}_{v-}\mathscr{B}_{v+}$ in solution. Simultaneous solution of Eqs. (11.36), (11.37), and (11.38) provides the solution composition.

Further considerations involving ion pairing and solvation start with an ion-molecule[8] that dissociates into v ions each of which associates with a number of solvent molecules. If the solute molality is $\mathfrak{m}$ and the solvent is water there are $55.51 - n\mathfrak{m}$ moles of free water if n moles of water are associated with each mole of solute. The true molality of the solution is then $\mathfrak{m}'$ where

$$\mathfrak{m}' = \frac{\mathfrak{m}}{(55.51 - n\mathfrak{m})(0.018)} = \frac{\mathfrak{m}(55.51)}{(55.51 - n\mathfrak{m})} = \frac{\mathfrak{m}}{1 - 0.018n\mathfrak{m}}. \quad (11.39)$$

In any solution, identifying the solvent as component 1, the chemical potential of the solvent is

$$\mu_1 = \mu_1^0 + \not{k}\mathrm{T} \ln a_1 \quad (11.40)$$

and of the solute

$$\mu_2 = \mu_2^0 + \not{k}\mathrm{T} \ln \mathfrak{m}_2 + \not{k}\mathrm{T} \ln \gamma_2. \quad (11.41)$$

Recalling the Gibbs–Duhem relationship [Eq. (8.64)] rephrased as $\sum n_i d\mu_i = 0$ in molar terms, we write for 1000 g solvent

$$n_1 d\mu_1 = 55.51\not{k}\mathrm{T} d \ln a_1$$

and for solutes of molality $\mathfrak{m}$ and $\mathfrak{m}'$

$$v_{\mathfrak{m}} d\mu_2 = v_{\mathfrak{m}}\not{k}\mathrm{T} d \ln \mathfrak{m} + v_{\mathfrak{m}}\not{k}\mathrm{T} d \ln \gamma_2$$

$$v_{\mathfrak{m}'} d\mu_2 = v_{\mathfrak{m}'}\not{k}\mathrm{T} d \ln \mathfrak{m}' + v_{\mathfrak{m}'}\not{k}\mathrm{T} d \ln \gamma_2'$$

so that

$$\frac{55.51}{v_{\mathfrak{m}}} d \ln a_1 = -(d \ln \mathfrak{m} + d \ln \gamma_2) \quad (11.42)$$

$$\frac{55.51}{v_{m'}} d \ln a_1 = -d \ln \mathfrak{m}' + d \ln \gamma_2'. \quad (11.43)$$

From Eq. (11.39)

$$d \ln \mathfrak{m}' = d \ln \mathfrak{m} - d \ln(1 - 0.018n\mathfrak{m}) \quad (11.44)$$

[8] See reference 4 (Moelwyn–Hughes, 1961), p. 914ff for original references and further detail.

and combining Eqs. (11.40), (11.41), (11.42) we obtain

$$d \ln \gamma_2 = d \ln \gamma_2' = \frac{n}{\nu} d \ln a_1 - d \ln (1 - 0.018 n\mathrm{m}).$$

Assuming that we can integrate both sides of this expression between the molality limits of 0 and m and that the activity coefficient γ given by the Debye–Hückel expression is the mean rational activity coefficient $\gamma = \gamma_2'(1 - 0.018\nu_{\mathrm{m}})$ we obtain

$$\ln \gamma_2 = \ln \gamma - \ln[1 - 0.018(n - \nu)\mathrm{m}] - \frac{n}{\nu} \ln a_1 \tag{11.45}$$

with $\ln \gamma$ represented by Eq. (11.31b). This expression, with the second term sign depending on the hydration number, provides a framework for fitting observed behavior. Hydration numbers can be estimated from various transport data.

Although it is clear from all of the above argument that the complexities that we anticipated for ion–solvent interactions (disruption of solvent structure, ion solvation, the contribution of van der Waals interactions to thermodynamic properties, etc.) are poorly addressed or ignored even in the extended theory, it is not clear that address to any single such feature will greatly improve the model. Indeed, having introduced corrections for ion pairing and ion solvation that rely on curve-fit to obtain the necessary parameters, it becomes increasingly clear that several different, reasonable modelistic extensions can be supported by this approach and different combinations of these would serve equally well so that appeal to experimental data will not selectively identify a model. For our purposes it is most useful to confine our attention to the extended applications of the simple Debye model in limiting form to the variety of thermodynamic features in the dilute region where the theory appears to apply. The theory can then serve as the basis for our concept of ideal behavior when we establish a framework that is consistent with the thermodynamic structure that we have developed to this point. Also, for electrolyte systems where electrical features dominate solution properties, there is a vast data bank of electrical measurements that provides operational extension when theory and experiment are incorporated into a common framework.

In our considerations of nonelectrolyte systems we found that, in addition to free energies of mixing, enthalpies and entropies of mixing together with volume changes on mixing were identifiable features of ideal mixtures and their deviations from prescribed values, features of nonideality. With some adaptation to the unique nonideality of electrolyte solutions and to the form and source of available data, these same thermodynamic solution properties will be useful measures of nonideality in electrolyte systems. Since electrolyte solutions will always be dilute in terms of the broad range of system compositions we encountered with nonelectrolyte mixtures, we will always use expressions for the chemical potential of the form of Eq. (11.26)

$$\mu_i = \mu_i^\circ + kT \ln a_i = \mu_i^\circ + \nu_i kT \ln \mathrm{m}_i + \nu_i kT \ln \gamma_i. \tag{10.46}$$

When we associate, as in Eq. (11.26), the term $v_i \not{k}T \ln \gamma_i$ with $\mu_{i,el}$, we require that either term be zero for the ideal solution and $\mu_{i,el}$ will be zero only for the charged ion in an infinitely dilute solution where each ion behaves as if it were isolated. Deviation from ideality is then regarded in the Henry's Law sense as deviation from the requirements of the linear Debye Limiting Law. Since the ideal solution with respect to both electrical and nonelectrical features is the infinitely dilute solution, deviations from ideality at moderate concentrations will be best approached through dilution studies in which the solution of molal concentation $\mathfrak{m}$ is diluted with a large amount of solvent and some observable related to nonideality measured.

The measured volume changes in constructing or diluting electrolyte solutions would be addressed in the Debye framework as follows. For the solute species (2) we write

$$\begin{aligned} \mu_2 &= \mu_2^\circ + \not{k}T \ln a_2 \\ &= \mu_2^\circ + v\not{k}T \ln \mathfrak{m}_\pm + v\not{k}T \ln \gamma_\pm \end{aligned} \qquad (10.47)$$

for

$$v = v_+ + v_-$$

and for

$$\gamma_\pm = \gamma_{el}\gamma_{NE}$$

where γ_{NE} are the nonelectrolyte contributions to the activity coefficient. These effects would be evident at infinite dilution where the electrical effects would be negligible. Since $(\partial\mu/\partial P)_{T,N} = \bar{V}$ where $\bar{V}$ is the partial molecular volume we can write

$$\left(\frac{\partial \mu_2}{\partial P}\right)_T = v\not{k}T\left[\left(\frac{\partial \ln \gamma_{el}}{\partial P}\right)_T + \left(\frac{\partial \ln \gamma_{NE}}{\partial P}\right)_T\right] = \bar{V}_2$$

since μ_2° is a function of T only. In the dilution limit $\mathfrak{m} \to 0$, $\ln \gamma_{el}$ will vanish and

$$v\not{k}T\left(\frac{\partial \ln \gamma_{NE}}{\partial P}\right)_T = \bar{V}_2^0$$

where $\bar{V}_2^0$ is the partial molecular volume of the solute at infinite dilution. Then in the region of finite concentration where the Debye–Hückel theory is valid

$$\left(\frac{\partial \mu_{2,el}}{\partial P}\right)_T = \bar{V}_2 - \bar{V}_2^0 = v\not{k}T\left(\frac{\partial \ln \gamma_{el}}{\partial P}\right)_T. \qquad (11.48)$$

From Eq. (11.29) we have

$$\mu_{i,el} = -\frac{z_i^2 \mathscr{q}_e^2 \kappa}{2D} \qquad (11.49)$$

and $\mu_{2,\text{el}} = \sum \mu_{i,\text{el}}$ so that

$$\mu_{2,\text{el}} = \frac{\sum \nu_i z_i^2 q_e^2}{2}\left[\frac{4\pi}{kT}\sum N_i z_i^2 q_e^2\right]^{1/2} D^{-3/2} V^{-1/2}. \tag{11.50}$$

Carrying out the differentiation required by Eq. (11.48), with $\bar{V}_2$ and $\bar{V}_2^0$ representing partial molar volumes,

$$\bar{V}_2 - \bar{V}_2^0 = \tfrac{3}{2} A D^{-3/2} V^{-1/2}\left[\left(\frac{\partial \ln D}{\partial P}\right)_T + \frac{1}{3}\left(\frac{\partial \ln V}{\partial P}\right)_T\right] \tag{11.51}$$

where A is the collection of pressure independent terms. Making the substitution $\sum (N_i/V)^{1/2} = (N_A c/1000)^{1/2}$ in the term $AV^{1/2}$ we obtain a final form relating $\bar{V}_2 - \bar{V}_2^0$ to concentration

$$\bar{V}_2 - \bar{V}_2^0 = \frac{3}{2}\left(\frac{8\pi N_A}{1000kT}\right)^{1/2}\left[\frac{q_e^2}{D}\sum(\nu_i z_i^2)\right]^{3/2}\left[\left(\frac{\partial \ln D}{\partial P}\right)_T + \frac{1}{3}\left(\frac{\partial \ln \bar{V}}{\partial P}\right)_T\right]\sqrt{c}$$

$$= \frac{3.70 \times 10^{14}}{D^{3/2}T^{1/2}}\left(\sum \nu_i z_i^2\right)^{3/2}\left[\left(\frac{\partial \ln D}{\partial P}\right)_T + \frac{1}{3}\left(\frac{\partial \ln \bar{V}}{\partial P}\right)_P\right]\sqrt{c}. \tag{11.52}$$

Where $\sqrt{c} \cong \sqrt{m}$ and for aqueous solutions, taking the compressibility of the solution to be that of water for which $(\partial \ln V/\partial P)_T = (1/V)(\partial V/\partial P)_T = \alpha_{T,H_2D}$ $45 \times 10^{-12}\,\text{dyne}^{-1}\,\text{cm}^2$ and $(\partial D_{H_2O}/\partial P)/D_{H_2O} = 47 \times 10^{-12}\,\text{dyne cm}^{-1}$

$$\bar{V}_2 = \bar{V}_2^0 + 0.99\left(\sum \nu_i z_i^2\right)^{3/2}\sqrt{m}\,\text{cm}^3\,\text{mol}^{-1}. \tag{11.53}$$

The resolution of volume changes in these systems is most appropriately accomplished by introducing the apparent partial molar property. We define the apparent partial molar volume[9] ϕ_V from

$$V = n_1\bar{V}_1 + n_2\bar{V}_2$$

as

$$\phi_V = \frac{V - n_1\tilde{V}_1^0}{n_2} \tag{11.54}$$

where $\tilde{V}_1^0$ is the molar volume of pure solvent. Then

$$\bar{V}_2 = \left(\frac{\partial V}{\partial n_2}\right)_{T,P,n_1} = n_2\frac{\partial \phi_V}{\partial n_2} + \phi_V$$

[9] This definition of an apparent partial molal property is equivalent to $n_2\phi_v = n_1(\bar{v}_1 - \tilde{v}_1^0) + n_2 v_2$ which as $m \to 0$ and $\bar{v}_1 - \tilde{v}_1^0 \to 0$ reduces to $\phi_v = \bar{v}_2$.

or, with n_1 constant and n_2 equivalent to molality

$$\bar{V}_2 = \mathfrak{m}\left(\frac{\partial \phi_V}{\partial \mathfrak{m}}\right)_{T,P} + \phi_V = \frac{d\mathfrak{m}\phi_V}{d\mathfrak{m}}$$

$$\phi_V - \phi_V^0 = \frac{1}{\mathfrak{m}}\int_0^{\mathfrak{m}} \bar{V}_2 d\mathfrak{m} \tag{11.55}$$

and, from Eq. (11.53)

$$\phi_V = \phi_V^0 + \tfrac{2}{3}\left[.99(\textstyle\sum v_i z_i^2)^{3/2}\sqrt{\mathfrak{m}}\right] \tag{11.56}$$

so a plot of ϕ_v vs $\mathfrak{m}^{1/2}$ should be linear in the Debye region with slope of 0.66 $(\sum v_i z_i^2)^{3/2}$. The linearity is confirmed by experiment but the slope is apparently not the simple function of charge type represented by $(\sum v_i z_i^2)^{3/2}$. Variation in hydration numbers for ions of different charge could be responsible.

If enthalpy changes (through calorimetrically measured heats of dilution) are the observable of interest and the integral heat of solution associated with constructing a solution of concentration $\mathfrak{m}$ (containing n_2 moles of solute) is

$$\begin{aligned}(\Delta H_{soln})_{\mathfrak{m}} &= n_2(\bar{H}_2 - \tilde{H}_2) + n_1(\bar{H}_1 - \tilde{H}_1^0)\\ &= (n_2\bar{H}_2 + n_1\bar{H}_1)_{\mathfrak{m}} - (n_2\tilde{H}_{2,\mathfrak{m}} + n_1\tilde{H}_1^0)\\ &= H_{soln,\mathfrak{m}} - (n_2\tilde{H}_{2,\mathfrak{m}} + n_2\tilde{H}_1^0)\end{aligned} \tag{11.57}$$

and the final solution is one of infinite dilution[10] for which

$$\begin{aligned}(\Delta H_{soln})_{\mathfrak{m}\to 0} &= [n_2(H_2^0 - \tilde{H}_2) + n_1(\bar{H}_1 - \tilde{H}_1^0)]_{\mathfrak{m}\to 0}\\ &= n_2[\bar{H}_2^0 - \tilde{H}_2]_{\mathfrak{m}\to 0}\end{aligned} \tag{11.58}$$

(since $\bar{H}_1 \to \tilde{H}_1^0$ as $\mathfrak{m} \to 0$), the heat of dilution from $\mathfrak{m}$ to $\mathfrak{m} \to 0$ is

$$\Delta H_D(\mathfrak{m}\to 0) = n_2(\bar{H}_2^\circ - \tilde{H}_2)_{\mathfrak{m}\to 0} - n_2(\bar{H}_2 - \tilde{H}_2)_{\mathfrak{m}} - n_1(\bar{H}_1 - \tilde{H}_1^0)_{\mathfrak{m}}$$

$$\begin{aligned}(-\Delta H_D)_{\mathfrak{m}\to 0} &= n_2(\bar{H}_{2,\mathfrak{m}} - \bar{H}_2^\circ) + n_1(\bar{H}_{1,\mathfrak{m}} - \tilde{H}_1^0)\\ &= H_{soln} - (n_2\bar{H}_2^\circ + n_1\tilde{H}_1^0).\end{aligned} \tag{11.59}$$

The relationship between the various thermal quantities is illustrated in Figure 11.4. The apparent partial molar enthalpy,[11] constructed as the Eq. (11.54)

[10] The n_1 moles of solvent is still the original n_1 since the added pure solvent is present in the infinitely dilute solution as (effectively) pure solvent.

[11] As we move toward infinite dilution $\bar{H}_{1,\mathfrak{m}} \to \tilde{H}_1^0$ and $(n_2\phi_H)_{\mathfrak{m}\to 0} \to n_2\bar{H}_2$ so $\phi^0 = \bar{H}_2^\circ$. The notation here and in Eq. (11.59)($\bar{H}_2^\circ$) designates the solute partial molar enthalpy in the infinitely dilute solution whereas $\tilde{H}_1^0$ refers to the pure solvent. Since the solvent is (effectively) pure in the infinitely dilute solution the notation does have some consistent basis.

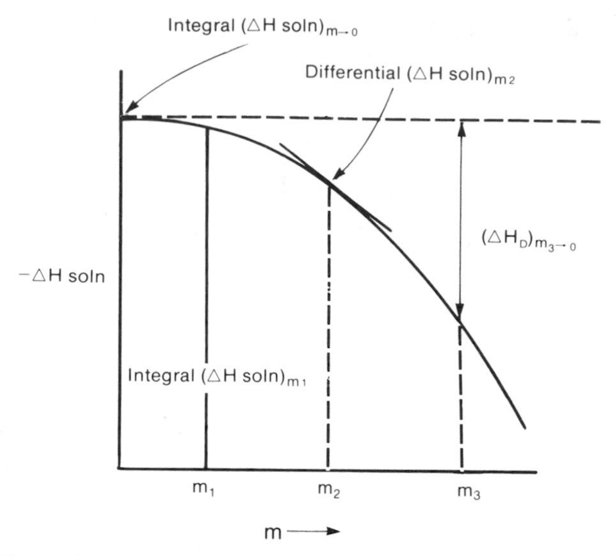

Figure 11.4. Illustration of the relationship between various solution enthalpy terms obtained by calorimetric measurements.

analog is

$$\phi_{\mathrm{H}} = \frac{\mathrm{H_{soln}} - n_1 \tilde{\mathrm{H}}_1^0}{n_2} = -(\Delta \mathrm{H_D} - n_2 \bar{\mathrm{H}}_{2,m\to 0}). \tag{11.60}$$

If we define $\bar{\mathrm{L}}_2 = \bar{\mathrm{H}}_2 - \bar{\mathrm{H}}_2^\circ$ and $\bar{\mathrm{L}}_1 = \bar{\mathrm{H}}_1 - \tilde{\mathrm{H}}_1^0$ as the relative partial molar enthalpies then

$$\begin{aligned} n_1 \bar{\mathrm{L}}_1 + n_2 \bar{\mathrm{L}}_2 &= \mathrm{H_{soln}} - n_1 \tilde{\mathrm{H}}_1^0 - n_2 \bar{\mathrm{H}}_2^\circ \\ n_2 \bar{\mathrm{L}}_2 &= n_2 \phi_{\mathrm{H}} - n_2 \bar{\mathrm{H}}_2^\circ \\ &= n_2 \phi_{\mathrm{L}} = -\Delta \mathrm{H_D} \end{aligned} \tag{11.61}$$

which relates the enthalpy change on dilution to the relative apparent partial molar enthalpy. It is this term which we can pursue in the Debye framework since

$$\frac{\mu_2}{\mathrm{T}} - \frac{\mu_2^0}{\mathrm{T}} = \nu k \ln \mathrm{m}_\pm + \nu k \ln \gamma_\pm$$

and

$$\left(\frac{\partial \tilde{\mu}_2/\mathrm{T}}{\partial \mathrm{T}}\right)_{\mathrm{P}} - \left(\frac{\partial \tilde{\mu}_2^0/\mathrm{T}}{\partial \mathrm{T}}\right)_{\mathrm{P}} = -\left(\frac{\mathrm{H}_2}{\mathrm{T}^2} - \frac{\bar{\mathrm{H}}_2^\circ}{\mathrm{T}^2}\right) = -\mathrm{T}^2 \bar{\mathrm{L}}_2. \tag{11.62}$$

We would relate $\bar{\mathrm{L}}_2$ to $k(\partial \nu k \ln \gamma_\pm/\partial \mathrm{T})_{\mathrm{P}}$ through arithmetic that closely parallels that set down in Eqs. (11.48) through (11.56) so here we set down only the result

$$\bar{L}_2 = -\frac{3.70 \times 10^{14}}{D^{3/2}T^{3/2}}(\sum \nu_i z_i^2)^{3/2}\left[\left(\frac{\partial \ln D}{\partial T}\right)_P + \frac{1}{3}\left(\frac{\partial \ln V}{\partial T}\right)_P + \frac{1}{T}\right]\sqrt{c}\ \text{erg mole}^{-1}.$$

For water as solvent with $(\partial D/\partial T)_P/D = -0.3613\ \text{deg}^{-1}$ and $(\partial V/\partial T)_P/V = -2.58 \times 10^{-4}\ \text{deg}^{-1}$ and 4.187 cal erg^{-1} and on a molar basis (molal equivalent)

$$\bar{L}_2 = 255(\sum \nu_i z_i^2)^{3/2}\sqrt{m} \tag{11.63}$$

and

$$\Delta H_D = -n_2\phi_L = -\frac{1}{m}\int_0^m \bar{L}_2\, dm. \tag{11.64}$$

Combining Eqs. (11.63) and (11.64) we obtain

$$\Delta H_D = -170(\sum \nu_i z_i^2)^{3/2}\sqrt{m}$$

and a plot of ΔH_D vs $\sqrt{m}$ should be linear in the Debye region with slope of $-170\,(\sum \nu_i z_i^2)^{3/2}$ for aqueous solutions at 25°C. The observed behavior is shown in Figure 11.5

In addition to the machinery of solution thermodynamics discussed in Chapter 10, all of which are applicable to electrolyte solutions when γ is replaced by $\gamma_\pm$ and m by $m_\pm$, and in the excess volume and enthalpy considerations

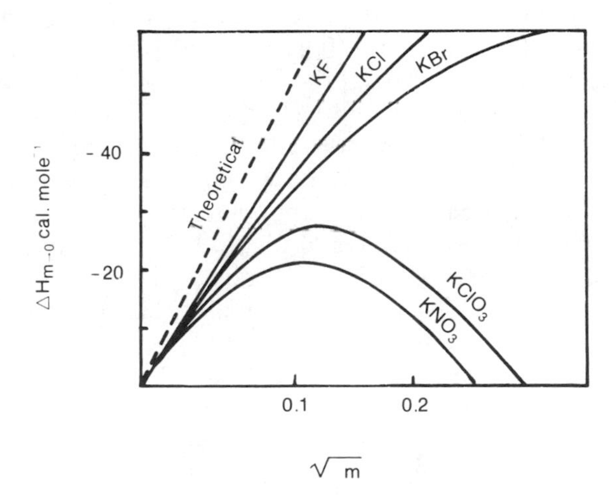

Figure 11.5. Comparison of heats of infinite dilution for various electrolytes with the Debye–Hückel limiting law requirement. From *Thermodynamics for Chemists* by Samuel Glasstone. Copyright(c) 1947 by D. Van Nostrand Company, Inc. Renewed (c) 1974 by Litton Educational Publishing Company, Inc. Reprinted by permission of Wadsworth, Inc.

applied to electrolyte solutions in this chapter, there is a vast body of electrical measrurement that is uniquely applicable to electrolyte solutions. A large part of this experimental practice relates to transport properties (electrical conductance, mobility, etc.) and provides much information about solute features such as hydration numbers, ion pairing, and degree of dissociation of weak electrolytes. It is not possible to cover this material here. However, that part of electrochemical measurement that relates directly to thermodynamic properties of the solutions should be addressed. We will confine our attention to electrochemical cells where the measured difference in electrical potential between phases can be related to chemical potentials and therefore to equilibrium properties.

An electrostatic potential $\Phi(=q/4\pi\epsilon_0 r)$ exists at any distance r from a point charge q in vacuum.[12] Our systems will not be point charges in vacuum but ions in which a unit charge $(\pm q_e)$ or multiple of the unit charge $(\pm zq_e)$ is associated with a real dimension (the ion radius) under conditions in which many such charges are closely spaced in a dielectric fluid. The ions are polarizable, they are mobile, and they exhibit moderately long-range, specific interactions with the dielectric. The configuration of charges at any point in a phase of interest would be perturbed by any direct effort to probe the field of charges to establish the electrostatic potential at that point. This does not mean that meaningful measurements cannot be accomplished but they must be carefully conceived.

Consider that a homogeneous phase ϑ, which could be a metal consisting of positive ions and electrons, is placed in a vacuum. The most energetic electrons will tend to escape from the metal and will, in effect, be regarded as an electron film at the surface. There will be a compensating positive charge excess in the immediately adjacent domain of the solid. This circumstance can be referred to as the formation of a double layer. If we locate a positive test charge at some remote point in the vacuum (phase ζ) the potential at that point established by the metal ion–electron collection will be zero and will increase in proportion to $1/r$ as the test charge is moved toward the interface between phases. The various regions of potential encountered would be as shown in Figure 11.6. The domain characterized by constant potential ψ reflects short-range interactions between the test charge and phase ϑ. The region bounded by the dashed lines reflects interactions within the double layer characterized overall by the potential χ and $\Phi = \psi + \chi$ is the potential within the body ϑ. Neither Φ nor χ is measureable by the test charge.

The circumstance of interest to us is one in which phase ζ is a solution in which ions of the metal body are present together with charge balancing anions that are not of present concern. The structure of the solution, particularly if the solvent is polar, and the structure of the double layer will be highly complex. Even though it is not possible to measure meaningful potentials within the body of phase ζ by the test charge (or equivalent) technique, an electrical potential difference will certainly exist between the body of phases ϑ and ζ due to the

[12] See Appendix D. The term ϵ_0 is called the vacuum permittivity.

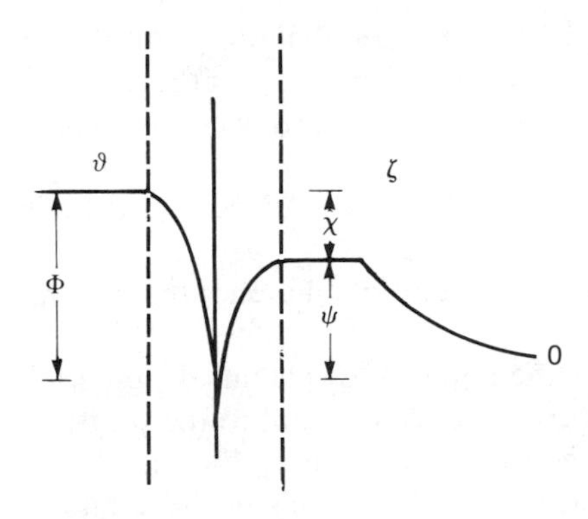

Figure 11.6. Illustration of the potential difference between a solid body (ϑ) and vacuum (ζ) as measured by a positive test charge.

differences in charge density that prevail in the two phases. If we designate this potential difference as $\Delta\Phi = \Phi_\vartheta - \Phi_\zeta$ we have defined a quantity that reflects the electrical work required to transfer charge from one phase to the other. By our definition each potential is positive and the sign of $\Delta\Phi$ depends on the direction of positive charge transfer and the magnitude of the potentials Φ_ϑ and Φ_ζ. Positive charge is transferred by the process (or the reverse process)

$$M^{z+}\,(\text{soln}) \rightarrow M\,(\text{metal})$$

and the electrical work is $Nze\Delta\Phi$.

When a metal (M) consisting of metal ions M^+ and electrons is immersed in a solution containing ions M^+, X^- we construct expressions for the chemical potentials of concern as

$$\mu^{\ddagger}(\text{metal}) = \mu^{\ddagger}_{M^+}(\text{metal}) + \mu^{\ddagger}_{e^-}(\text{metal})$$

$$\mu^{\ddagger}(\text{soln}) = \mu^{\ddagger}_{M^+}(\text{soln}) + \mu^{\ddagger}_{X^-}(\text{soln}).$$

The notation ‡ designates the electrochemical potential that is the sum of the normal chemical potential and the electrical work required to charge the ion in the phase of concern. If we take the metal/solution interface to be permeable to M^+ ions but not to X^- ions or to electrons and the equilibrium condition to be the equality of electrochemical potential, the establishment of equilibrium will require that M^+ ions move from the metal to the solution phase or from the solution phase to the metal phase until

$$\mu^{\ddagger}_{M^+}(\text{soln}) = \mu^{\ddagger}_{M^+}(\text{metal}).$$

In terms of the expressions for the normal chemical potential

$$\mu^{\pm}_{M^+}(\text{soln}) = \mu^{\circ}_{M^+}(\text{soln}) + kT \ln a_{M^+}(\text{soln}) + z_{M^+} q_e \Phi(\text{soln})$$

$$\mu^{\ddagger}_{M^+}(\text{metal}) = \mu^{\circ}_{M^+}(\text{metal}) + z_{M^+} q_e \Phi(\text{metal}).$$

Clearly, the process as outlined would produce a charge unbalance both in the metal electrode and in the solution and in the course of the process the original electrostatic potentials would change substantially. An equilibrium condition is

$$\mu^{\circ}_{M^+}(\text{metal}) - \mu^{\circ}_{M^+}(\text{soln}) - \textit{k}T \ln a_{M^+}(\text{soln}) = z_{M^+} \mathscr{q}_e \Delta\Phi(\text{eq}) \qquad (11.65)$$

where $\Delta\Phi(\text{eq}) = \Phi(\text{metal}) - \Phi(\text{soln})$ represents the potential difference that would prevail in this isolated system.

If we imagine that in order to establish the equality of chemical potential described by Eq. (11.65), M^+ ions are transferred from the metal to the solution leaving behind an equivalent number of electrons in the metal, the metal will now have a negative charge and the concentration of M^+ ions in the solution would increase. Suppose that the metal contains 1 mole of material and that as little as 10^{-15} mole of metal ions were transferred to the solution leaving behind 10^{-15} mole of electrons. We have a metal sphere which might be (0) 4 cm in radius having an excess charge of $10^{-15}\mathscr{F}$ $(= 9.648 \times 10^4 \times 10^{-15})$[13] that will generate a potential amounting to 8.7 volt at a point 10 cm from the center of the sphere in vacuum or (0) 0.1 volt at a point 10 cm from the center of the sphere in a homogeneous medium of dielectric constant (0) 100. This is certainly a significant potential although the concentration change is completely undetectable. In real ionic systems although the dielectric is not microscopically homogeneous, potentials of measureable magnitude should exist but we can expect to encounter great difficulty in obtaining absolute values for $\Delta\Phi_{eq}$ terms and for electrochemical potential since we have introduced a complex variety of charged domains.

To avoid phase boundary problems and deal with measurements that can be accomplished readily and with minimum ambiguity we consider a pair of electrodes immersed in separate solutions containing the metal ions at different concentrations. The solutions are in contact through a porous barrier or other conducting mechanism that affords electrical continuity but prevents mixing. Each electrode can establish the equilibrium represented by Eq. (11.64). The chemical potential of the (excess) electrons in the two metals will be different in view of the different concentrations of M^+ ions in the two solutions. That difference in electron chemical potential cannot be maintained if there is an electron path between the electrodes such as would be afforded by an external connecting wire. The equilibrium state for the two electrode system, equality of chemical potential for each species in the phases in which they appear, will be realized only when the M^+ concentrations have equalized through a solution process $M^+(\text{metal}) \rightarrow M^+(\text{soln})$ at one electrode and a deposition process $M^+(\text{soln}) \rightarrow M^+(\text{metal})$ at the other electrode. Under this circumstance the

[13] Avogadros number of electron charges is called the Faraday and designated by $\mathscr{F}$. The Faraday value is 9.648×10^4 coulombs.

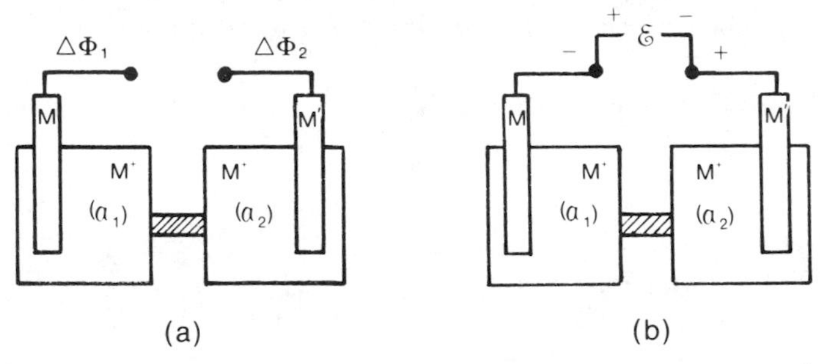

Figure 11.7. Schematic representation of (a) an open circuit electrolytic cell consisting of separate but same metal electrodes immersed in solutions containing the metal ion at different concentrations and (b) the same cell in a closed circuit including a variable voltage source of opposing polarity.

electron chemical potential will also be the same in each electrode. Figure 11.7a illustrates the initial state in which each electrode has established the internal equilibrium between metal ions in the metal and those in the solution resulting in the single electrode potentials $\Delta\Phi_1$ and $\Delta\Phi_2$. For $M^+(a_1) > M^+(a_2)$ the excess electron density is greater in the M electrode than in the M′ electrode. The M′ electrode is positive with respect to the M electrode and if the open circuit were closed the flow of current by convention would be from the positive to the negative electrodes. Electron flow would be in the opposite direction. It is not, however, the intention to cause this process to occur but rather to maintain a reversible state in which the internal equilibrium of each electrode system is undisturbed while a meaningful measurement relating to $\Delta\Phi_1$ and $\Delta\Phi_2$ is accomplished. This can be done as shown in Figure 11.7b where a counter potential $\mathscr{E}$ is applied that is of appropriate polarity and just sufficient in magnitude to prevent any electron transport. This value of $\mathscr{E}$ has established the value of the difference between the two electrode potentials under equilibrium conditions. We will call $-(\Delta\Phi_1 - \Delta\Phi_2)$ the cell potential ($\mathscr{E}_{\text{cell}}$). To the extent that any current is permitted to flow the spontaneous cell process would consist of a solution half cell (oxidation) and a deposition half cell (reduction) and, following convention, we write $\mathscr{E}_{\text{cell}} = \mathscr{E}_{1/2}(\text{ox}) + \mathscr{E}_{1/2}(\text{red})$ but recognizing the origins of the half cell potentials in the half cell potential difference terms $\Delta\Phi$.

For any infinitesimal step in the reversible process

$$M^+(a_2) \rightleftharpoons M^+(a_1)$$

which could be accomplished by making (Figure 11.7b) $\mathscr{E}$ infinitesimally less (or greater) than the cell potential, there is an infinitesimal change in the chemical potentials for the internal process. Since we have recognized that no discernible concentration changes accompanied the establishment of the individual half cell equilibria, normal chemical potential expressions apply and

$$\Delta\mu(\text{process}) = \mu^{\circ}_{M^+}(\text{metal}) + \mu^{\circ}_{M^+}(\text{soln 2}) - \mu^{\circ}_{M^+}(\text{metal}) - \mu^{\circ}_{M^+}(\text{soln 1}) + \ell T \ln\frac{a_2}{a_1} = \ell T \ln\frac{a_2}{a_1}. \tag{11.66}$$

In molar free energy terms

$$\Delta\tilde{G} = N_A \Delta\mu = RT \ln\frac{a_2}{a_1}. \tag{11.67a}$$

The molar free energy change is also given by the electrical work associated with the transfer of charge from one domain of the electrostatic potential to another. In this process, without quantifying the individual electrostatic potentials, we have obtained the electrical work as $-N_A z_+ q e \mathscr{E}_{\text{cell}}$ for the transfer of electrons in the external circuit, $\Delta\tilde{G} = -N_A z_+ q e \mathscr{E}_{\text{cell}}$ and

$$N_A z_+ q e \mathscr{E}_{\text{cell}} = -RT \ln\frac{a_2}{a_1}. \tag{11.67b}$$

The utility of this approach can be extended to any process that features the exchange of electrons

$$\Delta\tilde{G}(\text{process}) = -N_A z q e \mathscr{E}_{\text{cell}} \tag{11.67c}$$

in general. It only remains to construct cells that feature the reversibility requirements to quantify ion equilibria through cell voltage measurements.

If the two metal elctrodes are not the same substance, the half cell potentials will reflect the unique metal ion–electron character of each electrode as well as the positive ion activity in each half cell and the change in chemical potential associated with each half cell process and with the overall process will contain terms in $\Delta\mu^{\circ}$. The difference in half cell potentials still provides the driving force for electron transfer in the external circuit and Eq. (11.67) will apply. The directionality of the spontaneous process ($\Delta G < 0$) is that for which the measured $\mathscr{E}_{\text{cell}} > 0$ but we have only vague guidance, such as the magnitude of ionization potentials and possibly some understanding of solution structure, which could provide an awareness of the magnitude of the $\Delta\Phi$ terms. We will rely on the experimental measurement. If the first choice of polarity is not correct the current cannot be reduced to zero over any range in counter voltage $\mathscr{E}$ (Figure 11.7b). When the proper choice has been made we can construct the equilibrium expression as (taking metal 2/solution as having the more negative value of $\Delta\Phi$)

$$M_1^+(\text{metal } 1) + M_2^+(\text{soln } 2) \rightleftharpoons M_1^+(\text{soln } 1) + M_2^+(\text{metal})$$

$$\Delta\mu(\text{process}) = \Delta\mu^0(T) + kT \ln \frac{a_{M_1^+}}{a_{M_2^+}} \tag{11.68}$$

where all μ° terms have been collected into the single $\Delta\mu^\circ(T, P)$ and

$$N_A \Delta\mu = \Delta\tilde{G} = -N_A z_+ q_e \mathscr{E}_{\text{cell}} = \Delta G^\circ + RT \ln \frac{a_{M_1^+}}{a_{M_2^+}}. \tag{11.69}$$

The μ^0 terms for the different metals and the different solution species do not cancel in this, more common, case. We note that if the ion activities are each equal to 1

$$N_A z_+ q_e \mathscr{E}_{\text{cell}} = z_+ \mathscr{F} \mathscr{E}^e_{\text{cell}} \tag{11.70}$$

where we write $N_A q_e = \mathscr{F}$, calling this quantity of charge the Faraday. This observation can be generalized to any cell process thus extending the standard state concept to cell measurements.

There are other variations on the metal electrode–metal ion cell equilibria that extend the electrochemical measurements to other processes. No new features are introduced by these techniques. The two half cells can contain metal ions in different oxidation states and the half cell potentials will then reflect both the activity and the oxidation state. For an overall cell process involving a one electron change

$$M^{z_1+} \rightarrow M^{z_2+}$$

$$\Delta\mu(\text{process}) = \Delta\mu^\circ(T, P) + kT \ln \frac{a_{M^{z_2+}}}{a_{M^{z_1+}}}$$

$$\mathscr{F}\mathscr{E}_{\text{cell}} = \mathscr{F}\mathscr{E}^0_{\text{cell}} + RT \ln \frac{a_{M^{z_2+}}}{a_{M^{z_1+}}}. \tag{11.71}$$

Throughout this discussion of metal electrode–solution equilibria the metal chemical potential has been assumed to be given by the Eq. (11.55) expression, which is equivalent to assigning an invariant activity of one to the metal electrode. This is not required since it is possible to construct metal solutions in which the active metal activity is less than one. For such systems there will be an additional activity ratio term since the internal reversible cell process will now be

$$\text{Metal}(a_1) + M^+ \text{soln}(a_1') \rightarrow \text{Metal}(a_2) + M^+ \text{soln}(a_2').$$

Commonly the metal solutions are amalgams of different metal concentrations and the two electrodes are immersed in the same metal ion solution so the process is that of transferring metal (ions) from one solid solution to the other.

A different class of electrochemical cells can be constructed by several techniques in which the electrode–solution interface is permeable to electrons. A typical half cell of this nature would involve an inert metal surrounded by a gas phase in contact with a solution containing ions appropriate for electron transfer, such as chlorine gas in contact with a platinum electrode and a solution containing chloride ion. The platinum serves the dual purpose of establishing a potential with respect to the solution and catalyzing the dissociation of chlorine gas. The overall electrode–solution potential is taken to derive from electron transfer via atomic chlorine from the metal to the solution (or reverse). Arguments parallel those for active metal electrodes and result in the conclusion that an electron deficiency (excess) will develop in the platinum and we anticipate that this half cell, coupled with any metal–metal ion half cell, will produce the overall cell process

$$M^+(\text{metal}) + \tfrac{1}{2}Cl_2(\text{gas}) \rightarrow M^+(a_{M^+}) + Cl^-(a_{Cl^-})$$

and

$$\mathscr{F}\mathscr{E}_{\text{cell}} = \mathscr{F}\mathscr{E}^\circ_{\text{cell}} - RT \ln a_{M^+} a_{Cl^-} + \frac{RT}{2} \ln a_{Cl_2}. \tag{11.72}$$

It is clear that, in this case, a single cell with the electrolyte MCl in the solution phase will eliminate any phase boundaries, simplify the electrochemical measurement, and improve its accuracy.

Since we cannot, by unambiguous experiment or meaningful calculation, obtain absolute values for individual half cell potentials, a convention has been established for tabulation purposes. If we construct a hydrogen ion–hydrogen gas electrode according to the prescription discussed above for reversible gas electrodes, we can couple this half cell with any other oxidation–reduction half cell to obtain a cell voltage. If hydrogen ion and hydrogen gas are at unit activity, the half cell voltage for this electrode is $\mathscr{E}^\circ_{1/2}(H^+, H_2)$ and the half process is either an oxidation $\frac{1}{2}H_2 \rightarrow H^+ + e^-$ when coupled with a reduction half process or a reduction $H^+ + e^- \rightarrow \frac{1}{2}H_2$ when coupled with an oxidation half process. The cell voltage will always be positive for the spontaneous overall process. In terms of the absolute values of the half cell voltages $\mathscr{E}^\circ_{1/2}(\text{ox})_{\text{ab}}$ the cell voltage, which we will refer to as the reference, is

$$(\mathscr{E}^\circ_{\text{cell}})_{\text{ref}} = \pm[\mathscr{E}^\circ_{1/2}(\text{ox})_{\text{ab}} + \mathscr{E}^\circ_{1/2}(H^+ + \tfrac{1}{2}H_2)_{\text{ab}}] \tag{11.73}$$

where we recognize that the negative sign changes $\mathscr{E}_{1/2}(\text{ox})_{\text{ab}}$ to $\mathscr{E}_{1/2}(\text{red})_{\text{ab}}$ and $\mathscr{E}_{1/2}(H^+ \rightarrow \frac{1}{2}H_2)_{\text{ab}}$ to $\mathscr{E}_{1/2}(\frac{1}{2}H_2 \rightarrow H^+)_{\text{ab}}$ through their association with $\Delta\Phi$ and that the $\pm$ sign will be + or − as required for $\mathscr{E}_{\text{cell}} > 0$. Any two such processes can be added using the appropriate choice of sign to accomplish the objective of pairing a reduction and an oxidation half cell under the constraint that $\mathscr{E}_{\text{cell}} > 0$. The hydrogen gas–hydrogen ion absolute half cell voltage is eliminated in such a sum and we can be indifferent to its value, which can be set equal to zero for operational purposes. By this convention we have set $\mathscr{E}^\circ_{\text{cell}}$ in Eq. (11.73) equal to $\mathscr{E}^\circ_{1/2}(\text{ox})$ for the half cell process with which the standard hydrogen gas–hydrogen ion half cell was paired,

$$\pm[\mathscr{E}^\circ_{1/2}(\text{ox})]_{\text{ab}} = [\mathscr{E}^\circ_{\text{cell}}]_{\text{ref}} + [\mathscr{E}^\circ_{1/2}(\text{H}^+ + \tfrac{1}{2}\text{H}_2)]_{\text{ab}} \tag{11.74}$$

$$\pm[\mathscr{E}^\circ_{1/2}(\text{ox})]_{\text{op}} = [\mathscr{E}^\circ_{\text{cell}}]_{\text{ref}} > 0. \tag{11.75}$$

From a complete spectrum of cell measurements we construct a table of $[\mathscr{E}^\circ_{1/2}(\text{ox})]_{\text{op}}$ values that can be used, under the sign constraint of +, − pairs to define the standard cell voltage for arbitrary oxidation–reduction processes. Where

$$-[\mathscr{E}^\circ_{1/2}(\text{ox})]_{\text{op}} = [\mathscr{E}^\circ_{\text{cell}}]_{\text{ref}} > 0 \tag{11.76}$$

it is, of course, the reduction process in conjunction with the hydrogen gas–hydrogen gas electrode that is spontaneous. It has become the normal practice to list reduction potentials and a selected set of standard half cell reduction potentials $\mathscr{E}^\circ_{1/2}(\text{red})$ is collected in Table 11.2.

The utility of electrochemical cell voltages, of course, lies in the Eqs. (11.70), (11.71), and (11.72) analogs, or in general terms[14]

$$\mathscr{E}_{\text{cell}} = \mathscr{E}^\circ_{\text{cell}} - \frac{\text{N}_\text{A}k\text{T}}{n\mathscr{F}} \ln \prod \frac{(\text{a}_i^{\nu_i})_{\text{products}}}{(\text{a}_i^{\nu_i})_{\text{reactants}}} \tag{11.77}$$

for the cell process involving the exchange of n moles of electrons

$$\nu_{\mathscr{A}}\mathscr{A} + \nu_{\mathscr{B}}\mathscr{B} + \cdots \rightarrow \nu_{\mathscr{C}}\mathscr{C} + \nu_{\mathscr{D}}\mathscr{D} + \cdots \tag{11.78}$$

where each solution species is present at activity $\text{a}_{\mathscr{A}}, \text{a}_{\mathscr{B}}, \ldots, n = \sum(\nu_i z_i)_+ = \sum(\nu_i z_i)_-$ for all charged species and in the relationship

$$\Delta\text{G}_{\text{process}} = -n\mathscr{F}\mathscr{E}_{\text{cell}}, \qquad \Delta\text{G}^\circ_{\text{process}} = -n\mathscr{F}\mathscr{E}^\circ_{\text{cell}}.$$

If the activities are such that Eq. (11.78) can be expressed as an equilibrium state

$$\text{Reactants} \rightleftharpoons \text{Products} \tag{11.79}$$

[14]This expression is known as the Nernst equation.

TABLE 11.2. Selected Standard Electrode Potentials Based on the Hydrogen Gas (1 atm)/Hydrogen Ion (Unit Activity) Zero Convention

Electrode	$\mathscr{E}^\circ$	Half Cell Reaction
$Li^+ \mid Li$	−3.045	$Li^+ + e^- = Li$
$Na^+ \mid Na$	−2.714	$Na + e^- = Na$
$Mg^{2+} \mid Mg$	−2.37	$\frac{1}{2}Mg^{2+} + e^- = \frac{1}{2}Mg$
$OH^- \mid H_2 \mid Pt$[b]	−0.828	$H_2O + e^- = \frac{1}{2}H_2 + OH^-$
$Zn^2 \mid Zn$	−0.763	$\frac{1}{2}Zn^{2+} + e^- = \frac{1}{2}Zn$
$Fe^{2+} \mid Fe$	−0.440	$\frac{1}{2}Fe^{2+} + e^- = \frac{1}{2}Fe$
$Cd^{2+} \mid Cd$	−0.403	$\frac{1}{2}Cd^{2+} + e^- = \frac{1}{2}Cd$
$Ni^{2+} \mid Ni$	−0.250	$\frac{1}{2}Ni^{2+} + e^- = \frac{1}{2}Ni$
$Sn^{2+} \mid Sn$	−0.140	$\frac{1}{2}Sn^{2+} + e^- = \frac{1}{2}Sn$
$Pb^{2+} \mid Pb$	−0.126	$\frac{1}{2}Pb^{2+} + e^- = \frac{1}{2}Pb$
$D^+ \mid D_2 \mid Pt$	−0.0034	$D^+ + e^- = \frac{1}{2}D_2$
$H^+ \mid H_2 \mid Pt$	0.0000	$H^+ + e^- = \frac{1}{2}H_2$
$Cu^{2+}, Cu^+ \mid Pt$[c]	0.153	$Cu^{2+} + e^- = Cu^+$
$Cl^- \mid AgCl(cr) \mid Ag$	0.2224	$AgCl + e^- = Ag + Cl^-$
$Cl^- \mid Hg_2Cl_2(cr) \mid Hg$[d]	0.268	$\frac{1}{2}Hg_2Cl_2 + e^- = Hg + Cl^-$
$Cu^{2+} \mid Cu$	0.337	$\frac{1}{2}Cu^2 + e^- = \frac{1}{2}Cu$
$OH^- \mid O_2 \mid Pt$[e]	0.401	$\frac{1}{4}O_2 + \frac{1}{2}H_2O + e^- = OH^-$
$Cu^+ \mid Cu$	0.521	$Cu^+ + e^- = Cu$
$I^- \mid I_2(cr) \mid Pt$	0.5355	$\frac{1}{2}I_2 + e^- = I^-$
$Fe^{3+}, Fe^{2+} \mid Pt$	0.771	$Fe^{3+} + e^- = Fe^{2+}$
$Ag^+ \mid Ag$	0.7991	$Ag^+ + e^- = Ag$
$H^+ \mid O_2 \mid Pt$	1.229	$H^+ + \frac{1}{4}O_2 + e^- = \frac{1}{2}H_2O$
$Cl^- \mid Cl_2(g) \mid Pt$	1.3604	$\frac{1}{2}Cl_2(g) + e^- = Cl^-$
$Pb^{2+} \mid PbO_2 \mid Pb$	1.455	$\frac{1}{2}PbO_2 + 2H^+ + e^- = \frac{1}{2}Pb^{2+} + H_2O$
$Au^{3+} \mid Au$	1.50	$\frac{1}{3}Au^{3+} + e^- = \frac{1}{3}Au$
$F^- \mid F_2(g) \mid Pt$	2.87	$\frac{1}{2}F_2(g) + e^- = F^-$

Taken from R. A. Alberty, *Physical Chemistry*, 6th ed. Wiley, New York, 1983. Reprinted with permission by John Wiley and Sons Inc. Also see end of chapter sources.

the product term is the equilibrium constant and both $\Delta G_{process}$ and $\mathscr{E}_{cell}$ are zero

$$\mathscr{E}^\circ_{cell} = \frac{RT}{n\mathscr{F}} \ln K_{eq}. \tag{11.80}$$

All general thermodynamic relationships applying to ΔG and ΔG° will extend to $\mathscr{E}_{cell}$ and $\mathscr{E}^\circ_{cell}$ without detailed argument

$$\left(\frac{\partial \Delta G^\circ/T}{\partial T}\right)_P = -\frac{\Delta H^\circ}{T^2} = -n\mathscr{F}\left(\frac{\partial \mathscr{E}^\circ_{cell}/T}{\partial T}\right)_P \tag{11.81}$$

$$\left(\frac{\partial \Delta G^\circ}{\partial T}\right)_P = -\Delta S^\circ = -n\mathscr{F}\left(\frac{\partial \mathscr{E}^\circ_{cell}}{\partial T}\right)_P \tag{11.82}$$

$$\left(\frac{\partial \Delta G^\circ}{\partial P}\right)_T = \Delta V_{\text{process}} = -n\mathscr{F}\left(\frac{\partial \mathscr{E}^\circ_{\text{cell}}}{\partial P}\right)_T. \tag{11.83}$$

From Eq. (11.65) we have

$$\mu_{M^+}(\text{soln}) = \mu^\circ_{M^+}(\text{metal}) = \mu^\circ_{M^+}(\text{soln}) + kT \ln a_{M^+}(\text{soln}) + z_{M^+}q_e \Delta\Phi_{eq}$$

where $\mu^\circ_{M^+}(\text{soln})$ is defined in terms of the infinitely dilute solution and a proper procedure for obtaining standard ($\mathscr{E}^\circ_{1/2}$) cell potentials would recognize this constraint. Using a particular example, for the overall cell process

$$\tfrac{1}{2}H_2(1\text{ atm}) + \tfrac{1}{2}Cl_2(1\text{ atm}) = H^+(m) + Cl^-(m)$$

$$\mathscr{E}_{\text{cell}} = \mathscr{E}^\circ_{\text{cell}} - \frac{RT}{\mathscr{F}} \ln \frac{a_{H^+}a_{Cl^-}}{a_{H_2}^{1/2}a_{Cl_2}^{1/2}} = \mathscr{E}^\circ - \frac{RT}{\mathscr{F}}(\ln m_+m_- + \ln \gamma_+\gamma_-) \tag{11.84}$$

since we would assume ideal behavior for the gaseous components and $\nu_+ = \nu_- = 1$. Then from Eqs. (11.28), $m_+m_- = m_\pm^2 = m^2$; $\gamma_+\gamma_- = \gamma_\pm^2$

$$\mathscr{E}_{\text{cell}} = \mathscr{E}^\circ_{\text{cell}} - \frac{2RT}{\mathscr{F}} \ln(m_\pm\gamma_\pm). \tag{11.85}$$

Knowing through the Debye–Hückel result that in the limit $m \to 0$, $\ln \gamma_\pm$ assumes a linear form when plotted against $m^{1/2}$, $\ln \gamma_\pm = -A\sqrt{m} + Cm$, we rewrite Eq. (4.85) in the form

$$\mathscr{E}_{\text{cell}} + \frac{2RT}{\mathscr{F}}(-A\sqrt{m} + \ln m) = \mathscr{E}^\circ - \frac{2RT}{\mathscr{F}}Cm \tag{11.86}$$

and plot the left hand side as obtained from cell voltage measurements over a range of concentrations vs m. The intercept ($m = 0$) is $\mathscr{E}^\circ$ for the cell. On the basis of our hydrogen half cell convention this is $\mathscr{E}_{1/2}(Cl^-, Cl_2)$. Having $\mathscr{E}^\circ_{\text{cell}}$ we also obtain from the original data $\gamma_\pm$ at any concentration within the range of the experimental data from Eq. (11.85) independent of the linear constraint.

The standard half cell potentials will yield free energies of ion formation since half cell processes are consistent with our assignment of zero free energy of formation to elements in their standard state. As example half cell reactions

$$M(\text{metal}) \to M^{z+} + z_+e, \qquad \Delta\tilde{G} = -z_+N_Aq_e\mathscr{E}_{1/2}(M, M^{z+})$$

or

$$X_2 + z_-e \to 2X^{z-}, \qquad \Delta\tilde{G} = -z_-N_Aq_e\mathscr{E}_{1/2}(X_2, X^-).$$

The $\Delta\tilde{G}$ value relates to ion formation from the element. If all species are in appropriate standard states $\mathscr{E}_{1/2} = \mathscr{E}^{\circ}_{1/2}$ and $\Delta\tilde{G}$ is ΔG°_{f}. Free energies of ion formation can be tabulated and used in combination with standard free energies for molecular species even though we have introduced a further convention

$$\tfrac{1}{2}H_2(g) \rightarrow H^+ + e, \qquad \Delta\tilde{G} = 0 = N_A \mathscr{q}_e \mathscr{E}_{1/2}(H_2, H^+). \tag{11.87}$$

This is clear without arithmetic since, if an absolute half cell potential were available for the hydrogen gas-hydrogen ion half cell, the term $N_A\mathscr{q}_e[\mathscr{E}_{1/2}(H^+, H_2)]_{ab}$ would be added to (subtracted from) any reduction (oxidation) half process and the half processes must always combine as oxidation–reduction pairs in any overall process. Clearly, however, the free energies of ion formation obtained under this convention are not absolute values and have no significance by themselves.

We can also obtain standard enthalpies and standard entropies from standard half cell potential information using the relationship

$$\Delta G = \Delta H - T\Delta S = \Delta H + T\left(\frac{\partial \Delta G}{\partial T}\right)_P$$

$$-\Delta H = n\mathscr{F}\left[\mathscr{E} + T\left(\frac{\partial \mathscr{E}}{\partial T}\right)_P\right]. \tag{11.88}$$

The standard enthalpy of ion formation, however, is most commonly obtained from calorimetric integral heat of solution and heat of infinite dilution data. As an example the integral heat of solution $(\Delta H\,\mathrm{soln})_{m\rightarrow 0}$ accompanying the process

$$MX(\mathrm{solid}) \rightarrow [M^+(\mathrm{aq}) + X^-(\mathrm{aq})]_{m\rightarrow 0}$$

relates to the enthalpy of ion formation as

$$(\Delta H_{\mathrm{soln}})_{m\rightarrow 0} = \Delta H^{\circ}_{f}(\text{final species}) - \Delta H^{\circ}_{f}(\text{initial species})$$

$$\sum \Delta H^{\circ}_{f}(\text{ions}) = (\Delta H_{\mathrm{soln}})_{m\rightarrow 0} + \Delta H^{\circ}_{f}(\text{solid}). \tag{11.89}$$

Standard state conditions require 1 atm pressure and a stated temperature (usually 298 K) for the solid and infinite dilution for the ion system in accordance with our conventions. We must, however, attach the independent condition for ion formation

$$\tfrac{1}{2}H_2(\text{gas, 1 atm}) \rightarrow (H^+_{\mathrm{aq}})_{m\rightarrow 0}, \qquad \Delta H = 0 \text{ for all T} \tag{11.90}$$

to be consistent with the Eq. (11.87) convention. Under this convention we obtain single ion formation enthalpies starting with the process

$$(\Delta H_{soln}, HCl)_{m \to 0} = \Delta H_f^\circ(H_{aq}^+ + Cl_{aq}^-)_{m \to 0} + \Delta H_f^\circ(HCl)$$
$$\Delta H_f^\circ(H_{aq}^+ + Cl_{aq}^-)_{m \to 0} = \Delta H_f^\circ(Cl_{aq}^-)_{m \to 0} = (\Delta H_{soln}, HCl)_{m \to 0} - \Delta H_f^\circ(HCl) \quad (11.91)$$

from which, with a complete data set, other single ion enthalpies of formation follow.

Standard entropies of ion formation follow either from $(\partial \mathscr{E}^\circ_{1/2}/\partial T)_P$ or more commonly from $\mathscr{E}^\circ_{1/2}$ and calorimetric data for enthalpies. For the solution. process we take Eq. (11.89) and the analog

$$\Delta G_f^{\circ\prime}(\text{ions}) - \Delta G_f^\circ(\text{solid}) = [\Delta G^\circ(\text{soln})]_{m \to 0} \quad (11.92)$$

to obtain

$$\frac{(\Delta G^\circ_{soln} - \Delta H^\circ_{soln})_{m \to 0}}{T} = \Delta S^\circ_{soln} = S^\circ_{solid} - \sum S^\circ(\text{ions}). \quad (11.93)$$

Again we must introduce the convention

$$\tfrac{1}{2}H_2(\text{gas, 1 atm}) \to (H_{aq}^+ + e)_{m \to 0}, \qquad \Delta S^\circ = 0 \text{ for all T}. \quad (11.94)$$

Standard free energies, enthalpies and entropies of ion formation are listed in Table 11.3 for some common ions.

TABLE 11.3. Gibbs Free Energies, Enthalpies, and Entropies of Formation of Selected Ions in Water at 298 K

Ion	ΔG_f° (kJ mol^{-1})	ΔH_f° (kJ mol^{-1})	ΔS_f° (kJ mol^{-1} K^{-1})
H^+	0	0	0
OH^-	−157.3	−230.0	−10.5
Cl^-	−131.2	−167.4	+55.2
Br^-	−102.8	−120.9	+80.7
I^-	−51.7	−55.9	+109.4
NO_3^-	−110.6	−206.6	+125.0
NH_4^+	−79.5	−132.8	+112.8
LI^+	−293.8	−278.4	+14.0
Na^+	−261.9	−239.7	+60.2
K^+	−282.3	−251.2	+103.0
Ag^+	+77.1	+105.9	+73.9
Mg^{2+}	−456.0	−462.0	−118.0
Ca^{2+}	−553.0	−543.0	−55.2

Taken from a larger collection in R. S. Berry, S. A. Rice, and J. Ross, *Physical Chemistry*. Wiley, New York, 1980. Used with permission by John Wiley and Sons Inc.

PROBLEMS

11.1. Calculate the mean activity coefficient for $BaCl_2$ on the basis of the Debye limiting law for concentrations of 0.001, 0.005, and 0.01 M solutions. Plot these results vs $M^{1/2}$ to confirm the validity of the limiting law behavior at low concentrations.

11.2. Using the relationship developed as Eq. (11.10) and the following freezing point depression data,

$-\Delta T_f$	0.119	0.233	0.461	0.691
mole $BaCl_2$/liter	0.024	0.048	0.098	0.148

calculate the mean activity coefficient for $BaCl_2$ in a 0.10 molar solution.

11.3. The solubility limit of an electrolyte in any solvent reflects an equilibrium state

$$MX(\text{solid}) \rightleftharpoons M^+(\text{soln}) + X^-(\text{soln})$$

to which an equilibrium constant (the solubility product) must apply. Construct the form of this equilibrium constant in terms of ion concentrations for the saturated aqueous solution of $PbCl_2$. Using appropriate formation free energies, calculate the value of the solubility product and each ion concentration at saturation.

11.4. Cell voltage data are useful for purposes ranging from the calculation of equilibrium constants to answering questions related to process thermodynamics and activity coefficients. The values are of particular importance. From the following information for the cell,

$$Pt|H_2(1\text{ atm})|H^+(\mathfrak{m})Cl^-(\mathfrak{m})|AgCl|Ag$$

$\mathfrak{m}$(HCl)	0.1238	0.02563	0.00914	0.005619
$\mathscr{E}_{cell}$(volt)	0.34199	0.41824	0.46860	0.49257

find $\mathscr{E}^\circ$ for the cell. What is $\mathscr{E}^\circ_{1/2}$ for the silver–silver chloride half cell?

11.5. What is the equilibrium constant and standard free energy change for the reaction

$$Ag(s) + Fe^{3+}(aq) \rightleftharpoons Ag^+(aq) + Fe^{2+}(aq)$$

11.6. What is the mean activity coefficient of HCl at 25°C in a solution that, if used in the problem 11.4 cell, would produce a cell voltage of 0.3031 at 25°C?

11.7. Show, given the two cells,

$$\mathrm{Pt}|\mathrm{H_2}(1\ \mathrm{atm})|\mathrm{HCl}(m_1)\mathrm{MCl}(m_2)|\mathrm{AgCl}|\mathrm{Ag}$$

$$\mathrm{Pt}|\mathrm{H_2}(1\ \mathrm{atm})|\mathrm{MOH}(m_1')\mathrm{MCl}(m_2')|\mathrm{AgCl}|\mathrm{Ag}$$

and making the assumption that the ratio of activity coefficients is 1, that the ion product of water can be obtained from the two cell voltages.

11.8. We did not consider transport properties of ions in our discussion but electrical conductivity is clearly dependent on ion concentration. For a weak electrolyte the incomplete dissociation would provide a much reduced conductivity when compared to the same concentration of a strong electrolyte. If we compare the conductivity (Λ) at various concentrations with the limiting value at infinite dilution (Λ^0), the ratio $\Lambda/\Lambda_0(=\alpha)$ is the degree of dissociation since dissociation is taken to be complete in the infinitely dilute solution. Construct an expression for the equilibrium constant for dissociation of a general weak acid in terms of the degree of dissociation.

11.9. Use the problem 11.8 expression, together with the relationship $\Lambda/\Lambda_0 = \alpha$ to obtain

$$c\Lambda_0 = -\mathrm{K}\Lambda_0 + \frac{\mathrm{K}(\Lambda_0)^2}{\Lambda_c}.$$

Using the following data for acetic acid solutions,

c (mole liter^{-1}) $\times 10^5$	2.80	11.14	15.32	21.84	102.83	136.34
Λ ohm^{-1} cm^2 equiv^{-1}	210.32	127.71	112.02	96.47	48.13	42.22

obtain Λ_0 and the equilibrium constant for acetic acid dissociation from a plot of $c\Lambda_c$ vs $1/\Lambda_c$.

11.10. If water is added to a solution of given concentration the process is dilution. At any overall concentration the differential heat of solution is

$$\frac{\partial \Delta \mathrm{H_{soln}}}{\partial n_1} = \frac{\partial \mathrm{H_{soln}}}{\partial n_1} - \tilde{\mathrm{H}}_1 = \bar{\mathrm{H}}_1 - \mathrm{H}_1^\circ = \bar{\mathrm{L}}_1$$

where $\Delta \mathrm{H_{soln}} = \mathrm{H_{soln}} - (n_1\tilde{\mathrm{H}}_1 - n_2\tilde{\mathrm{H}}_2)$. Construct the argument that if we measure $\Delta\mathrm{H}$ for the addition of measured amounts (n_1 moles) of water of a particular solution, and extropolate the values $\Delta\mathrm{H}/n_1$ to zero added water, we obtain $\bar{\mathrm{L}}_1$ from this value. From the following data for the dilution of a strontium chloride solution,

n_1(moles)	37.2	30.0	27.7	20.0	10.0
$\Delta\mathrm{H_{soln}}$(cal)	−129.8	−113.3	−109.2	−86.3	−48.5

obtain $\bar{\mathrm{L}}_1$ for the 2.9 molar strontium chloride solution.

CHAPTER CONTENT SOURCES AND FURTHER READING

1. H. Eyring, ed., *Annual Review of Physical Chemistry*, Volumes 31 and 32. Annual Reviews, Palo Alto, CA.
2. I. Klotz, *Chemical Thermodynamics*, Chapters 21 and 22. W. A. Benjamin, New York, 1964.
3. R. S. Berry, S. A. Rice, and J. Ross, *Physical Chemistry*, Chapter 26. Wiley, New York, 1980.
4. E. A. Moelwyn-Hughes, *Physical Chemistry*, 2nd ed., Chapter XVIII. Pergamon Press, London, 1961.
5. H. S. Harned and B. B. Owen, *Physical Chemistry of Electrolytic Solutions*, 3rd ed. Van Nostrand, New York, 1958.
6. N. Davidson, *Statistical Mechanics*, Chapter 21. McGraw-Hill, New York, 1962.

DATA SOURCE

A. J. Bard, A. Parsons, and J. Jordan, *Standard Potentials in Aqueous Solution.* Marcel Dekker, New York, 1985.

SYMBOLS: CHAPTER 11

New Symbols

$\mathscr{z}_i, V_i, N_i, X_i \cdots$ = notation for a general solution component (subscript i rather than K as a reminder that special electrical considerations apply)

$\mathscr{q}, \mathscr{q}_e, z\mathscr{q}_e$ = charge, magnitude of charge on the electron, a number of electron charges

w_e = electrical work

D = dielectric constant

$f, \mathfrak{f}, f_x, f_y, f_z$ = force, force vector, force components

∇ = Laplacian operator

$\mathscr{V}(r)$ = potential energy in a system of charges

$\mathbf{F}$ = electric field

φ = electrical potential

$E_{\mathscr{A}\mathscr{B}}$ = interaction energy between charged species $\mathscr{A}$ and $\mathscr{B}$

κ = a collection of constants in the Debye-Hückel equations

b = distance of closest approach

ξ = a charging parameter

$\mu_{i,el}$ = electrical contribution to the chemical potential

$c, c_i \cdots$ = molar concentration

$\mathfrak{m}, \mathfrak{m}_i$ = molal concentration

ν_+, ν_- = number of positive or negative ions (in a charge neutral collection)

$\gamma_\pm, \mathfrak{m}_\pm$ = mean activity coefficient, mean molality of a species that dissociates into ions in solution

$u_{\mathscr{A}\mathscr{B}}$ = energy of ion pair formation (one time use)

$y, Q(y)$ = a function of energy in a system of charges (one time use)

$\bar{\mathrm{v}}_1, \bar{\mathrm{v}}_2$ = partial molecular property for a solution component

$\phi_\mathrm{V}, \phi_\mathrm{H}, \phi_\mathrm{L} \cdots$ = apparent partial molar property

$\bar{\mathrm{L}}_1, \bar{\mathrm{L}}_2$ = partial molar heat content for a solution component

$\Delta \mathrm{H_D}$ = heat of dilution

Φ = electrostatic potential

ψ, χ = components of electrostatic potential

$\Delta\Phi$ = difference in electrostatic potentials between two phases

$\mathscr{E}_\mathrm{cell}$ = difference between two $\Delta\Phi$ values when the two systems are components of an electrochemical cell

$\mathscr{E}_{1/2}$ = division of $\mathscr{E}_\mathrm{cell}$ into components by imposing a convention

$\mathscr{E}^\circ_\mathrm{cell}, \mathscr{E}^\circ_{1/2}$ = cell and half cell voltages under defined standard conditions

$\mathscr{F}, n\mathscr{F}$ = 1 mole of charge, a number of moles of charge

Carryover Symbols

$\mathscr{k}$, T, $\mathscr{z}$, $\mathscr{Z}$, E, V, P, T, K, K_eq, G, S, H, a, $\Delta \mathrm{G}^\circ_\mathrm{f}$, $\Delta \mathrm{H}^\circ_\mathrm{f}$, $\Delta \mathrm{S}^\circ_\mathrm{f}$, ϑ, ζ.

APPENDIX A

ENERGY

The concept of energy derives from Newton's laws of motion, phrased in terms of forces, and related statements. Lagranges and Hamiltons reformulation of the laws of motion in terms of energies associated with motion and position constitute the body of classical mechanics. These formulations have the advantage of separability of individual particle energies in a collection of particles and various modes of motion for a single particle. The Hamiltonian is an expression in momentum and positional coordinates for the total energy (motion and position related). The momentum is considered to be a continuous variable. Particles of small mass confined to motion in spaces of small dimension do not exhibit this property. Energy and momentum are quantized, i.e., only certain discrete values are permitted. These considerations are generally not important in center of mass related motions of atoms and molecules. They are important in the internal motions of atom and molecule components. The details associated with this summary are as follows.

Newton's first law of motion states that a particle free of external forces will move in a straight line at constant velocity. This is expressed as

$$\frac{d\mathbf{v}}{dt} = 0 \tag{A.1}$$

where $\mathbf{v}$ is the vector representing particle velocity. More often this statement is phrased in terms of momentum, $\mathbf{p} = m\mathbf{v}$, where m is the (constant in nonrelativistic mechanics) particle mass and

$$\frac{d\mathbf{p}}{dt} = 0 = m\frac{d\mathbf{v}}{dt}. \tag{A.2}$$

In integrated form $\mathbf{p} = \text{constant}$. This is summarized by the statement that momentum is conserved (constant) for a particle free from the influence of external forces.

Newton's second law governs the behavior of the particle when it is subject to external forces,

$$\frac{d\mathbf{p}}{dt} = \mathbf{f}_{ex} = m\frac{d\mathbf{v}}{dt} \tag{A.3}$$

for constant external force $\mathbf{f}_{ex}$. The first law is a special case of the second law for $\mathbf{f}_{ex} = 0$.

Newton's third law states that the agency causing the force experiences an equal and oppositely directed force

$$\mathbf{f}_{ij} = \mathbf{f}_{ji}. \tag{A.4}$$

The concept of energy derives from consideration of the rate of particle displacement when a force is applied. We define work as

$$w = \mathbf{f}\cdot d\mathbf{r} = f_x dx + f_y dy + f_2 dz \tag{A.5}$$

and consider the time derivative of this term

$$\frac{dw}{dt} = \mathbf{f}\cdot\frac{d\mathbf{r}}{dt} = \mathbf{f}\cdot\mathbf{v} \tag{A.6}$$

$d\mathbf{r}/dt$ is the velocity associated with particle motion under the applied force.

We recognize two circumstances:

1. The force acting on the particle is measured by the motion of the particle,

$$\mathbf{f} = m\frac{d\mathbf{v}}{dt}\frac{dw}{dt} = m\frac{d\mathbf{v}}{dt}\cdot\mathbf{v}$$

$$\frac{dw}{dt} = m\frac{dv_x}{dt}v_x + m\frac{dv_y}{dt}v_y + m\frac{dv_2}{dt}v_z$$

$$\frac{dw}{dt} = \frac{d}{dt}\left(\tfrac{1}{2}mv_x^2 + \tfrac{1}{2}mv_y^2 + \tfrac{1}{2}mv_y^2\right) = \frac{d}{dt}\left(\tfrac{1}{2}mv^2\right). \tag{A.7}$$

The work term can be represented by $\frac{1}{2}mv^2$ and is called the kinetic energy.

2. The force acting on the particle is measured by the negative gradient of a function of the particle position,

$$\mathbf{f} = -\Delta U(\mathbf{r}) \tag{A.8}$$

$$\frac{dw}{dt} = \mathbf{f}\cdot\frac{d\mathbf{r}}{dt} = \left[\frac{\partial U(\mathbf{r})}{\partial x}\left(\frac{\partial x}{\partial t}\right) + \frac{\partial U(\mathbf{r})}{\partial y}\left(\frac{\partial y}{\partial t}\right) + \frac{\partial U(\mathbf{r})}{\partial z}\left(\frac{\partial z}{\partial t}\right)\right. \tag{A.9}$$

Such forces are termed conservative. The term $U(\mathbf{r})$ is called the potential energy. Work done on the particle appears as a decrease in potential energy. The total energy for the particle is defined as the sum of kinetic and potential energies,

$$\epsilon = \tfrac{1}{2}mv^2 + U(\mathbf{r}) = \frac{p^2}{2m} + U(\mathbf{r}).$$

For an N particle system

$$E = \sum^{N} \tfrac{1}{2}m_i v_i^2 + U(\mathbf{r}_1, \ldots, \mathbf{r}_N) \tag{A.10}$$

and for a conservative many particle system the total energy is constant.

For one particle (in Cartesian coordinates) Newton's second law produces, in the absence of external forces,

$$m\ddot{x} + \partial U(x, y, z)/\partial x = 0$$

$$m\ddot{y} + \partial U(x, y, z)/\partial y = 0$$

$$m\ddot{z} + \partial U(x, y, z)/\partial z = 0$$

where we have introduced the dot notation often encountered in classical mechanical arguments [i.e., $(\cdot)$ specifies first, $(\cdot\cdot)$ specifies second, time derivatives]. These relationships may be referred to as equations of motion. For N particles there are 3N simultaneous equations of motion for motion in three dimensions summarized as

$$m_i\ddot{x}_i + \frac{\partial U(x_1, \ldots, x_{3N})}{\partial x_i} = 0, \qquad i = 1, 2, \ldots, 3N \tag{A.11}$$

where the three Cartesian coordinates of particle 1 are represented by x_1, x_2, x_3, those of particle 2 are represented by x_4, x_5, x_6, etc. for notational convenience. Also

$$E = \sum^{3N} \tfrac{1}{2}m\dot{x}_i + U(x_1, \ldots, x_{3N}). \tag{A.12}$$

The potential energy of the particle system cannot be represented as the sum of 3N terms, it appears in each of the 3N equations. Solution of these coupled equations is not possible unless the potential energy can be made separable. This can often be accomplished by transformation to a set of generalized coordinates.

Designate the total kinetic energy by T

$$T(\dot{x}_1, \ldots, \dot{x}_{3N}) = \sum^{3N} m\dot{x}_i.$$

If the velocity components can be treated as independent variables

$$m_i\dot{x}_i = \frac{\partial T(\dot{x}_1, \ldots, \dot{x}_{3N})}{\partial \dot{x}_i} \tag{A.13}$$

so that Newton's equations of motion become

$$\frac{d}{dt}\left(\frac{\partial T}{\partial \dot{x}}\right)_i + \frac{\partial U(x_1, \ldots, x_{3N})}{\partial x_i} = 0. \tag{A.14}$$

Now 3N generalized coordinates are introduced. These are defined in such manner that each generalized coordinate is a function of the 3N Cartesian coordinates,

$$q_K = q_K(x_1, \ldots, x_{3N}).$$

Conversely,

$$x_i = q_i(q_1, \ldots, q_{3N}). \tag{A.15}$$

Generalized coordinates will be chosen on the basis of the problem at hand. They can be Cartesian displacements, distances along curved paths, angular relationships, etc. For each generalized coordinate there is a generalized velocity, $\dot{q}_K = dq_K/dt$. The potential energy reexpressed is $U(q, \ldots, q_{3N})$ and the kinetic energy reexpressed is $T(q_i, \ldots, q_{3N}, \dot{q}_1, \ldots, \dot{q}_{3N})$. The analog of Eq. (A.14) is

$$\frac{d}{dt}\frac{\partial T}{\partial \dot{q}_K} - \frac{\partial T}{\partial q_K} + \frac{\partial U}{\partial q_K} = 0 \tag{A.16}$$

on the basis of which the Lagrangian function

$$L(q_1, \ldots, q_{3N}, \dot{q}_1, \ldots, q_{3N}) = T(q, \ldots, q_{3N}, \dot{q}, \ldots, \dot{q}_{3N}) - U(q_1, \ldots, q_{3N})$$

is defined so that

$$\frac{d}{dt}\left(\frac{\partial L}{\partial \dot{q}_K}\right) - \frac{\partial L}{\partial q_K} = 0. \tag{A.17}$$

The 3N second-order differential equations are the Lagrangian equations of motion for a system requiring 3N generalized coordinates for its complete description. These may be referred to as degrees of freedom or modes of motion.

In analogy to Eq. (A.13), a generalized momentum can be defined

$$\frac{\partial L(q_1,\ldots,q_{3N},\dot{q},\ldots,\dot{q}_{3N})}{\partial q_J} = p_J \tag{A.18}$$

the total derivative of L is

$$dL = \sum^{3N}\left[\left(\frac{\partial L}{\partial q_J}\right)dq_J + \left(\frac{\partial L}{\partial \dot{q}_J}\right)d\dot{q}_J\right] \tag{A.19}$$

so that, with Eq. (A.18),

$$dL = \sum^{3N}(\dot{p}_J dq_J + p_J d\dot{q}_J). \tag{A.20}$$

A transformation of variables from $q, \dot{q}$ to q, p produces

$$-H(q,\ldots,q_{3N},p_1,\ldots,p_{3N}) = L(q_1,\ldots,q_{3N},\dot{q}_1,\ldots,\dot{q}_{3N}) - \sum^{3N} p_J\dot{q}_J$$

and

$$-dH = dL - \sum^{3N} d(p_J\dot{q}_J)$$

or

$$-\sum^{3N}\left(\frac{\partial H}{\partial q_J}dq_J + \frac{\partial H}{\partial p_J}dp_J\right) = \sum^{3N}\dot{p}_J dq_J + \sum^{3N} p_J d\dot{q}_J - \sum^{3N} p_J d\dot{q}_J - \sum^{3N}\dot{q}_J dp_J$$

so that

$$dH = \sum^{3N}(-p_y dq_J + q_J dp_J)$$

or

$$-\frac{\partial H}{\partial q_J} = \dot{p}_J \qquad \text{and} \qquad \frac{\partial H}{\partial p_J} = \dot{q}_J \qquad J = 1,\ldots,3N. \tag{A.21}$$

These are Hamilton's equations of motion. Clearly, referring to Eq. (A.18), the Hamiltonian is the total energy of the system

$$H = T + U = E. \tag{A.22}$$

For a single particle in linear motion

$$\epsilon = \tfrac{1}{2}mv^2 + U(\mathrm{q}) = p^2/2m + U(\mathrm{q}). \tag{A.23}$$

In simple cases $U(\mathrm{q})$ may be a constant independent of position and in the simplest case it may be zero. Cartesian coordinates will then be most convenient. From Newton's second law

$$\mathrm{F(x)} = ma = m\ddot{\mathrm{x}}$$

$$p = m\dot{\mathrm{x}}, \qquad \dot{p} = m\ddot{\mathrm{x}} = F(\mathrm{x})$$

and a particle initially at rest subjected to a constant force F for a time t will have kinetic energy $\epsilon = F^2t^2/2m$. For arbitrary F and t, ϵ can assume any value.

Rotation of a particle with respect to an axial system is a common motion. The linear momentum p is replaced by the angular momentum $J = I\omega$ for moment of inertia I and angular velocity ω. Newton's equation of motion is $\dot{J}\mathfrak{T}$ for torque $\mathfrak{T}$. If constant torque $\mathfrak{T}$ is applied for time t to a particle initially at rest $\epsilon = \mathfrak{T}^2/2I$ and can be any value determined by arbitrary $\mathfrak{T}$ and t.

For vibration (oscillation) the restoring force of the oscillator is proportional to the displacement ξ

$$\mathrm{F}(\xi) = -\mathrm{K}\xi = m\frac{d^2\xi}{d\mathrm{t}^2}.$$

The force is also the negative gradient of the potential

$$\mathrm{F}(\xi) = -\frac{dU}{d\xi} = -\mathrm{K}\xi$$

so that

$$U = \frac{\mathrm{K}\xi^2}{2} \qquad \text{and} \qquad \mathrm{E} = \frac{p^2}{2m} + \frac{\mathrm{K}\xi^2}{2}. \tag{A.24}$$

A solution for the differential equation

$$\frac{d^2\xi}{d\mathrm{t}^2} + \frac{\mathrm{K}}{m}\xi = 0$$

is

$$\xi(\mathrm{t}) = A\sin\left(\frac{\mathrm{K}}{m}\right)^{1/2}\mathrm{t}.$$

A complete period of oscillation occurs for $(\mathrm{K}/m)^{1/2}\mathrm{t} = 2\pi$, or in the time period

$t = 2\pi(m/K)^{1/2}$. The oscillator frequency ν is

$$\frac{1}{t} = \frac{1}{2\pi}\left(\frac{K}{m}\right)^{1/2}. \qquad (A.25)$$

The energy is

$$\begin{aligned} E &= \frac{m}{2}\left[\left(\frac{K}{m}\right)^{1/2} A\cos\left(\frac{K}{m}\right)^{1/2} t\right]^2 + \frac{K}{2} A^2 \sin^2\left(\frac{K}{m}\right)^{1/2} t \\ &= \frac{K}{2} A^2 \cos^2\left(\frac{K}{m}\right)^{1/2} t + \frac{K}{2} A^2 \sin^2\left(\frac{K}{m}\right)^{1/2} t = \frac{K}{2} A^2. \end{aligned} \qquad (A.26)$$

An impulse provided to the oscillator at rest (potential energy zero) produces kinetic energy of $KA^2/2$, which remains constant after excitation and is, through the force constant K, proportional to the square of the amplitude A, which can be any value for given oscillator characteristics (K, m) by control of the initial impulse.

In summary, on the basis of classical mechanics governed by Newton's laws of motion, a particle can be given any energy through control of the appropriate forces that initiate motion. In the absence of dissipative forces (e.g., friction) the particle will continue to exhibit the motion without further action; its energy is conserved. These results are applicable under the circumstances of massive particles moving in macroscopic dimensions. When dealing with atomic masses and atomic dimensions, a more general mechanics is required. The basic feature of this more general mechanics is the quantization of energy. Particles subject to this (quantum) mechanics cannot assume any energy value in their various motions but only certain discrete values. The most compelling argument for accepting this premise arises from spectroscopy, i.e., the coupling of energy states with the absorption or emission of particular wave lengths of light by atoms and molecules in radiation related energy exchanges.

Electromagnetic radiation, for many purposes conveniently characterized by the wave features of amplitude (A), frequency (ν), and velocity (c), may also be characterized for other purposes in terms of the energy associated with a particle system of photons where the energy of each photon is $h\nu$ (h = Plancks constant = 6.62×10^{27} erg sec = 6.62×10^{20} J sec). If each photon emitted or absorbed by an atom or molecule has energy $\epsilon = h\nu = hc/\lambda$ where λ is the wavelength, then a change $-\Delta\epsilon(+\Delta\epsilon)$ has occurred in the energy of the emitting (absorbing) atom or molecule. Since only particular wavelengths are involved, only particular changes in energy can be acceptable and, it would follow, only particular energy states exist such that $\epsilon_2 - \epsilon_1 = hc/\lambda$.

The utility of the photon concept in many questions relating to radiation, which is normally considered in the wave framework, suggests that particles in motion having momentum p can be characterized by a wave length $\lambda = h/p$. This is an experimentally verifiable assertion (electrons are diffracted by crystal planes in the same manner that X-rays are diffracted). Consideration of the motion of atoms, molecules, and their constituent parts in a wave mechanical framework provides the quantized energy states we require.

Wave motion can be described by a function $\Psi(x, y, z)$ which relates amplitude to time and space coordinates

$$\nabla\Psi - \frac{1}{c^2}\frac{\partial^2\Psi}{\partial t^2} = 0 \tag{A.27}$$

where ∇^2 is the Laplacian operator $(\partial^2/\partial x^2) + (\partial^2/\partial y^2) + (\partial^2/\partial z^2)$ and c is the velocity of wave propagation. Stationary waves have the property that the wave function is factorable into separate functions of time and space

$$\Psi(x, y, z, \mathrm{t}) = \Psi(x, y, z)T(\mathrm{t})$$

so that

$$T\nabla^2\Psi = \frac{\Psi\ddot{T}}{c^2}$$

or

$$\frac{\nabla^2\Psi}{\Psi} = \frac{1}{c^2}\frac{\ddot{T}}{T}. \tag{A.28}$$

The left-hand side of Eq. (A.28) depends only on x, y, and z while the right-hand side depends only on t. This can be true for all x, y, z, and t only if the terms are separately equal to a constant ($-\mathrm{K}^2$ for arithmetic convenience) so that

$$\ddot{T} + c^2\mathrm{K}^2\mathrm{t} = 0$$

which has a general solution

$$T(\mathrm{t}) = a\sin c\mathrm{Kt} + b\cos c\mathrm{Kt} = a\sin\omega\mathrm{t} + b\cos\omega\mathrm{t}$$

where we identify $\omega = c\mathrm{K}$ as angular frequency. The corresponding linear frequency is $\nu = \omega/2\pi$ and the wavelength is $\lambda = 2\pi/\mathrm{K}$. Equation (A.28) then becomes

$$\nabla^2\Psi + \frac{4\pi^2}{\lambda}\Psi = 0 \tag{A.29}$$

which describes the space dependence of standing waves. It is known as the Helmholz equation.

If Eq. (A.29) with λ interpreted as the momentum equivalent wavelength $(=h/p)$ can be applied to matter waves then, with $p^2 = 2m[\mathrm{E} - U(\mathbf{r})]$ and $1/\lambda^2 = p^2/h^2 = 2m(\mathrm{E} - U(\mathbf{r}))/h^2$, the applicable wave equation would be

$$\nabla^2\Psi(\mathbf{r}) + \frac{8\pi^2 m}{h^2}(\mathrm{E} - U(\mathbf{r}))\Psi(\mathbf{r}) = 0. \qquad \text{(A.30)}$$

This equation, known as the Schrödinger equation, rearranges to

$$-\frac{h^2}{8\pi^2 m}\nabla^2\Psi(\mathbf{r}) + U(\mathbf{r})\Psi(\mathbf{r}) = \mathrm{E}\Psi(\mathbf{r}) \qquad \text{(A.31)}$$

or $\mathscr{H}\Psi = \mathrm{E}\Psi$ with $\mathscr{H}$ being the Hamiltonian operator

$$\mathscr{H} = -\frac{h^2}{8\pi^2 m}\nabla^2 + \Psi(\mathbf{r}).$$

The correspondence between the Hamiltonian operator and the Hamiltonian function permits the Schrödinger equation to be constructed by prescription for any dynamic system. The kinetic energy component of the Hamiltonian clearly transforms as

$$T = \frac{p^2}{2m} \rightarrow \frac{\hbar}{2m}\nabla^2 \qquad \text{(A.32)}$$

where $\hbar = h/2\pi$ is conventional notation. The individual momentum components in Cartesian coordinates transform as

$$p(x) = i\hbar\frac{\partial}{\partial x}$$

$$p(y) = i\hbar\frac{\partial}{\partial y}$$

$$p(z) = i\hbar\frac{\partial}{\partial z}. \qquad \text{(A.33)}$$

This transformation is not accomplished so directly when the momentum components are in terms of other coordinate systems.

Equation (A.31) is applicable to a single particle moving in three dimensions. It is a second-order differential equation with an infinity of solutions. Physical reality imposes some constraints that permit rejection of most solutions. For example, if a material particle is to be regarded as having wave properties, the classical localized position must be distributed over a region of space according to the amplitude and waveform of the wave. The intensity of electromagnetic radiation is given by the square of the amplitude. If $\Psi(\mathbf{r})$ is to be regarded as the amplitude of a matter wave, then its square is matter density. The square of the wave function at any point (or $\Psi\Psi^*$ if complex) is taken to be the probability that the particle is to be found there. The probability that the particle is somewhere requires that

$$\int_{-\infty}^{\infty} \Psi^*(\mathbf{r})\Psi(\mathbf{r})d\tau = 1$$

where $d\tau$ is a volume element. In addition to this normalization requirement the wave function should be finite, single valued, and continuous in order that the probability density be well behaved. There is a further feature of associating Ψ^2 with a probability. This feature is known as the Heisenberg uncertainty principle and can be phrased as $\Delta x \Delta p \geqslant h$, i.e., both position and momentum cannot be precisely known. This statement provides a physical meaning for the constant h and imposes constraints on the wave equation solutions.

Consider Eq. (A.31) applied to a particle moving in one dimension under circumstances in which $U(x) = 0$ in the region $0 < x < \mathscr{L}$ and $U(x) = \infty$ outside that region. The problem is appropriate for the description of translational motion in one dimension of any particle moving in a square well potential or of an isolated molecule moving in a bounded container, i.e., an ideal gas phase molecule moving in one dimension. The solution for

$$\frac{h^2}{2\pi}\frac{\partial^2\Psi(x)}{\partial x} + (\mathrm{E} - U(x))\Psi(x) = 0$$

under these conditions is

$$\Psi(x) = B\cos \mathrm{K}x + \mathrm{A}\sin \mathrm{K}x$$

with $\mathrm{K} = (2m\mathrm{E})^{1/2}/\hbar^2$ so that $\mathrm{E} = \mathrm{K}^2\hbar^2/2m$. At $x = 0$, $\cos \mathrm{K}x = 1$, therefore B must be 0 and $\Psi(x) = \mathrm{A}\sin \mathrm{K}x = 0$; at $x = \mathscr{L}$, for $\mathrm{A}\sin x$ to again be 0, $\mathrm{K}x$ must be an integral multiple of π, i.e., $\mathrm{K}x = n\pi$ and $\mathrm{K} = n\pi/\mathscr{L}$ for which $\mathrm{E} = n^2\hbar^2\pi^2/2m\mathscr{L}^2$ for $n = 1, 2, \ldots$. Only certain energy levels are acceptable. The absence of the $n = 0$ value produces a minimum (zero point) energy of $h^2/8m\mathscr{L}$ and successive levels are separated by $n^2(2n+1)/8m\mathscr{L}^2$. The three-dimensional problem is quite straightforward. The potential energy is zero within the box and

$$\mathrm{E} = \frac{n^2}{8m}\left(\frac{n_x^2}{a^2} + \frac{n_y^2}{b^2} + \frac{n_z^2}{c^2}\right) \qquad \text{(A.34)}$$

where a, b, c are linear dimensions. This treatment serves as a model for the translational motion of an ideal gas molecule.

Other modes of motion of interest include rotation and vibration. The construction of the Schrödinger equation from the classical Hamiltonian for an appropriate model of the motion and the wave equation solutions will provide the quantized energy states associated with that motion. We first consider those motions of electrons in atoms that can be treated as rotations. It is easiest to approach this problem in two steps: first rotation of a particle in a plane and second three-dimensional rotation. For the first case it is probably most constructive to arrive at the energy levels by intuitive argument that goes as follows.

Consider a particle of mass m moving around a circular path of radius r. The particle has momentum and this we associate with a wavelength $\lambda = h/p$. We can convert the linear momentum p to angular momentum $J = pr$ so that the energy given by $p^2/2m$ becomes $J^2/2mr^2$. We call mr^2 the moment of inertia (I) and write $\mathrm{E} = J^2/2I$. The momentum is quantized which implies that only certain values of λ are permissible. The permissible values of λ are those that lead to superposition of nodes and antinodes on successive circuits (otherwise the wave would cancel ultimately leading to zero probability of finding the particle on the ring). The circumference of the circle ($2\pi r$) must be an integral number of wavelengths and $\lambda = 2\pi r/h, n = 1, 2, \ldots$. The magnitude of linear momentum is therefore limited to $n\hbar/r$, the magnitude of angular momentum is limited to $J = n\hbar$, and the magnitude of energies of $\mathrm{E} = n^2\hbar^2/2I$ for $n = 0, 1, \ldots$. A particular momentum value can be achieved, however, by rotation in either the clockwise or counterclockwise direction so that all momentum values are included for $p = m_r\hbar/r$ and $J = m_r\hbar$ where $m_r = 0, \pm 1, \pm 2, \ldots$. Since $\mathrm{E}_{\mathrm{rot}} = m_r^2\hbar^2/2I$ the values are independent of the sign as opposed to the values of p^2 or J^2. The momentum vector **J** is taken to be positive for counterclockwise rotation and negative for clockwise rotation and in either case is perpendicular to the plane of rotation.

When rotation is extended to three dimensions the particle can be regarded as moving over the surface of a sphere and the wave function must match for a polar path. The problem is more complicated and solutions must be obtained from the wave equation that we set down as follows. The kinetic energy is $T = (p_x^2 + p_y^2 + p_z^2)2m$ in Cartesian coordinates and transforms into

$$T = \frac{1}{2m}\left(p_r^2 + \frac{p_\theta^2}{r^2} + \frac{p_\phi}{r^2 \sin 3\theta}\right)$$

in spherical polar coordinates that are more appropriate to rotation. The Hamiltonian operator is obtained from Eq. (A.32) and is

$$\mathscr{H} = -\frac{\hbar^2}{2m}\left[\frac{1}{r^2}\frac{\partial}{\partial r}r^2\frac{\partial}{\partial r} + \frac{1}{r^2 \sin\theta}\frac{\partial}{\partial\theta}\sin\theta\frac{\partial}{\partial\theta} + \frac{1}{r^2\sin^2\theta}\frac{\partial^2}{\partial\theta^2}\right] + U(r).$$

Only the results of the solution can be set down here. Angular momentum about the z axis is quantized and limited to $J = m_l\hbar, m_l = 0, \pm 1, \pm 2, \ldots, l$. There is now a limit to the absolute values of m_r, i.e., m_l is confined to $2l+1$ values. The magnitude of the angular momentum is quantized for all directions in space and is confined to the values $J=[l(l+1)]^{1/2}\hbar$ for $l=0,1,2,\ldots$. Both l and m_l are quantum numbers, which together specify the magnitude of angular momentum and the number of orientations that a particular angular momentum vector can assume. It should be particularly noted that the magnitude of angular momentum does not increase in steps of $l\hbar$ but in steps of $[l(l+1)]^{1/2}\hbar$. The reasons for this are related to the uncertainty principle. The occurrence of a product of this form is quite general and will be encountered in other circumstances.

The specification of the quantum number associated with projection of the orbital angular momentum on the z axis as m_l suggests that other rotation related momenta are sometimes present. This is true; fundamental particles (electrons, protons, neutrons, photons) have spin about their own axes. In analogy to the orbital angular momentum and its quantum number we designate a spin angular momentum and a spin quantum number m_s. The magnitude of the spin angular momentum is confined to $[s(s+1)]^{1/2}\hbar$ and its projection on the z axis to integral multiples of $\hbar, m_s\hbar$ with $m_s=s_1(s-1),\ldots,s$. There are only two spin possibilities, clockwise and counterclockwise, and the associated s values are $+1/2$, $-1/2$ so that m_s is $+1/2$ or $-1/2$.

Restriction of momentum quantization to a single axis in the case of both orbital and spin angular momentum derives from the Heisenberg uncertainty principle. The uncertainty principle would predict that, if the momentum J_z is precisely known, nothing can be said about the vector components J_x and J_y. Angular momentum is often usefully presented as a vector model as in Figure A.1. The circular surface of radius $[j(j+1)]^{1/2}$ and the side of the cone $[j(j+1)]^{1/2}$ are particular values of l drawn here for $j=2$. The z axis projection is given by the five m values -2, -1, 0, $+2$, $+1$ and the momentum vector is oriented somewhere in the surface of the cone.

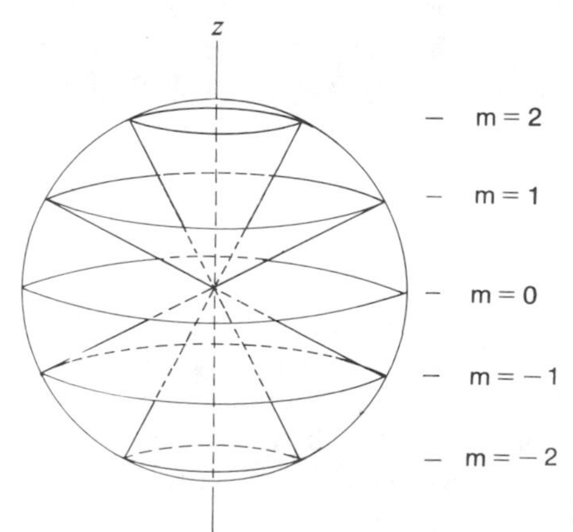

Figure A.1. The vector model for allowed values of angular momentum.

These considerations relating to rotation are of interest in that they are important features of the treatment of atomic orbitals that in turn provide the electronic structure of atoms and determine ground electronic and excited electronic state energies that will be of importance in statistical mechanical calculations. Much of the rotational treatment structure will also be applicable to our considerations of molecular rotation. We conclude the remarks on atomic structure with the observation that the Schrödinger equation for a single electron of charge $-\mathscr{q}_e$ in the vicinity of the nucleus of charge $+\mathscr{q}_e$ is

$$\frac{-h^2}{2m}\left(\frac{\partial^2}{\partial x^2}+\frac{\partial^2}{\partial y^2}+\frac{\partial^2}{\partial z^2}\right)\Psi-\left(\frac{\mathscr{q}_e^2}{4\pi\epsilon_0 r}\right)\Psi=\mathrm{E}\phi$$

where ϵ_0 is the electrical permitivity. Solving this equation with the required boundary conditions yields three quantum numbers, m_l and m_s as before and n, which regulates the distance from the nucleus. For the hydrogen atom the energies are related only to n,

$$\mathrm{E}_n=-\frac{m\mathscr{q}_e^4}{32\pi\epsilon_0\hbar^2}\left(\frac{1}{n^2}\right).$$

The Schrödinger equation cannot be solved in closed form for atoms other than hydrogen but the general form for the hydrogen orbitals can be utilized in the description of electronic configurations. The general procedure is to place a maximum of two (spin paired electrons) in each orbital. This follows from the requirement that no two electrons have identical quantum number descriptions (the Pauli exclusion principle). Orbitals associated with $n > 1$ will not interact with the nucleus in a simple Coulombic potential and the energies will depend on the n, l, and m quantum numbers. General rules will suffice to determine the electronic configuration of any atom. The values of l permitted are $0, 1, 2, \ldots, (n-1)$, the values of m_l permitted are $-l \cdots 0 \cdots +l$. We associate a particular l value with particular orbital forms (i.e., particular probability density patterns), which retain their specific form except for expansion as n increases. Since given values of the orbital angular momentum produce fixed values of E, the several permitted projections of orbital angular momentum all lead to the same value of E. Each orbital associated with a given l value is m_l-fold degenerate, i.e., can occur m_l-fold times with the same energy. Since the actual energies cannot be calculated, it will be necessary to rely on spectroscopic data to obtain energies of excited states.

We complete the problem of rotation by considering the rotation of molecules in free space. It is most practical to assume that rigid bonds exist in the molecule, to obtain the energy states on this basis and then consider whether or not refinement is required. Actually rotational energies will almost always be obtained from spectroscopy so that precise calculation is not required. The framework is, however, important for our understanding.

For a body rotating about some axis x,

$$E_r = \tfrac{1}{2} I_x \omega_x^2$$

where I is the moment of inertia and ω is the angular velocity. Free rotation about three axes is

$$E_r = \tfrac{1}{2} I_x \omega_x^2 + \tfrac{1}{2} I_y \omega_y^2 + \tfrac{1}{2} I_z \omega_z^2 \tag{A.35}$$

or, in terms of angular momentum

$$E_r = \frac{J_x^2}{2I_x} + \frac{J_y^2}{2I_y} + \frac{J_z^2}{2I_z}. \tag{A.36}$$

Moment of inertia and angular velocity are formal analogs of mass and linear velocity. It is required that I_x, I_y, and I_z be so-called principal moments of inertia in order that the stationary state corresponds to a motion with no moments acting on the center of mass.

This general rotation about the principal axes is characterized by the elipsoid of rotation, which can be visualized for a molecule as follows. Take any axis passing through the center of mass and, defining the moment of inertia as

$$I = \sum m_i x_i^2 \tag{A.37}$$

construct the moment of inertia about that axis by measuring the perpendicular distance (x_i) from each mass point (m_i) in the molecule to the axis. Evaluate I from the x_i and m_i. Construct the vector $\boldsymbol{\rho} = \mathbf{n}/\sqrt{I}$ where n is a convenint scaling factor and $\mathbf{n}$ is directed along the axis of rotation. Repeat the process for all possible axes. The terminus of all vectors is an elipsoid, the axes of which are inversely proportional to the square root of the three principal moments.

The choice of this axial set has the advantage of permitting the Hamiltonian for rotational energy to be written

$$H_r = \frac{J_x^2}{2I_x} + \frac{J_y^2}{2I_y} + \frac{J_z^2}{2I_z}$$

where the J_x, etc. are angular momentum vectors perpendicular to the corresponding principle axis for which the principle moment of inertia is I_x, etc. If other axes were used there would be cross-product terms. These particular rotations are independent.

When the quantum prescription is applied to this rotation two quantum numbers emerge as in the rotating electron case. These we refer to as j and m. The j quantum number expresses the requirement that the angular momentum be quantized. The $m = -j \cdots 0 \cdots j$ quantum number requires that only $2j + 1$ quantized projections of J onto the z axis be permitted where $J = [j(j+1)]^{1/2}\hbar$.

Molecule structures may have symmetry elements that reduce the number of independent moments that must be considered. For a molecule such as methane (called a spherical top) $I_x = I_y = I_z$. For a molecule such as CH_3I and NH_3 (symmetric tops) $I_x = I_y \neq I_z$, and for linear molecules in general $I_x = I_y$, $I_z = 0$. Moments of inertia for these general molecule forms are given in Table 5.3. Inertial constants are commonly obtained from spectroscopic data. Spectroscopists usually report rotational constant $B = \hbar/2I$, which has units of frequency. The rotational constant is, however, usually reported in wave numbers.

A complete dynamic description of a molecule cannot be accomplished in terms of fewer than $3n$ coordinate specifications where n is the number of atoms in the molecule. The coordinate specifications do not, however, have to be those of the mass centers of the n constituent atoms. In the gas phase, all molecules will have translation and rotation. These motions are not readily described in terms of the individual atom coordinates. Regardless of whether we ultimately apply classical or quantum considerations it will be necessary, in principle, to construct the Hamiltonian for these motions. Translation will require the specification of three coordinates for the center of mass whether they be Cartesian or some generalized coordinate system. Rotation is characterized by either two (linear) or three (nonlinear) principal moments of inertia, which are in turn defined by either two or three coordinate specifications. The remaining $3n - 5$ or $3n - 6$ coordinate specifications can be utilized in describing internal motions.

The internal motions in polyatomic molecules can almost always be regarded as oscillations to which the harmonic oscillator approximation can be applied if the complex motions in the molecule can be resolved into independent modes. This is accomplished by a technique called normal mode analysis, which we will not cover here but only recognize that techniques are available that, when used to interpret spectroscopic data, will provide the $3n - 5$ or $3n - 6$ independent vibrational modes.

The classical (mass on a spring) approach to vibration was summarized by Eqs. (A.24) through (A.26). When the oscillator is two atomic masses vibrating with respect to the center of mass and the restoring force is the chemical bond stiffness a somewhat different approach is required.

For vibration, the oscillator can have only one frequency (ν) and therefore one value of K, m, and also ω. The value of K clearly derives from the strength of the chemical bond; the mass value is not obvious. The center of mass of an atom (homonuclear or heteronuclear) pair is obtained by equalizing the moments m_1R_1, and m_2R_2 where R_1 and R_2 are measured from the center of mass. So, for $m_1R_1 = m_2R_2$ and $R_1 + R_2 = R$ where R is the bond length, a vibration may be regarded as a change in R from a rest position R_0 and we require an expression for change in R. Adding R_2m_1 to each side of the equal moments expression we obtain

$$m_1R_1 + m_1R_2 = m_2R_2 + m_1R_2$$

$$\frac{m_1R_1}{m_1 + m_2} = R_2$$

and similarly

$$\frac{m_2 R}{m_1 + m_2} = R_1.$$

The restoring force for each atom must be proportional to the total bond stretch $R - R_0$, so

$$m_1 \frac{d^2 R_1}{dt} = \mathrm{K}(R - R_0) = m_2 \frac{d^2 R_2}{dt}$$

$$\frac{m_1 m_2}{m_1 + m_2} \frac{d^2 R}{dt^2} = \mathrm{K}(R - R_0).$$

This is the classical expression for a particle having mass $m_1 m_2 / m_1 + m_2$ under displacement $R - R_0$ with restoring force constant K. This mass term is referred to as the reduced mass and given the symbol μ_{R}.

The displacements are small fractions of the bond length and μ_{R} is order of atomic mass so that quantum considerations must be applied. The Schröedinger equation is

$$-\frac{\hbar^2}{2\mu_n} \frac{\partial}{\partial R} \Psi(x) + \frac{\mathrm{K}}{2}(R - R_0)\Psi(x) = \mathrm{E}\Psi(x). \tag{A.38}$$

The potential energy increases with displacement and the wave function approaches 0 for large displacement. Solution of the wave equation with these boundary conditions produces

$$\mathrm{E}_{\mathrm{v}} = (n + \tfrac{1}{2}) \frac{\mathrm{K}}{\mu_{\mathrm{R}}} \hbar = (n + \tfrac{1}{2}) \hbar\omega \tag{A.39}$$

where ω is an angular frequency. The expression is often written

$$\mathrm{E}_{\mathrm{v}} = (n + \tfrac{1}{2}) h\nu$$

where ν is a linear frequency. Recalling that $\hbar = h/2\pi$; $\nu = \omega/2\pi$ with units for either quantity being sec^{-1}. In some compilations of spectroscopic data $\tilde{\nu}$ is used for the wave number defined as $1/\lambda$ so that $c\tilde{\nu}$ (wave number) $= \nu$, i.e., the dimension of $\tilde{\nu}$ (wave number) is cm^{-1} for c (velocity of light) $= 3 \times 10^{10}\,\mathrm{cm\,sec}^{-1}$.

Not all oscillations making up the $3n - 5$ or $3n - 6$ modes of motion are associated with the simplest form of vibration, which we could refer to as a symmetric stretch (both atoms simultaneously moving away from the center of mass of the molecule). There can also be asymmetric stretching of the bonds between atoms and various bending and torsional modes of motion. These usually can be treated as oscillations within the harmonic oscillator framework, and quite often the normal modes (those which are independent) are apparent. For example the linear molecule CO_2 has the following normal modes of motion

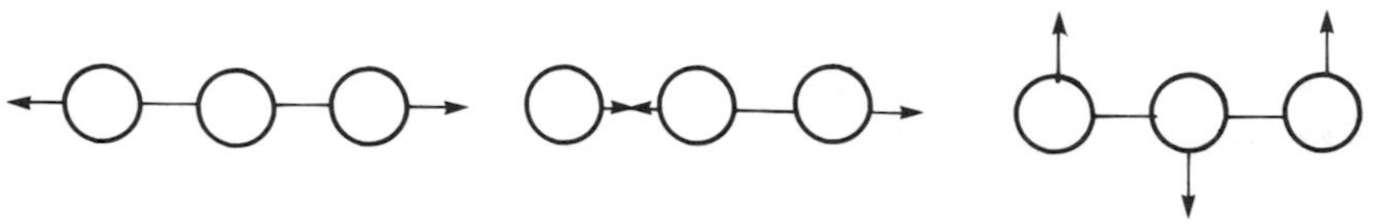

in which the motion of one atom pair does not in itself incite other motion. In a similar manner the nonlinear water molecule has normal modes, which can be represented

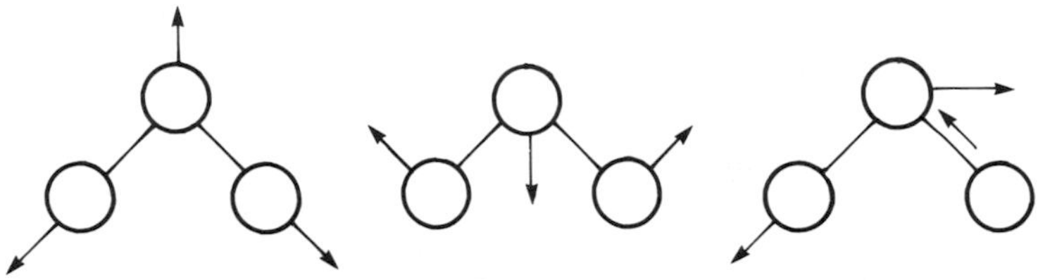

In molecules such as ethane there could be free rotation of the methyl groups about the carbon–carbon bond and in ethylene a torsional oscillation of the two methylene groups of the carbon–carbon double bond. Actually, in ethane rotation is only a high temperature feature. Consider the following sketch for staggered orientations of the methyl hydrogens.

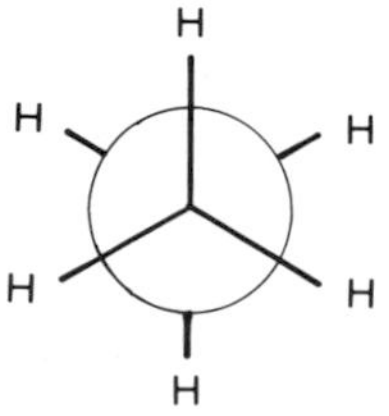

Free rotation requires that a potential energy barrier associated with the passage of two hydrogens be overcome. We describe the barrier in terms of the angle ϕ between hydrogens as

$$\phi = \tfrac{1}{2}U_0(1 - 3\cos 3\phi). \tag{A.40}$$

Free rotation will occur only if $kT \gg U_0$. Hindered rotation would occur if kT is order of U_0 and torsional oscillation if $kT < U_0$. In some cases (not ethane) these vibrations are spectroscopically active. They can be resolved, of course, from heat capacity measurements.

CONTENT SOURCES AND FURTHER READING

1. P. W. Atkins, *Physical Chemistry*, 2nd ed., Chapter 13. W. H. Freeman, San Francisco 1982.
2. S. M. Blinder, Advanced Physical Chemistry, Chapters 4 and 5, Macmillan Co., New York, 1969.
3. R. S. Berry, S. A. Rice, and J. Ross, *Physical Chemistry*, Chapters 3, 4, and 5. Wiley, New York, 1980.
4. E. A. Moelwyn-Hughes, *Physical Chemistry*, 2nd ed., Chapter IV. Pergamon Press, London, 1961.

APPENDIX B

MATHEMATICAL CONSIDERATIONS

DERIVATIVE FUNCTIONS

The derivative of a functional relationship $y = f(x)$ tells us how y changes as x changes. The derivative, written dy/dx, can be obtained by methods of the calculus from an analytic form, e.g., for $y = x^2$, $dy/dx = 2x$, or from the slope of the tangent to the plotted curve at any point. If y is a function of two or more variables it is necessary to introduce partial derivatives. For the two independent variable case the concept can be illustrated graphically. Let $z = f(x, y)$ be represented as a surface in the Cartesian coordinate system (Figure B.1). Any point P on the surface can be represented as the intersection of two lines which can be the intersections of planes representing constant values of any two of the three variables (constant x and constant y in the Figure B.1 example) with the surface. The slopes of the two curves, both taken at point P are each just like an ordinary derivative but expressed in a unique notation [$(\partial z/\partial y)_x$ and $(\partial z/\partial x)_y$ in the Figure B.1 example] indicating the variable held constant in the construct by a subscript. Clearly, the entire surface can be generated by a sequence of such constructs taken over all values of any two of the three variables x, y, z. Such partial derivatives can be obtained by geometric construct as illustrated or from the analytical expression defining the surface by differentiation if in any differentation the indicated variable is treated as a constant, e.g., for $z = x^2y^2$, $(\partial z/\partial x)_y = 2y^2x$ and $(\partial z/\partial y)_x = 2x^2y$.

The total differential of $y = x^2$ is written as $dy = 2xdx$, which can be read as $dy = (dy/dx)dx$. The total differential of a function in several variables substitutes

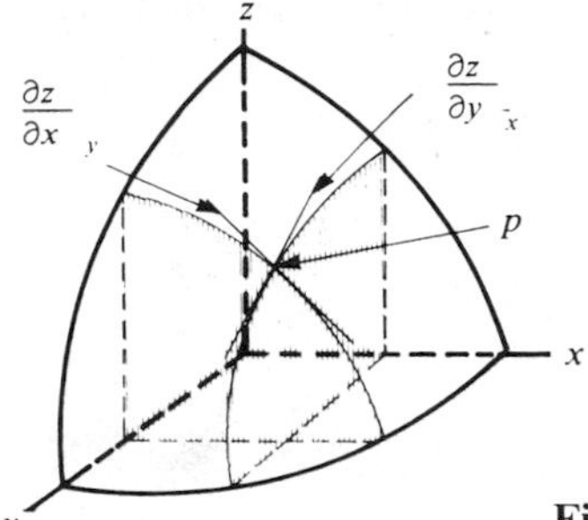

Figure B.1. Illustration of the partial derivative concept.

the partial derivatives for the ordinary derivative, e.g., for the two variable case,

$$dz = \left(\frac{\partial z}{\partial x}\right)_y dx + \left(\frac{\partial z}{\partial y}\right)_x dy \tag{B.1}$$

or more generally for $z = \mathrm{f}(x_1, \ldots, x_n)$,

$$dz = \sum^{\kappa} (\partial z/\partial x_\kappa)_{x_J} dx_\kappa, \qquad J \neq \kappa.$$

Often there are interrelationships between two sets of variables such that there are alternative choices of independent variables. If $z = \mathrm{f}(x, y)$ but $x = \mathrm{f}_x(u, v)$ and $y = \mathrm{f}_y(u, v)$ the total derivative dz [from Eq. (B.1)] may be partially differentiated by u or by v to give

$$\left(\frac{\partial z}{\partial u}\right)_v = \left(\frac{\partial z}{\partial x}\right)_y \left(\frac{\partial x}{\partial u}\right)_v + \left(\frac{\partial z}{\partial y}\right)_x \left(\frac{\partial y}{\partial u}\right)_v \tag{B.2}$$

$$\left(\frac{\partial z}{\partial v}\right)_u = \left(\frac{\partial z}{\partial x}\right)_y \left(\frac{\partial x}{\partial v}\right)_u + \left(\frac{\partial z}{\partial y}\right)_x \left(\frac{\partial y}{\partial v}\right)_u \tag{B.3}$$

also

$$dz = \left(\frac{\partial z}{\partial u}\right)_v du + \left(\frac{\partial z}{\partial v}\right)_u dv. \tag{B.4}$$

More generally for $z = \mathrm{f}(x_1, \ldots, x_n)$ where $(u_1, \ldots, u_n)$ are an alternative set of variables

$$\left(\frac{\partial z}{\partial u_r}\right)_s = \sum^{n,r} \left(\frac{\partial z}{\partial x_i}\right)_{x,j} \left(\frac{\partial x_i}{\partial u_r}\right)_{u,s}, \qquad j \neq i, s \neq r. \tag{B.5}$$

This relationship is referred to as the chain rule in mathematical texts.

Various differential expressions can be constructed with the useful identities

$$\left(\frac{\partial z}{\partial y}\right)_x = 1 \Big/ \left(\frac{\partial y}{\partial z}\right)_x, \qquad \left(\frac{\partial z}{\partial x}\right)_u \left(\frac{\partial x}{\partial y}\right)_u = \left(\frac{\partial z}{\partial y}\right)_u \tag{B.6}$$

e.g., from Eq. (B.3),

$$\left(\frac{\partial z}{\partial v}\right)_u \left(\frac{\partial v}{\partial x}\right)_u = \left(\frac{\partial z}{\partial x}\right)_y + \left(\frac{\partial z}{\partial y}\right)_x \left(\frac{\partial y}{\partial v}\right)_u \left(\frac{\partial v}{\partial x}\right)_u$$

$$\left(\frac{\partial z}{\partial x}\right)_u = \left(\frac{\partial z}{\partial x}\right)_y + \left(\frac{\partial z}{\partial y}\right)_x \left(\frac{\partial y}{\partial x}\right)_u \tag{B.7}$$

which is appropriate for the construction of expressions where, e.g., $z = f(x, y)$ and $z = g(x, u)$.

MAXIMA AND MINIMA

Any point on a curve $y = f(x)$ for which $dy/dx = 0$ is called a stationary point. Stationary points may be associated with maxima or minima or (certain) points of inflection. The maximum/minimum value of a function of any number of variables is similarly characterized. Consider the surface shown in Figure B-2a

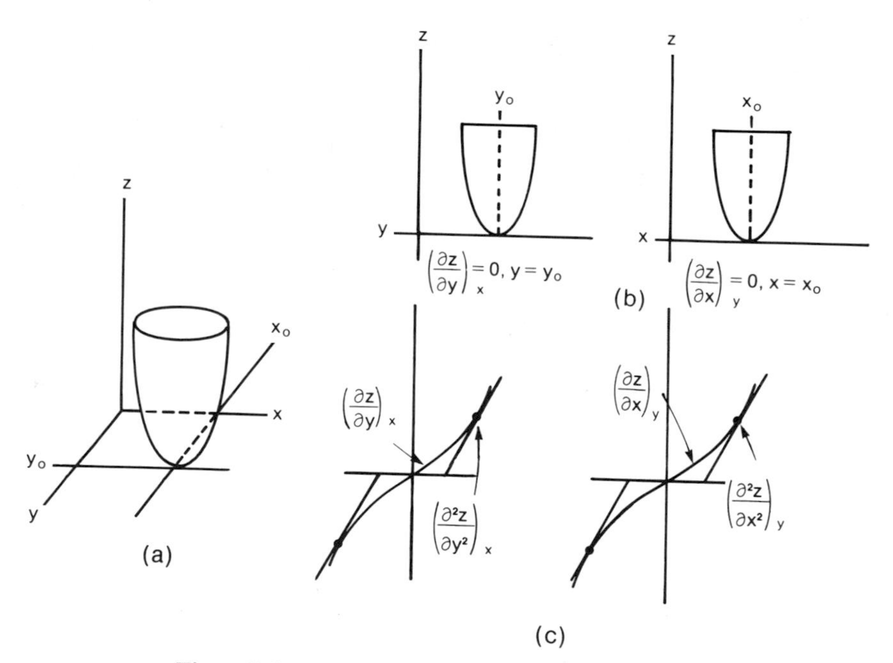

Figure B.2. Illustration of a minimum in a function.

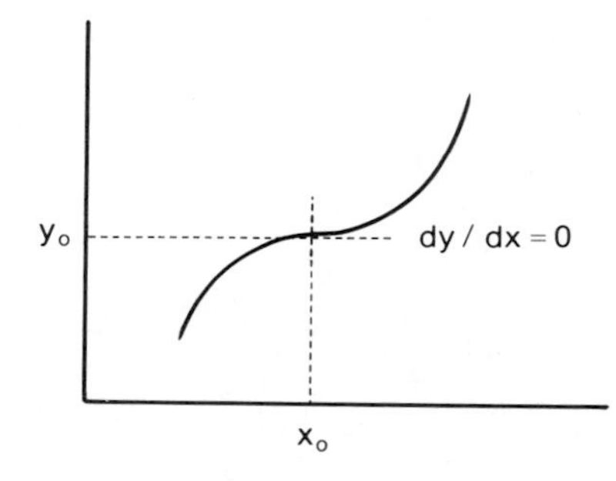

Figure B.3. Illustration of a point of inflection.

and the cross-sections (b, c) obtained by passing planes through the surface at $x = x_0$, $y = y_0$ where x_0, y_0, z_0 are the coordinates of the stationary point for which $(\partial z/\partial x)_y$ and $(\partial z/\partial y)_x$ equal zero. For the minimum illustrated, the tangent to the curve $(\partial z/\partial y)_x$ *vs* y [i.e., $(\partial^2 z/\partial y^2)_x$] is always an increasing function. An analogous stationary point associated with a maximum would again be characterized by $(\partial z/\partial x)_{y_0} = 0 = (\partial z/\partial y)_{x_0}$ but $(\partial^2 z/\partial^2 x)_y$ and $(\partial^2 z/\partial y^2)_x$ are both negative. A horizontal point of inflection is characterized by (two dimensions for clarity) $dy/dx = 0$ and $(d^2y/d^2x) = 0$, as shown in Figure B.3.

CONDITIONAL MAXIMA (MINIMA)

In many cases it is necessary to find the maximum (minimum) in a function subject to one or more conditions (constraints). As a simple example consider the maximum in f(x, y) subject to the constraint $g(x, y) = 0$. First we write

$$df = \left(\frac{\partial f}{\partial x}\right)_y dx + \left(\frac{\partial f}{\partial y}\right)_x dy$$

$$dg = \left(\frac{\partial g}{\partial x}\right)_y dx + \left(\frac{\partial g}{\partial y}\right)_x dy = 0 \qquad \text{(since } g \text{ is a constant)}$$

$$dy = -\left(\frac{\partial g}{\partial x}\right)_y \left(\frac{\partial y}{\partial g}\right)_x dx$$

$$df = \left[\left(\frac{\partial f}{\partial x}\right)_y - \left(\frac{\partial f}{\partial y}\right)_x \left(\frac{\partial g}{\partial x}\right)_y \left(\frac{\partial y}{\partial g}\right)_x\right] dx$$

$$\frac{df}{dx} = 0 = \left[\left(\frac{\partial f}{\partial x}\right)_y - \left(\frac{\partial f}{\partial g}\right)_x \left(\frac{\partial g}{\partial x}\right)_y\right] \qquad \text{at } x = x_0, y = y_0 \tag{B.8}$$

so that

$$\left(\frac{\partial f}{\partial x}\right)_y \Big/ \left(\frac{\partial f}{\partial y}\right)_x = \left(\frac{\partial g}{\partial x}\right)_y \Big/ \left(\frac{\partial g}{\partial y}\right)_x$$

or

$$\left(\frac{\partial \mathrm{f}}{\partial x}\right)_y \Big/ \left(\frac{\partial g}{\partial x}\right)_y = \left(\frac{\partial \mathrm{f}}{\partial y}\right)_x \Big/ \left(\frac{\partial g}{\partial y}\right)_x \equiv \lambda$$

$$\left(\frac{\partial \mathrm{f}}{\partial x}\right)_y = \left(\frac{\partial g}{\partial x}\right)_y \lambda \quad \text{or} \quad \left(\frac{\partial \mathrm{f}}{\partial x}\right)_y - \lambda\left(\frac{\partial g}{\partial x}\right)_y = 0 = \frac{\partial}{\partial x}(\mathrm{f} - \lambda g)$$

also

$$\left(\frac{\partial \mathrm{f}}{\partial y}\right)_x - \lambda\left(\frac{\partial g}{\partial x}\right)_y = 0 = \frac{\partial}{\partial y}(\mathrm{f} - \lambda g) \tag{B.9}$$

and these statements are true when Eq. (B.8) is true, i.e., for the conditional maximum. The conclusions embodied in Eqs. (B.9) can be phrased as, the conditional maximum of $\mathrm{f}(x, y)$ subject to the condition $g(x, y) = 0$ is equal to the unconditional maximization of $(\mathrm{f} - \lambda g)$ where λ may be referred to as an undetermined multiplier. These arguments can be extended to any number of variables and conditions, observing from the steps leading to Eqs. (B.9) that there must be at least one more independent variable than there are conditions. For $\mathrm{f} = \mathrm{f}(x_1, \ldots, x_n)$ subject to $g(x_1, \ldots, x_n) = 0$, $h(x_1, \ldots, x_n) = 0, \ldots$ produces

$$\begin{aligned} &\partial/\partial x_1(\mathrm{f} - \lambda g - \mu h \cdots) = 0 \\ &\quad\vdots \\ &\partial/\partial x_n(\mathrm{f} - \lambda g - \mu h \cdots) = 0. \end{aligned} \tag{B.10}$$

These expressions, together with $g(x_1, \ldots, x_n)$, $h(x_1, \ldots, x_n)$ are sufficient to determine the values of the variables $(x_1, \ldots, x_n)$ at the maximum and to determine the multipliers. The method of undetermined multipliers is due to Lagrange and $\lambda, \mu \cdots$ are referred to as Lagrangian multipliers.

DIFFERENTIAL EXPRESSIONS

The expression

$$dz(x, y) = \mathrm{M}(x, y)dx + \mathrm{N}(x, y)dy \tag{B.11}$$

is a linear differential form known as a Pfaff differential expression. It may be that the expression is obtained by the direct differentiation of some function $\mathrm{F}(x, y)$

$$d\mathrm{F}(x, y) = \left(\frac{\partial \mathrm{F}}{\partial x}\right)_y dx + \left(\frac{\partial \mathrm{F}}{\partial y}\right)_x dy$$

$$\left(\frac{\partial \mathrm{F}}{\partial x}\right)_y = \mathrm{M}, \qquad \left(\frac{\partial \mathrm{F}}{\partial y}\right)_x = \mathrm{N} \tag{B.12}$$

in which case dz is referred to as a perfect or exact differential. In case there is no such function $F(x, y)$ the term $dz(x, y)$, which represents the right-hand side of Eq. (B.11), is an imperfect or inexact differential. In this case dz is not to be regarded as the derivative of some function.

We first consider the case where $dz(x, y)$ is exact. On the basis of Eqs. (B.9)

$$F(x, y) = \int M(x, y)dx + K(y)$$

where $K(y)$ is independent of x. Then

$$\frac{\partial F(x, y)}{\partial y} = \frac{\partial}{\partial y}\int M(x, y)dx + \frac{dK}{dy} = N(x, y)$$

$$\frac{dK}{dy} = N(x, y) - \frac{\partial}{\partial y}\int M(x, y)dx$$

and

$$K_y = \int N(x, y) - \int \frac{\partial}{\partial y}\int M(x, y)dx$$

leading to

$$F(x, y) = \int M(x, y)dx + \int N(x, y)dy - \iint \frac{\partial M(x, y)}{\partial y} dx\,dy.$$

Now

$$\left(\frac{\partial F}{\partial y}\right)_x = \int M(x, y)dx + \int N(x, y)dy - \int M(x, y)dx$$

and

$$\left(\frac{\partial F}{\partial x}\right)_y = M(x, y) - \int \left(\frac{\partial N}{\partial x}\right)_y dy - \int \left(\frac{\partial M}{\partial y}\right)_x dy \tag{B.13}$$

so that $(\partial F/\partial x)_y = M(x, y)$ only if $(\partial N/\partial x)_y = (\partial M/\partial y)_x$. This is the condition for exactness (known as Euler's reciprocity relation). Any differential expression can be tested for exactness by this criterion. The form of the function $F(x, y)$ can be established by inspection in the simplest cases; in any event, knowing that a particular expression $M dx + N dy$ is exact, $F(x, y)$ can be obtained from Eqs. (B.13).

In the case where the expression $M dx + N dy$ is not exact (the reciprocity relationship fails), there may be an integrating factor that, when applied, renders the expression into an exact (integrable) form. We summarize this remark for the inexact expression

$$\left(\frac{\partial M}{\partial y}\right)_x \neq \left(\frac{\partial N}{\partial x}\right)_y$$

but

$$\left(\frac{\partial GM}{\partial y}\right)_x = \left(\frac{\partial GN}{\partial x}\right)_y \tag{B.14}$$

where G is the integrating factor.

There will always be an integrating factor for an expression in two independent variables. If there are more than two independent variables, the reciprocity relationship requires that $dz(x_1,\ldots,x_n)=\sum^n M_i(x_1,\ldots,x_n)dx_i$

$$\left(\frac{\partial M_i}{\partial x_j}\right) = \left(\frac{\partial M_j}{\partial x_i}\right)$$

for every possible variable pair [there are $n(n-1)/2$]. Failing this criterion the expression may have an integrating factor if (neglecting the background arithmetic)

$$M_i\left(\frac{\partial M_i}{\partial x_k} - \frac{\partial M_i}{\partial x_j}\right) + M_j\left(\frac{\partial M_i}{\partial x_k} - \frac{\partial M_k}{\partial x_i}\right) + M_k\left(\frac{\partial M_i}{\partial x_j} - \frac{\partial M_j}{\partial x_i}\right) = 0$$

for all triplets i, j, k.

The expression $dq(x, y)$ is a differential equation. For

$$M\mathrm{d}x + N dy = 0$$
$$dy/dx = -M(x, y)/N(x, y) \tag{B.15}$$

the general solution is $F(x, y) = \text{constant}$. When dq is exact $dF(x, y) = dq(x, y)$. The form of $F(x, y)$ can be obtained as described above and the family of solutions represented in the x, y plane.

In three variables, the differential equation (B.15) is now a partial differential equation and, if $dq(x, y, z)$ is exact or integrable, i.e.,

$$M(x, y, z)dx + N(x, y, z)dy + P(x, y, z)dz = 0,$$

has a family of solutions of the form

$$\int dq(x, y, z) = \int F(x, y, z) = 0. \tag{B16}$$

The solutions represent a family of surfaces that we will refer to as solution surfaces. Solution curves are particular x, y, z values that satisfy the Eq. (B.16) relationship and thus represent a path that lies entirely within a solution surface. If the differential expression is not exact, solution surfaces do not exist although solution curves that satisfy Eq. (B.16) can be obtained by numerical methods. In n variables the solution surfaces are n-dimensional hypersurfaces and $F(x_1,\ldots,x_n)=0$ when the differential expression is exact or integrable. Solution curves but not solution surfaces exist if dq is inexact.

These geometric properties of the differential expression can be stated: "If a Pfaff differential expression $dq=\sum^i M_i dx_i$ possesses an integrating factor, then in the neighborhood of every point $P(x_i,\ldots,x_n)$ there exist innumerable points which are inaccessible from P by paths that are solution curves of the differential equation $dq=0$." The converse of this statement is also true, i.e., if there are points in the neighborhood of $P(x_1,\ldots,x_n)$ that are inaccessible by a solution curve of the Pfaff differential equation $\sum^i M_i dx_i=0$ then the expression $dq=\sum^i M_i dx_i$ possesses an integrating factor.

LINE INTEGRALS

Line integrals can be defined as follows

$$\Delta q_c = {}_c\int_{x_0 y_0}^{x_1 y_1} [M(x,y)dx + N(x,y)dy] \tag{B.17}$$

for the case of two independent variables. They represent the continuous summation of a differential expression. In contrast to the ordinary Rieman integral for a function in one variable

$$\int_a^b F(x)dx = F(b) - F(a) \tag{B.18}$$

which can be represented as an area, as indicated in Figure B.4, there is no

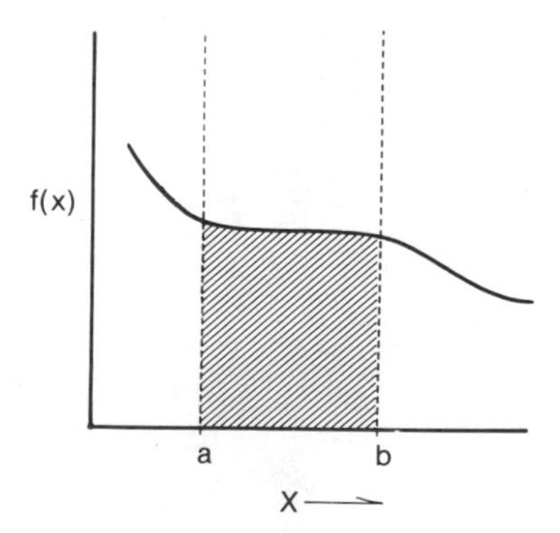

Figure B.4. Illustration of a Rieman integral.

simple way to represent the line integral. The line integral does, however, reduce to a Rieman integral when the path of integration is parallel to (in this case) either coordinate axis, e.g., $y = \text{constant}$, $dy = 0$

$$\Delta qc = \int_{x_0}^{x_1} M(x, y)dx.$$

In general, when the curve along which integration is carried out represents a relationship $y = g(x)$, y can be eliminated to yield a Rieman integral

$$\Delta qc = \int_{x_0}^{x_1} [M(x, g(x))]dx.$$

The line integral depends on the direction in which the curve is traced

$$\int_{c\ x_0 y_0}^{x_1 y_1} dq(x, y) = -\int_{c\ x_1 y_1}^{x_0 y_0} dq(x, y).$$

The integral around a closed path in which the initial and final points but no other points lie on the same curve is known as a cyclic integral. The value of the cyclic integral has important implications if the differential expression defining the curve is exact.

If the cyclic integral is taken to be a rectangle with sides parallel to the x, y coordinate axes

$$\begin{aligned}
&\oint [\mathrm{M}(x, y)dx + \mathrm{N}(x, y)dy \\
&= \int_{x_0}^{x_1} \mathrm{M}(x_1 y_0)dx + \int_{y_0}^{y_1} \mathrm{N}(x_1, y)dy \\
&\quad + \int_{x_1}^{x_0} \mathrm{M}(x, y_1)dx + \int_{y_1}^{y_0} \mathrm{N}(x_0, y)dy \\
&= \int_{y_0}^{y_1} [\mathrm{N}(x, y) - \mathrm{N}(x_0, y)]dy - \int_{x_0}^{x_1} [\mathrm{M}(x, y_1) - \mathrm{M}(x, y_0)]dx.
\end{aligned}$$

Also

$$\mathrm{N}(x, y) - \mathrm{N}(x_0, y) = \int_{x_0}^{x} \left(\frac{\partial \mathrm{N}}{\partial x}\right)_y dx$$

$$\mathrm{M}(x, y_1) - \mathrm{M}(x, y_0) = \int_{y_0}^{y} \left(\frac{\partial \mathrm{M}}{\partial y}\right)_x dy$$

and

$$\oint [\mathrm{M}dx + \mathrm{N}dy] = \int_{x_0}^{x_1} \int_{y_0}^{y_1} \left[\left(\frac{\partial \mathrm{N}}{\partial x} \right)_y - \left(\frac{\partial \mathrm{M}}{\partial y} \right)_x dx\,dy \right.$$

which vanishes if $[\mathrm{M}dx + \mathrm{N}dy]$ is a perfect differential. Since any closed curve can be considered to be made up of rectangles, the result is general. If the curve is irregular the rectangles may be shrunk to the smallest dimension. Integration in the clockwise direction cancels common boundaries. The converse of

$$\oint dq(x, y) = 0$$

for a perfect differential is that if the integration around a cycle is zero, $dq(x, y)$ is perfect and if nonzero then $dq(x, y)$ is imperfect.

HOMOGENEOUS FUNCTIONS

A function $\mathrm{f}(x_1, \ldots, x_n)$ is said to be homogeneous of degree n if

$$\mathrm{f}(\lambda x_1 + \lambda x_2, \ldots, \lambda x_n) = \lambda^n \mathrm{f}(x_1, \ldots, x_n) \tag{B.19}$$

where λ is an arbitrary number. If we define a new set of independent variables $u_i = \lambda x_i$ then $\mathrm{f}(u_1, \ldots, u_n) = \lambda^n \mathrm{f}(x_1, \ldots, x_n)$ and when differentiated with respect to λ

$$\frac{\partial \mathrm{f}}{\partial \lambda} = \sum^{i} \left(\frac{\partial \mathrm{f}}{\partial u_i} \right) \left(\frac{\partial u_i}{\partial \lambda} \right) = n\lambda^{n-1} \mathrm{f}(x_1, \ldots, x_n)$$

where $(\partial \mathrm{f}/\partial u)(\partial u/\partial \lambda) = (\partial \mathrm{f}/\partial x_i) x_i$. Then

$$\sum^{n} \left(\frac{\partial \mathrm{f}}{\partial x_i} \right) x_i = n\lambda^{n-1} \mathrm{f}(x_1, \ldots, x_n).$$

Homogeneous functions of degree 0 produce

$$\sum^{n} \left(\frac{\partial \mathrm{f}}{\partial x_i} \right) x_i = 0. \tag{B.20}$$

Homogeneous functions of degree 1 produce

$$\sum \left(\frac{\partial \mathrm{f}}{\partial x_i} \right) x_i = \mathrm{f}(x_1, \ldots, x_n). \tag{B.21}$$

These considerations are important in thermodynamics since thermodynamic functions can be divided into two classes: those that are independent of material quantity (T, P, etc.) and those that do depend on material quantity (E, V, etc.). These two classes of functions are designated as intensive and extensive properties of the system of N molecules, which may or may not be identical. A general intensive property (variable) of a system containing various molecular species α in various n_α mole numbers $(n = (N_\alpha)N_0)$

$$\mathrm{f}(\lambda n_1, \ldots, \lambda n_\alpha) = \mathrm{f}(n_1, \ldots, n_\alpha)$$

is homogeneous of degree 0. For a general extensive property (variable) g

$$g(\lambda n_1, \ldots, \lambda n_\alpha) = \lambda g(n_1, \ldots, n_\alpha)$$

i.e., the extensive property is homogeneous of degree 1 and

$$\sum\left(\frac{\partial g}{\partial n_\alpha}\right)n_\alpha = g(n_1, \ldots, n_\alpha).$$

The ratio of two homogeneous functions of order 1, i.e., extensive properties, is intensive

$$h(n_1, \ldots, n_\alpha) = \frac{g_1(n_1, \ldots, n_\alpha)}{g_2(n_1, \ldots, \mathrm{n}_\alpha)}$$

$$h(\lambda n_1, \ldots, \lambda n_\alpha) = \frac{g_1(\lambda n_1, \ldots, \lambda n_\alpha)}{g_2(\lambda n_1, \ldots, \lambda n_\alpha)} = \frac{\lambda g_1(n_1, \ldots, n_\alpha)}{\lambda g_2(n_1, \ldots, n_\alpha)} = h(n_1, \ldots, n_\alpha).$$

Any molar quantity [the ratio of an extensive variable E_α, V_α, etc. to the total quantity (n_α)] is intensive just as is the ratio of mass to volume (density). To summarize: intensive properties are uniform

$$\mathrm{f} = \mathrm{f}_1 = \mathrm{f}_2, \ldots, \mathrm{f}_\alpha$$

and extensive properties are additive

$$g = g_1 + g_2 + \cdots + g_\alpha.$$

LEGENDRE TRANSFORMATIONS

The independent variables of a differential expression may be changed by the following procedure. Consider

$$dF(x, y) = M(x, y)dx + N(x, y)dy \tag{B.22}$$

which is taken to be exact. Define the function

$$\begin{aligned}\Phi &= F - Mx \\ d\Phi &= dF - Mdx - xdM \\ &= Mdx + Ndy - Mdx - xdM = Ndy - xdM; \end{aligned} \tag{B.23}$$

this is the complete differential for a function

$$\Phi = \Phi(N, x).$$

FACTORIALS

The gamma function is useful in dealing with factorials. It is defined by the expression

$$\Gamma(z) = \lim_{n \to \infty} \frac{1 \cdot 2 \cdot 3 \cdots (n-1)}{z(z-1) \cdots (z-n-1)} n^2 \tag{B.24}$$

from which it follows that

$$\begin{aligned}\Gamma(1) &= \lim_{n \to \infty} \frac{n!}{n!} = 1 \\ \Gamma(z+1) &= z\Gamma(z) \\ \Gamma(z) &= (z-1)!. \end{aligned} \tag{B.25}$$

This definition can be shown to be identical to the infinite limit of the term

$$F(z, n) = \int_0^n \left(1 - \frac{t}{n}\right)^n t^{(z-1)} dt$$

where n is a positive integer. Then

$$\lim_{n \to \infty} F(z, n) = \Gamma(z) = \int_0^\infty e^{-t} t^{(z-1)} dt \tag{B.26}$$

where

$$\lim_{n \to \infty} \left(1 - \frac{t}{n}\right)^n \to e^{-t}.$$

Then

$$\frac{d}{dz} \ln \Gamma(z) = \int_0^\infty \left(\frac{e^{-t}}{t} - \frac{e^{-zt}}{1 - e^{-t}}\right) dt \tag{B.27}$$

$$\ln \Gamma(z) = (z - 1/2)\ln z - z + \tfrac{1}{2}\ln 2\pi + \text{terms in } \frac{1}{z}$$

$$\Gamma(z) = e^{-z} z^{(z-1/2)} 2\pi^{1/2}\left(1 + \frac{1}{12z} + \frac{1}{288z^2} + \cdots\right). \tag{B.28}$$

From Eqs. (B.25) and (B.28) $N\Gamma(N) = N!$ and if N is large

$$N! = N\Gamma(N) = e^{-N} N^N (2\pi N)^{1/2}(1 + \cdots)$$

$$\ln N! = N \ln N - N + \tfrac{1}{2}\ln(2\pi N) \cong N \ln N - N. \tag{B.29}$$

Equation (B.29) is known as the Stirling Approximation.

VECTORS

Vector quantities, as opposed to scalar quantities, which have only magnitude, have both magnitude and direction. Energy is a scalar. Force is a vector. Scalars combine in accordance with the ordinary rules of arithmetic. Vectors combine in accordance with special rules dictated by the laws of physics.

The addition rules for vectors are summarized in Figure B.5 where vectors **a** and **b** add to produce a resultant vector r.

For many operations it is convenient to resolve a vector into a set of component vectors. Consider any vector **a** as a displacement from the origin of a Cartesian coordinate system. The coordinate system may be rotated in any manner. Assume an orientation such that the vector lies in the positive quadrant and makes an angle ϕ with respect to the xz plane. The vector components are the projections of the vector onto the x, y; and z axes and are expressed in terms of scalar magnitudes a_x, a_y, and a_z and unit vectors **i**, **j** and **k**. The concept and common notation is illustrated in Figure B.6. Geometry provides the relationships $a_y = a \sin\phi$ and $a_x^2 + a_y^2 = (a\cos\phi)^2$ so that $a^2 = a_x^2 + a_y^2 + a_z^2$.

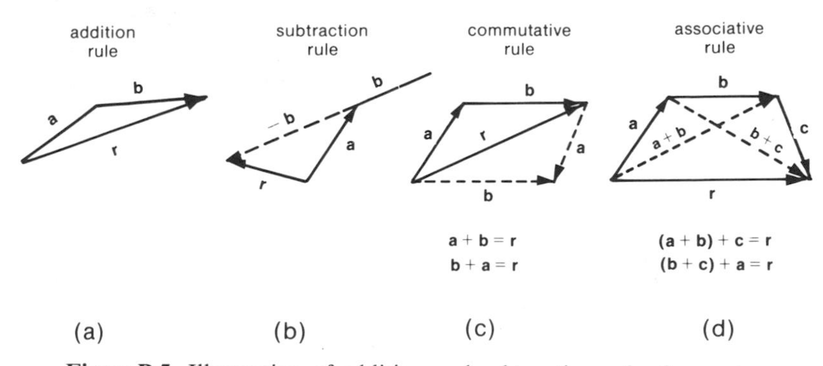

Figure B.5. Illustration of addition and subtraction rules for vectors.

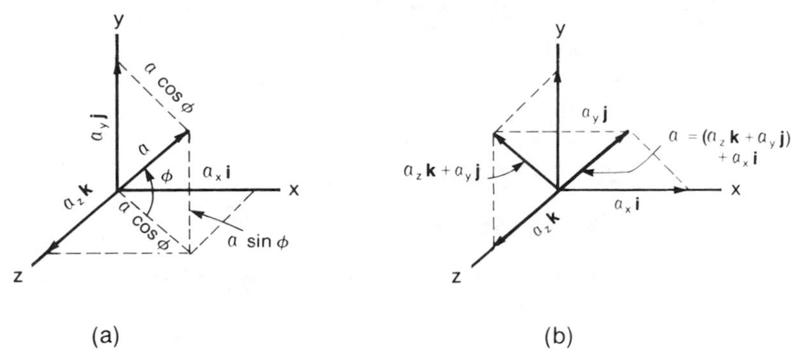

Figure B.6. Illustration of the unit vector concept for vector components.

Given the vector components the vector is resolved as the vector sum of the components

$$\mathbf{a} = a_x\mathbf{i} + a_y\mathbf{j} + a_z\mathbf{k}. \tag{B.30}$$

Two vectors expressed in terms of their vector components are added by adding their vector components

$$\mathbf{a} + \mathbf{b} = (a_x + b_x)\mathbf{i} + (a_y + b_y)\mathbf{j} + (a_z + b_z)\mathbf{k}$$
$$\mathbf{r} = r_x\mathbf{i} + r_y\mathbf{j} + r_2\mathbf{k}$$

and subtracted by subtracting their vector components

$$\mathbf{a} - \mathbf{b} = (a_x - b_x)\mathbf{i} + (a_y - b_y)\mathbf{j} + (a_z - b_z)\mathbf{k}$$
$$\mathbf{r} = r_x\mathbf{i} + r_y\mathbf{j} + r_z\mathbf{k}.$$

Geometrically $\mathbf{a} - \mathbf{b}$ is equivalent to reversing the direction of **b** and adding those vector components to the a components.

Multiplication rules provide for the multiplication of a vector (**a**) by a scalar (k) to produce the product $k\mathbf{a}$ in which the magnitude is k times the original magnitude with the direction unchanged (momentum is the product of mass, and the velocity vector, $\mathbf{p} = m\mathbf{v}$). The rules also provide for the multiplication of vector by vector to produce a scalar (work, a scalar, is the product of the force vector and a displacement vector) or to produce a vector (angular momentum, a vector, is the product of the radius vector and the linear momentum vector). Multiplication that produces a scalar is called the dot product and defined as

$$\mathbf{a}\cdot\mathbf{b} = ab\cos\phi = \mathbf{b}\cdot\mathbf{a} \tag{B.31}$$

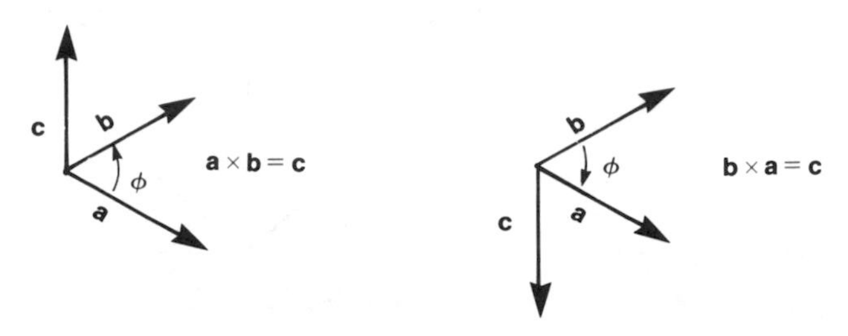

Figure B.7. Illustration of the vector cross-product.

where ϕ is the (smallest) angle between the two vectors. Multiplication to produce a vector is called the cross product and defined as

$$\mathbf{a} \times \mathbf{b} = \mathbf{c} \tag{B.32}$$

where the magnitude of c is $ab \sin \phi$ where ϕ is again the smallest angle between the two vectors. The direction of **c** is perpendicular to the plane of **a** and **b** pointing up or down depending on the order in the cross-product as indicated in Figure B.7.

In terms of unit vectors the multiplication rules can be summarized as

$$\mathbf{i}\cdot\mathbf{i} = \mathbf{j}\cdot\mathbf{j} = \mathbf{k}\cdot\mathbf{k} = 1 \qquad \mathbf{i}\times\mathbf{i} = \mathbf{j}\times\mathbf{j} = \mathbf{k}\times\mathbf{k} = 0$$

$$\mathbf{i}\cdot\mathbf{j} = \mathbf{j}\cdot\mathbf{k} = \mathbf{k}\cdot\mathbf{i} = 0 \qquad \mathbf{i}\times\mathbf{j} = \mathbf{k}, \quad \mathbf{j}\times\mathbf{k} = \mathbf{i}, \quad \mathbf{k}\times\mathbf{i} = \mathbf{j} \tag{B.33}$$

$$\mathbf{a} \times \mathbf{b} = -\mathbf{b} \times \mathbf{a} = (a_y b_x - b_y a_x)\mathbf{i} + (a_z b_x - b_z a_x)\mathbf{j} + (a_x b_y - a_y b_x)\mathbf{k}.$$

CONTENT SOURCES AND FURTHER READING

1. S. M. Blinder, *Advanced Physical Chemistry*, Chapters 1, 2 and 3. Macmillan, Toronto, 1969.
2. H. Morgenau and G. M. Murphy, *The Mathematics of Physics and Chemistry*, 2nd ed., Chapters 1 and 4. Van Nostrand, New York, 1956.

APPENDIX C

STATISTICAL CONSIDERATIONS

INTRODUCTION

If there is a property x that is common to each of N objects, the average value of the property taken over all of the N objects is

$$\bar{x} = \frac{1}{N} \sum_{i=1}^{N} x_i \tag{C.1}$$

where the sum is taken over the individual object property. Several or many of the objects may have a common value of the property in which case a more convenient expression would be

$$\bar{x} = \frac{1}{N} \sum^{j} N_j x_j \tag{C.2}$$

where the sum is taken over the values of the property.

The probability that an object has the particular property is

$$p_j = \frac{N_j}{N} \tag{C.3}$$

so that

$$\bar{x} = \frac{1}{N}\sum^{j} N p_j x_j = \sum^{j} p_j x_j \tag{C.4}$$

and

$$\sum p_j = 1.$$

If the property is not discrete but varies continuously over some range, the probability that it has some exact value is effectively zero. The probability is replaced by a probability density [$\rho(x)$] function such that $\rho(x)dx$ identifies the probability that an individual object have a value of the property that lies in the range x to $x + dx$, i.e., in the range of dx. The probability that the object has a value of the property lying within some finite range $x_1 \leqslant x \leqslant x_2$ is

$$p(x_1 \leqslant x \leqslant x_2) = \int_{x_1}^{x_2} \rho(x)dx \tag{C.5}$$

subject to the normalization requirement that

$$\int \rho(x)dx = 1$$

taken over the entire range of values of x. Assuming that the probability density function has been normalized, the average value of the property $\bar{x}$ is

$$\bar{x} = \int x\rho(x)dx.$$

The probability that a property x lies between x and $x + dx$ while simultaneously another independent property y lies between y and $y + dy$ is

$$\rho(x, y) = \rho(x)\rho(y).$$

The probability density function $\rho(x)$, independent of any value of y is

$$\rho(x) = \int \rho(x, y)dy$$

where the integral is taken over the range of y values. The normalization requirement is

$$\iint \rho(x, y)dx\,dy = 1.$$

The spread of values around the average value is a feature of the form of the distribution function and is commonly characterized by a term called the standard deviation defined as

$$\frac{1}{\sigma_x} = \overline{(x - \bar{x})^2} = \int (x - \bar{x})^2 \rho(x) dx$$

$$\frac{1}{\sigma_x} = \int x^2 \rho(x) dx - 2\bar{x} \int x \rho(x) dx + \bar{x}^2 \int \rho(x) dx = (\overline{x^2} - \bar{x}^2)^{1/2}. \tag{C.6}$$

The distribution function is often of the form

$$\rho(x) = \beta e^{-\alpha(x - x_0)^2}$$

which is normalized for $\beta = (\alpha/\pi)^{1/2}$. When α is related to the standard deviation we have

$$\sigma_x^2 = \int_{-\infty}^{\infty} x^2 \rho(x) dx - \left[\int_{-\infty}^{\infty} x \rho(x) dx \right]^2 = \frac{1}{2\alpha}$$

so that

$$\rho(x) = \frac{1}{(2\pi)^{1/2} \sigma_x} e^{-(x - x_0)/2\sigma^2}.$$

This is the Gaussian distribution which produces the familiar bell-shaped curve centered on x_0.

COMBINATORIAL CONSIDERATIONS

If a number n objects can be given n different assignments the number of ways in which the objects can be assigned follows from the argument that the first object can be given n different assignments, the second object $(n - 1)$ different assignments for each of the n assignments of the first object, the third $(n - 2)$, etc. In total there are $(n)(n - 1)(n - 2) \cdots (1) = n!$ ways to accomplish the assignments.

If a number n of one type of object and a number m of another type of object are to be given $(n + m)$ different assignments with no constraint on the assignments given to different objects, then, as before, the first object has $(n + m)$ possible assignments, the second $(n + m - 1)$, etc., there being $(n + m)!$ ways to accomplish the assignments.

In either case of n objects and n assignments or m objects and m assignments the total number of ways in which the assignments are made could be cataloged by accomplishing one random distribution then making all possible exchanges between object and assignment. These we refer to as permutations and there are $n!$ or $(n+m)!$ of these to accomplish. If, in the case of two identifiable object types, we are interested in the permutations among the n-type objects there are $n!$ of them and there are similarly $m!$ permutations of the m-type objects. If we wish to disregard the permutations among the n objects and those among the m objects, calling all of these indistinguishable permutations, there are $(n+m)!/n!m!$ distinguishable permutations, i.e., those involving exchanges between n and m type objects only.

Other forms of the combinatorial term will be required for other constraints. The number of ways in which m objects can be selected from a set of n objects is $n!/m!(n-m)!$. This same relationship results when we ask for the number of ways in which m indistinguishable objects can be given assignments from among a larger set of n assignments. The number of distinguishable assignments of the n objects among the $n, (n<m)$, assignment set is $n!/(n-m)!$. Other and more complicated distribution questions will lead to other forms of the combinatorial terms. Probabilities associated with postulated events will depend on the form of the combinatorial term involved.

Consider as an example a question related to the distribution of the n and m type (indistinguishable among each group) objects. We inquire into the probability, that in $(p+q)$ selections of an object from the $(m+n)$ set of objects, we obtain p of the n objects and q of the m objects. First to simplify notation, we make $n=m$ so that the probability of obtaining an n or an m object on any single selection is 1/2, i.e., on any selection there are two possibilities and in $(p+q)$ selections there are $2^{(p+q)}$ sequences of n and m occurrences in the selection process. The probability that any particular sequence is encountered is $(1/2)^{p+q}$. The number of those sequences that represent p of the n objects and q of the m objects is $(p+q)!/p!q!$ since we are not interested in the exact sequence. The probability then is

$$\wp(p, q)=\frac{(p+q)!}{p!q!}\left(\frac{1}{2}\right)^{p+q}.$$

This expression can be reduced to a more useful form by conversion to logarithmic form, application of Stirling's formula to the factorials, and expansion in a Taylor series. The arithmetic is not interesting. The result for $\eta=p+q$ is

$$\wp(p, q) \simeq\left(\frac{2}{\pi \eta}\right)^{1/2} e^{-2(p+\eta/2)^2/\eta}.$$

The distribution is Gaussian and centered on $\eta/2$ with standard deviation $\sqrt{\eta}/2$. The ratio of the standard deviation to the average value is

$$\frac{\sigma}{\eta/2}=\frac{\sqrt{\eta}/2}{\eta/2}=\frac{1}{\sqrt{\eta}}.$$

When η is very large the approximation becomes exact and fluctuations around the mean value become insignificant with respect to the mean value. The distribution becomes more and more sharply peaked as η increases.

APPLICATION TO MOLECULE SYSTEMS

It is a fundamental postulate of statistical mechanics that particles are distributed among energy states with equal a priori probability. However in a many particle system there are particular energetic distributions that can occur in so many ways that these distributions actually dominate the statistical features of the system. We are concerned then with the construction of combinatorial statements that reflect the number of distinguishable ways in which a particle collection can be distributed among energy states. Quantum arguments would appear to be required in that discrete energy states are a feature of that approach. Clearly classical arguments would require some arbitrary division of the energy continuum into discrete packages. This introduced puzzling contradictions in the prequantum development of the subject. We will phrase our arguments here in quantum mechanical terms that provide some unique constraints, resolving some classical problems and confirming some classical results (also see Appendix F on Classical Considerations).

In a many particle system the system wave function is the sum of the products of the individual particle wave function, e.g.,

$$\psi_{123} = \sum^{k} \sum^{l} \sum^{m} \psi_k(1)\psi_l(2)\psi_m(3).$$

Since the particles are not identifiable

$$\psi_{123} = \sum^{k} \sum^{l} \sum^{m} a_{klm}\psi_l(2)\psi_k(1)\psi_m(3)$$

and any other construct obtained by the interchange of particle coordinates is also acceptable. The system wave function is not any one of these but a linear combination of all of them. The wave function obtained from such linear combinations can be symmetric or antisymmetric, i.e., it may change sign or it may not, with respect to particle interchange. The fundamental particles themselves (electrons, protons, neutrons) are antisymmetric and require that the wave function describing their energy states be antisymmetric. Such particles and their odd-numbered collections in atoms and molecules cannot doubly occupy the same wave function, or their interchange would leave the wave function unchanged in sign (i.e., symmetric). For antisymmetric systems each wave function ψ_k of an energy level i is either empty or has in it one particle. The total number of particles N_i in an energy level is less than the degeneracy g_i of that level. There is no such restriction on the number of symmetric particles that can occupy a single wave function. This situation requires that we develop two separate statistical combinatorial statements; one for symmetric and one for antisymmetric particles.

For antisymmetric particles the combinatorial problem reduces to the question of how many different ways can N_i identical objects be distributed among g_i boxes ($g_i \neq N_i$) in such a manner that each box contains either 1 or 0 objects. This reduces to the number of permutations between N_i objects and ($g_i - N_i$) nonobjects with neither objects nor nonobjects being distinguishable. There are $g_i!$ ways of permuting the objects and nonobjects. Among these permutations those that involve only the interchange of objects or those that involve only the interchange of nonobjects are not distinguishably different if the objects and the nonobjects are indistinguishable within their own class. The g_i permutations must be first divided by $N_i!$ and then by $(g_i - N_i)!$. The number of ways of obtaining a given distribution among many energy levels is the product of all the individual, corrected permutations,

$$\Omega = \prod^{i} \frac{g_i!}{N_i!(g_i - N_i)!}. \tag{C.7}$$

This statistics is known as Fermi–Dirac (FD) statistics and particles obeying it are called fermions.

If there is no restriction on the number of particles that can occupy the same quantum state (wave function), the combinatorial problem is somewhat different. Consider the group of objects as being placed in line so that $(g_i - 1)$ walls can be placed in the collection creating g_i compartments. The result is $g_i + N_i - 1$ stations, each of which can be occupied by an object or a wall. The correction

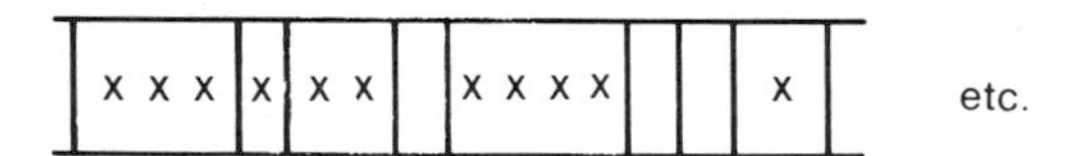

for indistinguishability among the total permutations is arrived at as before and

$$\Omega = \prod^{i} \frac{(g_i + N_i - 1)!}{N_i!(g_i - 1)!}. \tag{C.8}$$

This statistics is known as Bose–Einstein (BE) statistics and particles obeying it are called bosons.

Both FD and BE combinatorial statements recognize the indistinguishability of particles and the arguments are completely valid for the constraints on energy state occupancy for fermions and for bosons. If for the moment we do not recognize any constraints that relate to particle occupancy we obtain a still different statistical statement. For a total of N particles distributed among energy levels ϵ each having degeneracy g_i, there are N! permutations among the $g_i^{N_i}$ assignments of N_i particles to each state. Here we neglect consideration of empty quantum states. Permutations among the particles assigned to each state reduces the total number of permutations and the combinatorial expression is

$$\Omega = \mathrm{N}!\prod^{i} \frac{g_i^{\mathrm{N}_i}}{\mathrm{N}_i!}. \tag{C.9}$$

No correction for indistinguishability has yet been made. The $(\prod^{i} \mathrm{N}_i!)^{-1}$ term simply corrects for calling the assignment $^1\mathrm{N}$, $^2\mathrm{N}$, $^3\mathrm{N}$ (where the superscript identifies particular particles) to energy level ϵ_i a different assignment from $^3\mathrm{N}$, $^2\mathrm{N}$, $^1\mathrm{N}$ in that same level. A true indistinguishability correction does not recognize any interchange of particles, i.e., the assignment of $^4\mathrm{N}$, $^5\mathrm{N}$, $^6\mathrm{N}$ to energy level ϵ_i is not different from $^1\mathrm{N}$, $^2\mathrm{N}$, $^3\mathrm{N}$ in energy level ϵ_i. We note that division of Eq. (C.9) by N!, which is certainly a maximum correction for indistinguishability, produces the result

$$\Omega = \prod^{i} \frac{g_i^{\mathrm{N}_i}}{\mathrm{N}_i!} \tag{C.10}$$

or a product of terms $\omega_i = (g_i)^{\mathrm{N}_i}/\mathrm{N}_i!$.

This statistics, which predates quantum theory and which must now be regarded as applying to fictitious particles, is known as Maxwell–Boltzmann statistics and its particles are called boltzons.

For bosons, when each term ω_i in the product is expanded,

$$\omega_i = \frac{(g_i + \mathrm{N}_i - 1)!}{\mathrm{N}_i!(g_i - 1)!}$$

we have

$$\omega_i = \frac{(g_i + \mathrm{N}_i - 1)(g_i + \mathrm{N}_i - 2)\cdots(g_i + \mathrm{N}_i - \mathrm{N}_i)[(g_i - 1)\cdots 1]}{\mathrm{N}_i!(g_i - 1)!}$$

where the bracketed term in the numerator is $(g_i - 1)!$, which cancels the same term in the denominator. Then we note that each remaining term in the numerator is $\geqslant g_i$ and

$$(\omega_i)_{\mathrm{BE}} \geqslant \frac{g_i^{\mathrm{N}_i}}{\mathrm{N}_i!}.$$

For fermions, when each term in the product is expanded,

$$\omega_i = \frac{g_i!}{\mathrm{N}_i!(g_i - \mathrm{N}_i)!}$$

we have

$$(\omega_i) = \frac{g_i(g_i - 1)(g_i - 2)\cdots(g_i - N_i + 1)[(g_i - N_i)\cdots 1]}{N!(g_i - N_i)!}$$

where the bracketed term is $(g_i - N_i)!$, which again cancels the same term in the denominator leaving a product set in which each term is $\leqslant g_i$ and

$$(\omega)_{FD} \leqslant \frac{g_i^{N_i}}{N_i!}. \tag{C.11}$$

Combining Eqs. (C.9) and (C.11) we observe that

$$\omega_{FD} \leqslant \frac{g_i^{N_i}}{N_i!} \leqslant \omega_{BE} \tag{C.12}$$

where the limiting term for either statistics is Eq. (C.10). Maxwell–Boltzmann statistics, overcorrected for indistinguishability by the N! division, is a limiting case for either of the quantum statistical statements. Clearly the circumstances in which the Fermi–Dirac constraint set (no more than one particle in any quantum state) would be met by the Bose–Einstein statement only when so many states are available that no two particles are in the same state even though such multiple occupancy is not prohibited.

Equations (C.9), (C.10), and (C.11) represent, in each case, the number of ways in which a single unique distribution of N particles among all available energy states can be accomplished. Each unique distribution contributes one such term to the total number of ways in which all possible distributions can be achieved

$$\Omega = \sum t.$$

It is a feature of the statistics of very large numbers that particular distributions will dominate the sum and, in fact, there is a single term that effectively is the sum. We write $\Omega = t_{max}$ where t_{max} is the maximum valued term W for any of the three combinatorial expressions. The particular distribution N_i which leads to t_{max} is clearly *the* distribution which will occur, not because of a priori favor but because it can be achieved in so many ways compared to any other distribution. The favored distribution can be obtained by maximizing the expressions for Ω. It is more convenient, however, to maximize $\ln \Omega$ since factorials can then be accommodated ($\ln N! = N \ln N - N$) and, when t is maximum valued, $\ln t$ is also maximum valued.

We write

$$\ln \Omega_{FD} = \sum^{j} \{(g_i \ln g_i - g_i) - (N_i \ln N_i - N_i) - [(g_i - N_i)\ln(g_i - N_i) - (g_i - N_i)]$$

and consider the variation in $\ln W_{\mathrm{FD}}$ as the N_i vary

$$d\ln\Omega_{\mathrm{FD}} = \sum^{j}\{-\mathrm{N}_i d\ln\mathrm{N}_i - \ln\mathrm{N}_i d\mathrm{N}_i + d\mathrm{N}_i - (g_i-\mathrm{N}_i)d\ln(g_i-\mathrm{N}_i)$$
$$-\ln(g_i-\mathrm{N}_i)d\mathrm{N}_i - d\mathrm{N}_i\}$$
$$=\sum^{i}\{-d\mathrm{N}_i - \ln\mathrm{N}_i d\mathrm{N}_i + d\mathrm{N}_i - d(g_i-\mathrm{N}_i) + \ln(g_i-\mathrm{N}_i)d\mathrm{N}_i - d\mathrm{N}_i\}$$
$$=\sum\left[\ln\left(\frac{g_i-\mathrm{N}_i}{\mathrm{N}_i}\right)\right]d\mathrm{N}_i. \tag{C.13}$$

Without arithmetic

$$d\ln\Omega_{\mathrm{BE}} = \sum\left[\ln\left(\frac{g_i+\mathrm{N}_i-1}{\mathrm{N}_i}\right)\right]d\mathrm{N}_i \tag{C.14}$$

and

$$d\ln\Omega_{\mathrm{MB}} = \sum\left[\ln\frac{g_i}{\mathrm{N}_i}\right]d\mathrm{N}_i. \tag{C.15}$$

A general expression then is

$$d\ln\Omega = \sum\left[\ln\left(\frac{g_i+k\mathrm{N}_i}{\mathrm{N}_i}\right)\right]d\mathrm{N}_i \tag{C.16}$$

where k is -1, 0, $+1$ for fermions, boltzons, and bosons, respectively.

The distribution numbers N_i can be assigned in many ways but their assignments are subject to two constraints if we wish to associate the favored distribution with particular system properties. Among the constraints which we could impose, specified E and N are among the most convenient (see Appendix E on Ensembles). We express these constraints as $\sum d\mathrm{N}_i = 0$ for $\sum \mathrm{N}_i = \mathrm{N}$, and $\sum \epsilon_i d\mathrm{N}_i = 0$ for $\sum \epsilon_i \mathrm{N}_i = \mathrm{E}$. Equation (C.15) should now be maximized with respect to variations $d\mathrm{N}_i$ subject to the above constraints. This is a conditional maximization and the mathematical approach is set down in Appendix B on Mathematical Considerations. The procedure requires that we multiply each constraint condition by an undetermined multiplier and add these terms to the derivative expression setting the sum equal to zero. We obtain from that procedure with undetermined multipliers $-\alpha$ and $-\beta$,

$$-\alpha\sum d\mathrm{N}_i = 0$$
$$-\beta\sum \epsilon_i d\mathrm{N}_i = 0$$
$$\sum^{i}\left\{\ln\frac{g_i+k\mathrm{N}_i}{\mathrm{N}_i} - \alpha - \beta\epsilon_i\right\}d\mathrm{N}_i = 0.$$

Since the dN_i can be arbitrary variations, each term in the sum is separately equal to zero and then each bracketed term must also be separately equal to zero

$$\ln \frac{g_i + kN_i}{N_i} = \alpha + \beta\epsilon_i$$

$$N_i = \frac{g_i}{e^{\alpha} e^{\beta\epsilon_i} - k} \tag{C.17}$$

with $k = +1$ for fermions, -1 for bozons, and 0 for boltzons. For simplicity in arithmetic we consider the boltzon expression and write

$$N_i = g_i e^{-\alpha} e^{-\beta\epsilon_i}$$

then, summing over the N_i and dividing by that sum to eliminate $e^{-\alpha}$,

$$\frac{N_i}{\sum N_i} = \frac{g_i e^{-\beta\epsilon_i}}{\sum g_i e^{-\beta\epsilon_i}} = \frac{g_i e^{-\epsilon_i/kT}}{z} = \frac{N_i}{N}. \tag{C.18}$$

In Eq. (C.18), anticipating arguments developed in the text, we have set $\beta = 1/kT$ and $z = \sum^i g_i e^{-\epsilon_i/kT}$. We now construct a dilution ratio term

$$\frac{g_i}{N_i} = \frac{z}{N} e^{\epsilon_i/kT}$$

where g_i/N_i reflects the ratio of number of states at energy level i to the number of particles occupying them. If g_i/N_i is very large there will be no occasion for any state to be occupied by more than one particle. The distinction between fermions and bolsons will have vanished, leaving us with Maxwell–Boltzmann statistics.

Since ϵ_i/kT is a positive number and $e^{\epsilon_i/kT} \geqslant 1$, $g_i/N_i \geqslant z/N$. We obtain the magnitude of the term from the ideal gas partition function as developed in the text

$$\frac{z}{N} = \left(\frac{2\pi m kT}{h^2}\right)^{3/2} \frac{V}{N} = \frac{0.0257 M^{3/2} T^{3/2}}{P}$$

where M is conventional molecular weight (mN_A) and P is ideal gas pressure. For real gases which approximate the ideal gas the lightest gases will produce the smallest z/N value as will the lowest temperatures and the greatest pressures. For Helium z/N is (0) 10^5 at 300 K and one atmosphere pressure. For argon z/N is (0) 10^6 under the same conditions. At one atmosphere pressure the temperature at which $z/N = 10$ is 4 K for helium and 1 K for argon. This is at the boiling point for helium but far below the boiling point for argon (87 K) so z/N may not be always large for helium gas. Increase in temperature will increase the population in higher energy levels and increase z/N. Increase in pressure will increase the spacing between levels causing molecules to shift into lower energy levels. If the temperature is 300 K a z/N ratio of 10 is observed for helium at (0) 10^4 atm and for argon at (0) 10^6 atm. Lower pressures, increasing the ratio, are of ordinary concern. The dilution ratio is large under normal circumstances. The FD vs BE distinction vanishes and MB statistics is adequate.

It is necessary to establish that the procedure of maximizing the combinatorial term expression, replacing a sum by a single term, is legitimate. This can be most readily demonstrated by comparing a most probable distribution with an adjacent distribution. Consider three energy levels populated according to the most probable Maxwell–Boltzmann distribution. Promote one molecule in 10^6 from the middle level to the next higher and drop one molecule from the middle level to the next lower. The total energy is constant and this distribution is clearly an adjacent distribution with only two molecules in 10^6 displaced for the most probable distribution.

We have for

Level	**Start**	**Finish**
1	c	$c + \alpha b$
2	b	$b - 2\alpha b$
3	a	$a + \alpha b$

with the ratio of initial probability to final probability of

$$R = \frac{\Omega_{\text{initial}}}{\Omega_{\text{final}}} = \frac{(a + \alpha b)!(b - 2\alpha b)!(c + \alpha b)!}{a!b!c!}$$

and

$$\frac{(n + x)!}{n!} = \frac{(n + x)(n + x - 1)(n + x - 2)\cdots(n + x - x)!}{n!} = \left(n + \frac{x}{2}\right)^x$$

$$R = \left[\frac{(a + ab/2)(c + ab/2)}{(b - \alpha b)^2}\right]^{\alpha b}$$

$$\frac{c}{b} = \frac{N_3}{N_2} = \frac{Ne^{-\epsilon_3/kT}}{Ne^{-\epsilon_2/kT}} = e^{-(\epsilon_3 - \epsilon_2)/kT} = e^{-\Delta\epsilon/kT}$$

$$\frac{b}{a} = \frac{N_2}{N_1} = e^{-(\epsilon_2 - \epsilon_1)/kT} = e^{-\Delta\epsilon/kT}$$

$$c = be^{-\Delta\epsilon/kT}, \qquad a = be^{\Delta\epsilon/kT}$$

$$R = \left[\frac{(be^{\Delta\epsilon/kT} + \alpha b/2)(be^{-\Delta\epsilon/kT} + \alpha b/2)}{(b - \alpha b))^2}\right]^{\alpha b}$$

$$= \left[\frac{1 + (\alpha/2)e^{\Delta\epsilon/kT} + (\alpha/2)e^{-\Delta\epsilon/kT} + (\alpha/2)^2}{1 - 2\alpha + (\alpha/2)^2}\right]^{\alpha b}$$

where α is 10^{-6}. Therefore

$$R \cong \left[1 + \frac{\alpha}{2} e^{\Delta\epsilon/kT} + \frac{\alpha}{2} e^{-\Delta\epsilon/kT} \right]. \tag{C.19}$$

If we consider level 1 to be the ground state and level 2 to be the first excited state then b is the number of molecules in the first excited state. Using the classical limit vibration as an example

$$N_2 = b = \frac{N}{z} e^{-\epsilon_2/kT} = \frac{Nh\nu^0}{kT} e^{-\Delta\epsilon/kT}$$

with $h\nu^0/kT \simeq 15$

$$N_2 = N(15)e^{-15} \simeq 10^{18}$$

$$R = \left[1 + \frac{(10^{-6})(3 \times 10^6)}{2} + \frac{10^{-6}(3 \times 10^{-7})}{2} \right]^{10^{18}} \simeq 2^{10^{18}}. \tag{C.20}$$

CONTENT SOURCES AND FURTHER READING

1. R. E. Dickerson, *Molecular Thermodynamics*, Chapter 2, W. A. Benjamin, Menlo Park CA, 1969.
2. J. Kestin and J. R. Dorfman, *A Course in Statistical Thermodynamics*, Chapter 4, Academic Press, New York, 1971.

APPENDIX D

INTERMOLECULAR FORCES

From Newton's second law force is defined as the negative gradient of a potential. The term $-d\varphi/dr$, where φ is the potential energy and r is the distance of separation defines the force acting between two molecules. These forces arise from interactions between the separate electric fields and can, with the exception of London forces that derive from quantum effects, be treated in the classical framework using straightforward electrostatic principles.

The force between two charges separated by a distance r in vacuo is experimentally

$$f = \frac{q_1 q_2}{r^2} \tag{D.1}$$

where q_1 and q_2 are the magnitude of the two charges. In the cgs system the unit of charge is the electrostatic unit (esu) so defined that a force of 1 dyne acts between two charges each of magnitude 1 esu placed 1 cm apart. The unit of charge more commonly used is the coulomb (C), which derives from magnetic field argument. In many discussions the electron charge in esu or coulomb units is required. These charge units are related as follows.

In Ohms law argument a current of 1 ampere represents a charge flow of 1 coulomb per second. The electron charge in coulombs can be measured in terms of the current produced by counted electron flow, e.g., 9.65×10^4 coulombs/electron mole gives the electron charge as 1.60×10^{-19} C/e. The force acting between two electron charges separated by a distance of 1 cm can be measured. It is $4.33 \times 1D^{18}$ dynes so that the esu is 2.08×10^9 electron charges or 3.33×10^{-10} C.

In the SI system of units the charge unit is the coulomb, the distance of separation is expressed in meters, and the force is expressed in Newtons. Two charges of 1 coulomb each separated by 1 meter produce a force of κ Newtons where κ is a constant having units $C^2/N\,m^2$. Spherical field problems make it convenient to write the constant as $4\pi\epsilon^0$, calling ϵ^0 the vacuum permitivity. Then

$$4\pi\epsilon^0\,N = 4\pi\epsilon^0 \times 10^7\,\text{dynes} = 4\pi\epsilon^0 \times 10^7\,\text{esu}^2/\text{cm}^2$$

$$\epsilon^0 = \frac{10^7\,(3.33 \times 10^{-10})^2 C^2}{4\pi \qquad N\,10^4\,m^2} = 8.82 \times 10^{-10}\,C/N\,m^2.$$

We then write

$$f = \frac{q_1 q_2}{4\pi\epsilon^0 r^2} \tag{D.2}$$

where the charges are expressed in coulombs and the separation distance r in meters.

If the interacting charges are ions we can describe them as $z_A q_e$ and $z_B q_e$ where z_A, z_B identify the sign and multiplicity to be attached to unit electron charge q_e. The force acting between univalent ions of opposite charge separated by distance r/z in vacuum is

$$f = \frac{z_A z_B q_e}{4\pi\epsilon_0 r^2} = -\frac{q_e^2}{4\pi\epsilon_0 r^2} \tag{D.3}$$

and relates to a potential energy

$$\phi_r = -\int_r^\infty -\frac{q_e^2\,dr}{4\pi\epsilon_0 r^2} = -\frac{q_e^2}{4\pi\epsilon_0 r} \tag{D.4}$$

where ϕ_r is the potential energy at separation r with reference to the particles at infinite separation. A factor $1/D$, where D is known as the dielectric constant, is introduced when the charges are in a material medium.

The electric field associated with a charge q is defined as the force at any point in space exerted on a unit positive charge placed at that point. The field of charge q_2 at a point identified by $\mathbf{r}$ in Eq. (D.3) is $q_2/4\pi\epsilon_0\mathbf{r}^2$ taking q_1 to be positive and of magnitude 1 coulomb.

When two electric charges of magnitude q are fixed a distance ℓ apart the configuration is characterized by an electric moment $\mu = q\ell$. All asymmetric molecules possess such a moment arising from their electron distribution relative to the mass center. Absolute values of q and ℓ are not accessible. It is convenient to imagine, in a two charge center circumstance, that the charges reside at the

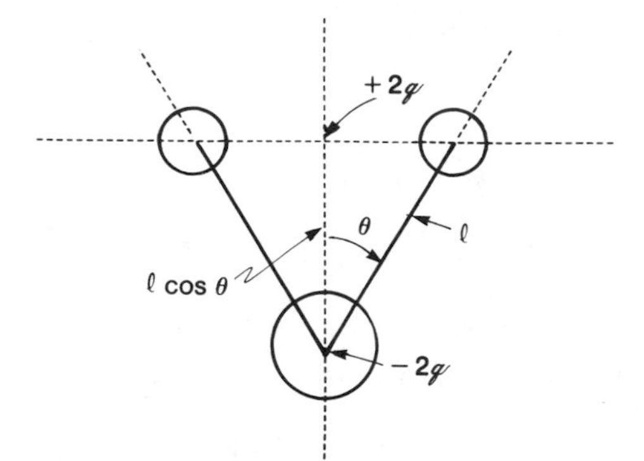

Figure D.1. Illustration of the dipole moment for a water molecule.

nuclei; ℓ is then the internuclear distance and q is some fraction of the electron charge. The quantity μ is referred to as the dipole moment. Any multiple charge distribution can be referred to this format, e.g., for the water molecule we consider a cylindrical axial system as in Figure D.1 with dipole moment $\mu = 2q\ell \cos\theta$, i.e., the molecule is represented by charge $+2q_e$ located a distance $\ell \cos\theta$ from the charge $-2q_e$ located at the oxygen nucleus.

Any multiple charge center molecule may also have a quadrupole moment defined at $Q = \sum q_i r_i^2$ (in analogy to the moment of inertia in a mechanical system) where r is a distance from the charge center to a reference axis. If we again consider the water molecule referenced to cylindrical coordinates (Figure D.1), each charge $+q$ resides a distance $\ell \sin\theta$ from the reference (dipole) axis while the charge $-2q$ lies on the reference axis. The quadrupole moment about the reference axis is $Q = 2q\ell^2 m^2\theta$. As a further example, a linear, symmetrical system of point charges such as indicated in Figure D.2 has both the center of negative and of positive charge coincident with the center of mass and has no dipole moment. There is a quadrupole moment $Q = 2q\ell^2$.

Molecules with dipole or quadrupole moments produce an electric field at any point in space. Referring to Figure D.3 it is clear that simple (but somewhat tedious) geometric argument will produce the force acting along the line of centers in terms of the forces acting between the reference charge and the dipole charges for any orientation of the dipole axis and for any distance r therefore defining the field at all points in space which is

$$F = \left[\frac{\mu^2}{r^6}(1 + 3\cos^2\theta)\right]^{1/2}. \qquad \text{(D.5)}$$

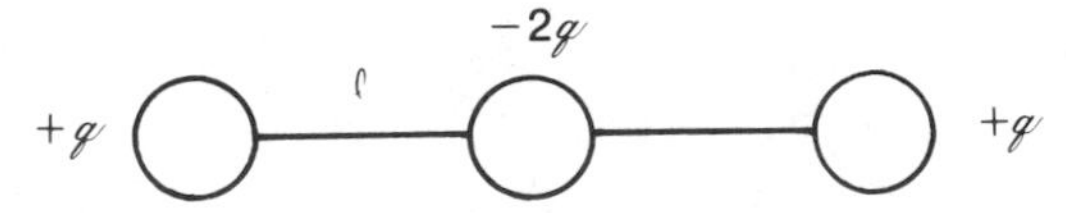

Figure D.2. Illustration of a symmetrical molecule with a quadrupole moment.

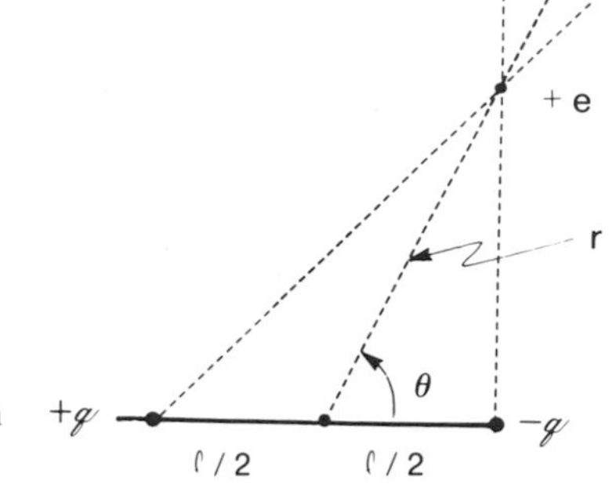

Figure D.3. Illustration of the field at a point in space arising from the remote presence of a dipole.

Definition of the field at any point leads to the potential energy of interaction between the dipole and a charge placed at that point. Further argument leads to the potential energy of interaction between two dipoles

$$\phi_{\mathrm{DD}} = \frac{\mu_{\mathrm{A}}\mu_{\mathrm{B}}}{r^3}[2\cos\theta_{\mathrm{A}}\cos\theta_{\mathrm{B}} - \sin\theta_{\mathrm{A}}\sin\theta_{\mathrm{B}}\cos\psi_{\mathrm{A}} - \cos\psi_{\mathrm{B}}]$$

where θ_A and θ_B are angles defined in Figure D.4 and the angles ψ_A, ψ_B reflect orientation in a plane perpendicular to the plane of the sketch. A similar but considerably more involved expression can be obtained for dipole–quadrupole interaction energies

$$\phi_{\mathrm{QD}} = -\frac{3}{2}\frac{\mu_{\mathrm{A}}Q_{\mathrm{B}}}{r^4}\,\mathrm{f}(\theta_{\mathrm{A}}\theta_{\mathrm{B}}\psi_{\mathrm{A}}\psi_{\mathrm{B}}).$$

When two dipoles interact there is maximum repulsion when the alignment is head to head or tail to tail ($\varphi_r = 2\mu_A\mu_B/4\pi\epsilon_0 r^3$) and maximum attraction ($\psi_a = -2\mu_A\mu_B/4\pi\epsilon_0 r^3$) when head to tail orientation occurs. Taking the dipolar charge to be (0) $0.2q_e$ and the distance of charge separation to be (0) 1 Å in

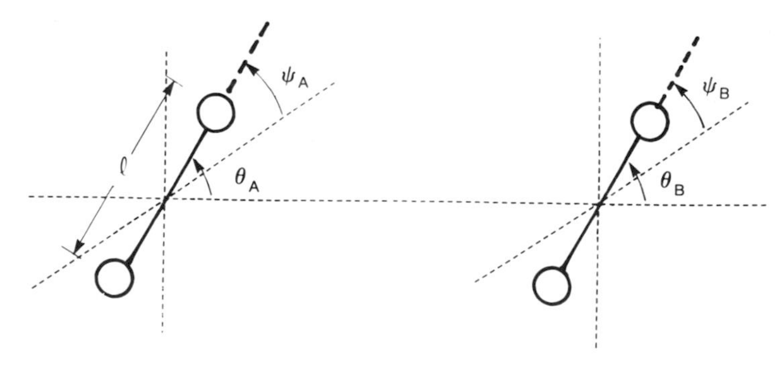

Figure D.4. Illustration of the interaction between two linear dipoles.

the dipole we find the maximum attractive interaction at an assumed 5 Å separation of the dipoles to be

$$\phi_a = \frac{2\left[\left(\frac{1.16}{5} \times 10^{-19}\right)(5 \times 10^{-10})\right]C^2m^2}{4\pi(8.82 \times 10^{-10})(5 \times 10^{-10})C^2\,m^3/N\,m^2} = 7.76 \times 10^{-20}\,J.$$

Thermal energies are order kT or about 10^{21} J at 300 K so that the attractive interactions are about the same order as thermal energies when the separation distance between dipoles is 5 Å. We would anticipate that there is little prevailing orientation under these circumstances. The dipole field averaged over all values of θ [Eq. (D.5)] is $F = (2\mu^2/r^6)^{1/2}$. This is the field that we associate with a freely rotating dipole.

Even if a molecule does not possess a permanent dipole, a charge separation may be induced by interaction with the field of an ion, a dipole, or a quadrupole. It may be assumed that the induced moment (μ_i) is uniquely proportional to the field strength, $\mu_i = \alpha F$. The constant of proportionality ($\propto$) is referred to as the polarizability of the molecule.

The charge separation associated with the induced moment requires electrical work,

$$\begin{aligned} w &= \int_0^l F q\,d\ell = \int_0^l \frac{\mu_i}{\alpha} q\,d\ell = -\frac{q^2}{\alpha}\int_0^l \ell\,d\ell \\ &= \frac{1}{2}\frac{q^2\ell^2}{\alpha} = \frac{\ell}{2}\frac{\mu_i^2}{\alpha} = \tfrac{1}{2}\alpha F. \end{aligned}$$

The attractive interaction of the dipole with the field which generated it is

$$\phi' = -\mu_i F = \alpha F^2$$

so that the net potential energy of interaction is

$$\phi_i = -\alpha F^2 + \frac{\alpha F^2}{2} = -\frac{\alpha F^2}{2}. \tag{D.6}$$

If the field is that of a rotating dipole

$$\phi_{BA} = -\frac{\alpha_B \mu_A^2}{4\pi\epsilon_0 r^6} \tag{D.7}$$

where B refers to the molecule with polarizability α_B in the average dipole field of the molecule with moment μ_A. The total pair interaction has two such terms

$$\phi_{BA,AB} = \frac{-\alpha_A\mu_B^2 + \alpha\beta\mu_A^2}{4\pi\epsilon_0 r^6}. \tag{D.8}$$

In the normal circumstance the molecules are of the same species and

$$\phi_{AA} = -\frac{2\alpha_A\mu_A^2}{4\pi\epsilon_0 r^6}. \tag{D.9}$$

Since permanent dipoles also have polarizability there is an additional, long range interaction term with r^{-6} dependence to be added to the r^{-3} dependence of potential energy between molecules with permanent moments.

Even the simplest particles do have oscillating electric fields that can couple to provide an interaction potential energy. The treatment involves quantum considerations. If two particles each have a single electron oscillating along a common axis, there will be electrostatic interaction terms associated with their charges and distances apart. These terms must be added to the normal displacement energies associated with the restoring force of the oscillator. The total potential energy term in the Schrödinger equation is

$$\mathscr{V} = \tfrac{1}{2}\left(fx_1^2 + \tfrac{1}{2}fx_2^2 + \frac{2q_e^2x_1x_2}{r^3}\right)$$

where f is the commonly valued restoring force for two like oscillators and x_1, x_2 are the displacements from rest when the two rest positions are separated by distance r. The classical expression for the total energy of two oscillators vibrating at frequency v when the momentum terms are written as $p_a = \frac{1}{2}(p_1 + p_2)$ and $p_s \sim \frac{1}{2}(p_1 - p_2)$ and the frequencies of vibration as $v_s = (1 + 2\alpha/r^3)^{1/2}$ and $v_a = (1 - 2\alpha/r^3)^{1/2}$ is $E = \frac{1}{2}[(p_s^2 + \frac{1}{2}fx_1) + (p_a^2 + \frac{1}{2}fx_2)]$. The Schröedinger equation in terms of these variables is

$$\frac{\partial^2\varphi}{\partial x_1^2} + \frac{\partial^2\psi}{\partial x_2^2} + \frac{8\pi^2 m}{h^2}(E - \tfrac{1}{2}fx_1^2 - \tfrac{1}{2}fx_2^2)\psi = 0$$

and the allowed energy levels, $E = (n_s + \frac{1}{2})hv_s + (n_a + \frac{1}{2})hv_a$, provide a ground state energy of $\frac{1}{2}(v_s - v_\alpha)$ while the ground state energy of two independent oscillators is hv. After algebra and expansion the difference is[1]

$$\phi_L = -\frac{hv\alpha^2}{2r^6}. \tag{D.10}$$

The r^{-6} dependence thus appears to be common to attractive interactions between like particles. Generally $\varphi = -B/r^m$ with $\varphi = -B/r^6$ being a particularly useful case. Repulsive interactions are much more obscure. Clearly, since they will derive from the interaction of shielded nuclei, they will fall off much more rapidly than attractive interactions. For mathematical convenience

[1] Forces that produce this interaction potential are called London forces.

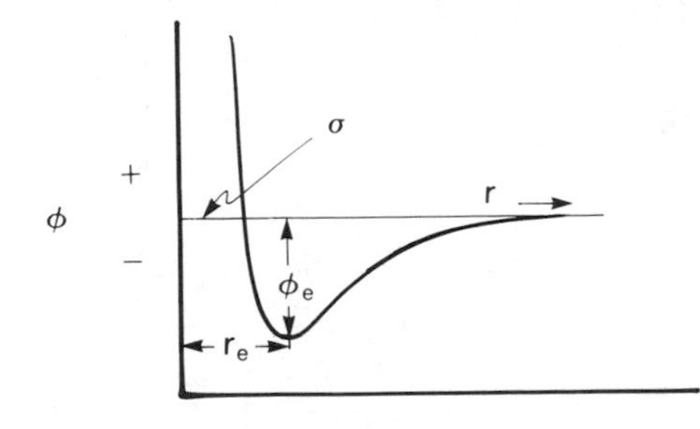

Figure D.5. Illustration of the general interaction energy between particles as a function of their distance of separation.

we can assign the same form to them writing,

$$\phi_\Sigma = Ar^{-n} - Br^{-m}, \qquad n > m.$$

A value of 9 or 12 for n is commonly utilized. The $12-6$ potential is widely used and is known as the Leonard–Jones potential

$$\phi_\Sigma = Ar^{-12} - Br^{-6}. \tag{D.11}$$

Taken over all values of separation Eq. (D.11) produces an interaction energy like that shown in Figure D.5.

The force between the pair is zero at the minimum, $-(d\varphi/dr)r_e = 0$ so that

$$d(A/r_\mathrm{e}^n - B/r_\mathrm{e}^m)/dr = 0$$

$$-(-nAr_\mathrm{e}^{n-1} + mBr_\mathrm{e}^{m-1}) = 0$$

$$r_\mathrm{e}^{n-m} = \frac{nA}{mB} \tag{D.12}$$

and A or B can be expressed in terms of the minimum value for $\phi(=\phi_\mathrm{e})$, i.e.,

$$A = \frac{m}{n} Br_\mathrm{e}^{n-m}$$

$$\varphi_\mathrm{e} = B\left(\frac{m}{n}\frac{r_\mathrm{e}^{n-m}}{r_\mathrm{e}^m} - \frac{1}{r_\mathrm{e}^m}\right) = Br_\mathrm{e}^m\left(\frac{m}{n} - 1\right)$$

and similarly

$$\phi_\mathrm{e} = Ar^{-n}\left(1 - \frac{n}{m}\right) \tag{D.13}$$

so that

$$B = \frac{n\phi_\mathrm{e} r_\mathrm{e}^m}{m-n}, \qquad A = \frac{m\varphi_\mathrm{e} r^n}{m-n}$$

and

$$\phi = \frac{\phi_e}{m-n}\left[m\left(\frac{r_e}{r}\right)^m - n\left(\frac{r_e}{r}\right)^n\right].$$

The distance of separation term σ (Figure 0.5) corresponds to $\varphi = 0$ and

$$m\left(\frac{r_e}{\sigma}\right)^m = n\left(\frac{r_e}{\sigma}\right)^n$$

so

$$r_e = \left(\frac{n}{m}\sigma^{m-n}\right)^{1/(m-n)} = \sigma\left(\frac{n}{m}\right)^{1/(m-n)}$$

In terms of σ and ϕ_e the potential energy at any distance of separation is

$$\phi = \frac{\phi_e}{n-m}\left(\frac{n^n}{m^m}\right)^{1/(n-m)}\left[\left(\frac{\sigma}{r}\right)^n - \left(\frac{\sigma}{r}\right)^m\right]. \quad \text{(D.14)}$$

The Leonard–Jones 12–6 potential becomes

$$\phi = 4\phi_e\left[\left(\frac{\sigma}{r}\right)^{12} - \left(\frac{\sigma}{r}\right)^6\right]. \quad \text{(D.15)}$$

This potential energy expression is suitable for many problems involving gas phases at moderately high density. Other potentials will be required for gases at high density and particularly for molecules with permanent dipoles. As interaction becomes more complex (orientation features, etc.) any equation having only a few (2, 3, 4) adjustable constants will fail to reproduce the interaction circumstances. Since, for our purposes, models that are relatively simple and that reflect general behavioral features are most appropriate, this discussion will not be carried further. Extensions will be found in cited discussions.[2]

The potential energy between a set of interacting particles is best reflected by the virial theorem of Clausius. Here the components, in Cartesian coordinates of the force acting on a particle in motion are

$$X = m\frac{d^2x}{dt^2}, \qquad Y = m\frac{d^2y}{dt^2}, \qquad Z = m\frac{d^2z}{dt^2}.$$

[2]see references 1, 2 (Moelwyn–Hughes, 1961; Mayer and Mayer, 1977).

Writing

$$xX + yY + zZ = m\left(x\frac{d^2x}{dt^2} + y\frac{d^2y}{dt^2} + z\frac{d^2z}{dt^2}\right)$$

and noting that

$$\frac{d}{dt}\left(x\frac{dx}{dt}\right) = x\frac{d^2x}{dt^2} + \left(\frac{dx}{dt}\right)^2$$

$$x\frac{d^2x}{dt^2} = \frac{d}{dt}\left(x\frac{dx}{dt}\right) - \left(\frac{dx}{dt}\right)^2, \quad \text{etc.}$$

$$xX + yY + zZ = m\frac{d}{dt}\left[x\left(\frac{dx}{dt}\right) + y\left(\frac{dy}{dt}\right) + z\left(\frac{dz}{dt}\right)\right]$$
$$- m\left[\left(\frac{dx}{dt}\right)^2 + \left(\frac{dy}{dt}\right)^2 + \left(\frac{dz}{dt}\right)^2\right].$$

Since dx/dt is the x component of velocity, etc. the final (bracketed) term is the squared velocity of the particle and, since, $x(dx/dt) = \frac{1}{2}(dx^2/dt)$, etc.

$$xX + yY + zZ = \frac{m}{2}\left[\frac{d}{dt}(x^2 + y^2 + z^2)\right] - mv^2. \qquad \text{(D.16)}$$

For a confined system of particles the bracketed term must vanish for any longterm average, otherwise the average distance between particles would change. Therefore

$$\tfrac{1}{2}\sum mv^2 = \mathrm{E}_t = \tfrac{1}{2}\sum(xX + yY + zZ) = \tfrac{3}{2}N\mathcal{k}\mathrm{T}$$

for a gas phase. The term $(xX + yY + zZ)$ is known as the virial of the forces.

There are two contributions to the forces acting on a confined system of particles. First, there is the force exerted by the confining walls and second, there is the force associated with molecular interactions. For one wall, say xy, the force exerted by molecules striking the wall is $P(xy)$ so the (equal and opposite) force exerted by the wall is $-P(xy)$. The distance of a molecule from the wall can be 0 to z or, on the average $z/2$. The force term for this one wall is $PV/2$. For six walls the contribution to the sum appearing in Eq. (D.11) due to the confining walls is $-\frac{3}{2}PV$. We write then

$$-\tfrac{3}{2}N\mathcal{k}T = \tfrac{1}{2}(3PV) + \tfrac{1}{2}\sum(xX + yY + zZ)$$

or

$$PV = N\mathcal{k}\mathrm{T} + \tfrac{1}{3}\sum(xX + yY + zZ) \qquad \text{(D.17)}$$

where the sum now relates to intermolecular forces only. In the absence of such interactions the ideal gas result emerges. The virial can be expressed in terms of an interaction energy between two particles. If the interaction energy is ϕ along the line of centers of two particles separated by distance a where the line of centers is oriented at angle θ to the axis, the force f is $-d\phi/da$ and the force component along the x axis is $f\cos\theta = f(x_2 - x_1)/a$ where x_1 and x_2 are the projections of the x coordinates of the two particles onto the x axis. Then

$$\sum xX = x_1X_1 + x_2X_2 = f\left(\frac{x_2 - x_1}{a}\right)x_1 + f\left(\frac{x_1 - x_2}{a}\right)x_2 = (x_2 - x_1)^2\frac{f}{a}$$

$$\sum(xX + yY + zZ) = \frac{f}{a} - [x_2 - x_1)^2 + (y_2 - y_1)^2 + (z_2 - z_1)^2] = f\,a$$

and

$$PV = N\mathscr{k}T + \tfrac{1}{3}\sum a\frac{d\phi}{da}. \tag{D.18}$$

The summation of $ad\phi/da$ for every molecule pair in the system can be approached as follows. The number of molecules in a spherical shell at distance a from a given molecule is $n_a 4\pi^2 a^2 da$ where n_a is the concentration at a. If this is governed by the Boltzmann distribution law

$$n_a = ne^{-\phi(a)/\mathscr{k}T} = \frac{N}{V}e^{-\phi(a)/\mathscr{k}T}. \tag{D.19}$$

The sum $ad\phi_1/d\phi$ for a single molecule over all molecules in the system is

$$\int_0^\infty \frac{N}{V}e^{-\phi/\mathscr{k}T}(a)\frac{d\phi}{da}(4\pi a)^2 da$$

and there are $n/2$ such terms so that

$$\sum a\frac{d\phi}{da} = \frac{2\pi N^2}{V}\int_0^\infty e^{-\phi/\mathscr{k}T}a^3 d\phi. \tag{D.20}$$

Integration by parts ($\int u dv = uv - \int v du$) produces

$$u = a^3, \qquad dv = e^{-\phi/\mathscr{k}T}d\phi$$

$$du = 3u^2 da, \qquad v = -\mathscr{k}Te^{-\phi/\mathscr{k}T}$$

$$uv - [\int v du = \mathscr{k}Ta^3 e^{-\phi/\mathscr{k}T}]_0^\infty + \int_0^\infty \mathscr{k}Te^{-\phi/\mathscr{k}T}3a^2 da. \tag{D.21}$$

If ϕ is taken to be zero at infinity, the first term in Eq. (D.21) becomes $-\mathscr{k}Ta^3$

and

$$-\ell T a^3 = \ell T \int 3a^2 da.$$

The resulting expression is

$$\sum a \frac{d\phi}{da} = \frac{6\pi N^2}{V} \ell T \int_0^\infty (1 - e^{-\phi/\ell T}) a^2 da$$

so that an equation of state in the form

$$PV = N\ell T + \frac{2\pi N^2 \ell T}{V} \int_0^\infty (1 - e^{-\phi/\ell T}) a^2 da \tag{D.22}$$

results. The treatment is due to Raleigh. A general expression of the form

$$PV = N\ell T + \frac{B(T)}{V} + \frac{C(T)}{V^2} + \cdots \tag{D.23}$$

is known as the virial equation of state with $B(T)$, $C(T)$,... known as the second, third,... virial coefficients.

The form of Eq. (D.23) is particularly well fitted to curve fit from experimental data but any appropriate potential function can be introduced into Eq. (D.22) and virial coefficients approximated by integration. For example, if the Raleigh equation integral is separated into two parts,

$$I = \int_0^\sigma (1 - e^{-\phi/\ell T}) a^2 da + \int_\sigma^\infty (1 - e^{-\phi/\ell T}) a^2 da \tag{D.24}$$

the first integral represents the repulsive region (see Figure D.5) and increases very rapidly with a; ϕ is also positive and $\exp(-\phi/\ell T) \to 0$ and may be ignored. The second integral can be reduced to

$$\int_\sigma^\infty \left[1 - \left(1 - \frac{\phi}{\ell T}\right)\right] a^2 da$$

when ϕ is small compared to ℓT. Then

$$I = \tfrac{1}{3}\sigma^3 + \int_\sigma^\infty \frac{\phi}{\ell T} a^2 da. \tag{D.25}$$

Writing Eq. (D.25) in the form

$$\phi = \frac{\gamma\phi_e}{n-m}\left[\left(\frac{\sigma}{r}\right)^n - \left(\frac{\phi}{r}\right)^m\right]$$

$$\gamma = \left(\frac{n^n}{m^m}\right)^{1/(n+m)}$$

$$I = \tfrac{1}{3}\sigma^3 + \frac{\gamma\phi_e}{kT}\left[\int_\sigma^\infty \sigma^n r^{(2-n)}\,dr + \int_\sigma^\infty \sigma^m r^{(2-m)}\,dr\right]$$

$$= \tfrac{1}{3}\sigma^3\left[1 - \frac{3\gamma}{(n-3)(m-3)}\frac{\phi_e}{kT}\right]. \tag{D.26}$$

The second virial coefficient is, then, using the Leonard–Jones values for n and m

$$B(\mathrm{T}) = \frac{2\pi \mathrm{N}^2\sigma^3}{3\mathrm{V}}\left[1 - 1.285\frac{\phi_e}{kT}\right].$$

There are other approaches that produce correction terms to ideal gas behavior that can be couched in the form of Eq. (D.23) and that will provide useful virial coefficient values.

A strictly statistical approach to the evaluation of second and higher order virial coefficients is based on the theory of cluster integrals. We begin with the introduction of the function[3]

$$e^{-u(r_{ij})/kT} = 1 + f(r_{ij}) \tag{D.27}$$

which has the behavior indicated in Figure D.6. so that $f(r_{ij})$ approaches 0 at large separation distances and 1 at small distances of separation. Only the first consideration is of importance; most intermolecular interactions do not extend beyond 20 Å so that particles must be within that distance else f_{ij} will vanish. The configurational integral becomes

$$Q_\mathrm{N} = \int \cdots \int \prod_{i>j} (1 + f_{ij})\,d\tau_1, \ldots, d\tau_\mathrm{N}. \tag{D.28}$$

Expansion of the product terms leads to a collection of terms some of which involve only pair functions while others involve pair function products

$$(1 + f_{ij})(1 + f_{kl})\cdots = 1 + f_{ij} + f_{kl} + f_{ij}f_{kl}\cdots.$$

[3] This treatment follows reference 3.

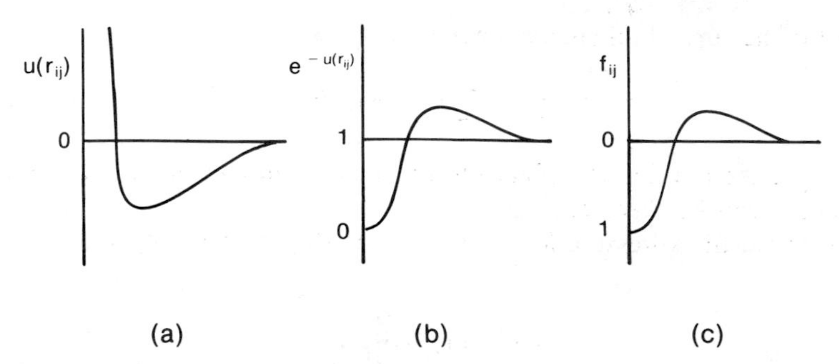

Figure D.6. Illustration of features of the Mayer function for constructing the configuration integral.

We can regard Eq. (D.28) as representing clustered molecule interactions with the terms unity representing single molecules, pair functions representing two molecule interactions, while pair product terms represent three, four, etc. molecule interactions. We shall refer to these two, three, four molecule interaction groupings as $1, 2, \ldots, \ell$ clusters and designate the number of each cluster present in the N particle system as N_ℓ, then $N = \sum^\ell lN_\ell$. The configurational integral can be expressed in terms of integrals taken over each cluster type.

The three cluster is sufficiently general to illustrate the manner in which this is accomplished. The three cluster involves all product terms with only three indices in common, e.g., $f_{ij}f_{jk}, f_{ik}f_{jk}, f_{ij}f_{ik}, f_{ij}f_{jl}, \ldots$. For any such three clusters we write $f(pqr)$ and express the three cluster component of the configurational integral as

$$\iiint \prod_{p>q,r} f(pqr)d\tau_1 d\tau_2 d\tau_3 = \int\left[\iint \prod_{p>q,r} f(pqr)d\tau, d\tau_2\right]d\tau_3 = VB_3 \quad \text{(D.29)}$$

where integration over one of the volume elements to produce V can be taken as a reflection of the independence of the three cluster integral (bracketed term) on any particular location of the molecule cluster within the total volume. Again the notation $p > q, r$ prevents multiple counting. Using Eq. (D.29) as a general form, we write

$$V\underbrace{\int \cdots \int}_{\ell-1} \prod_{p>q} f(pq)d\tau, \ldots, d\tau_{\ell-1} = VB_\ell \quad \text{(D.30)}$$

for any single ℓ cluster where the indices p, q now represent all of the ℓ indices involved in an ℓ-cluster. The term VB_ℓ will be identical for each of the ℓ-clusters of which there are N_ℓ. Any single assignment of the N molecules in the system to $1, 2, 3, \ldots, \ell$-clusters produces terms $\prod(VB_\ell)^{N_\ell}$ and the configurational integral is the sum of all such assignments

$$Q_N = \sum g_{N_\ell} \pi (VB_\ell)^{N_\ell} \tag{D.31}$$

where g_{N_ℓ} is the multiplicity of each term, i.e., the number of ways in which the distribution can be effected.

The statistical expression for g_{N_ℓ} is, for any distribution $(N_1, \ldots, N_\ell)$

$$g_{N_\ell} = \frac{N!}{\prod N_\ell! \prod (\ell!)^{N_\ell}} \tag{D.32}$$

arising from the consideration that the N particles can be ordered, e.g., in a straight line, in N! ways if they are distinguishable, then divided into $1, 2, 3, \ldots, \ell$-clusters by taking the first N_1 particles to be isolated, the next N_2 to be in two clusters, etc. for each permutation. We then divide $\prod N_\ell!$ since the $N_1, \ldots, N_\ell$ sequence is not important and for each N_ℓ by $\ell!$ since the i, j, k-sequence is not important. The sum of all distributions $N_1, \ldots, N_\ell$ consistent with $\Sigma \ell N_\ell = N$ is the sum of all such terms

$$g = \sum_{N,\ell} g_{N_\ell}$$

Equation (D.31) presents us with a now familiar problem. What distribution of the particles among clusters leads to the maximum term in g_{N_ℓ} and therefore to the maximum term in the configurational integral Q_N? It is of some mathematical convenience to note that carrying out the maximization of the partition function, in which Q_N appears as a multiplier, will accomplish the same result as maximizing the configurational integral. We then write for the indistinguishable particles which make up a single component gas phase

$$\mathscr{Z} = \left(\frac{2\pi m k T}{h^2}\right)^{3N/2} \frac{Q_N}{N!} = \sum^{N} \frac{1}{\prod N_\ell! \prod (\ell!)^{N_\ell}} \prod (VB_\ell)^{N_\ell} \left(\frac{2\pi m k T}{h^2}\right)^{3N/2}$$

or simplifying

$$\mathscr{Z} = \sum \frac{1}{\prod^{\ell} N_\ell!} \prod^{\ell} \left[\left(\frac{2\pi m k T}{h^2}\right)^{3\ell/2} V \frac{B_l}{\ell!}\right]^{N_\ell} = \sum_{N} \frac{1}{\prod^{\ell}_{N} N_\ell!} \prod^{\ell} (V q_\ell)^{N_\ell}$$

defining q_ℓ to include all terms containing ℓ in any direct manner

$$q_\ell = \left[\left(\frac{2\pi m k T}{h^2}\right) \frac{B_\ell}{\ell!}\right]. \tag{D.33}$$

If we now maximize $\ln \mathscr{Z}$ subject to the constraint $\sum \ell N_\ell = N$, we obtain

$$\ln \mathscr{Z} = (-\sum N_\ell \ln N_\ell + \sum N_\ell + \sum N_\ell V q_\ell)$$

$$d\ln \mathscr{Z} = (-\sum N_\ell d\ln N_\ell - \sum \ln N_\ell dN_\ell + \sum dN_\ell + \sum \ln Vq_\ell dN_\ell)$$

and for the term

$$d\ln \mathscr{Z}_{\max} + \alpha \sum \ell dN_\ell = (-\sum \ln N_\ell + \sum Vq_\ell + \alpha\ell) dN_\ell$$

or, from familiar argument,

$$N_\ell^* = e^{\alpha\ell} V q_\ell, \qquad N_\ell^* = \lambda^\ell V q_\ell \qquad \text{for } \lambda = e^\alpha \tag{D.34}$$

and $\ln N_\ell^* = \ell \ln \lambda + \ln Vq_\ell$. We now have

$$\ln \mathscr{Z}_{\max} = -\sum (N_\ell^* l \ln \lambda + N_\ell \ln q_\ell V) + \sum N_\ell^* + \sum N_\ell \ln Vq_\ell$$

or

$$\ln \mathscr{Z}_{\max} = \sum N_\ell^* - \sum \ell N_\ell \ln \lambda = \sum N_\ell^* - N \ln \lambda. \tag{D.35}$$

Since we are interested in following the consequences to the nonideal gas equation of state, we will follow the route

$$A = -kT \ln \mathscr{Z}, \qquad P = -\left(\frac{\partial A}{\partial V}\right)_{N,T}$$

so that

$$A = -kT(\sum N_\ell^* + NkT \ln \lambda) \tag{D.36}$$

$$P = kT \sum \left(\frac{\partial N_\ell^*}{\partial V}\right)_{N,T} - NkT \frac{\partial \ln \lambda}{\partial V}. \tag{D.37}$$

From Eq. (D.33)

$$\frac{\partial N_\ell^*}{\partial V} = \left(\frac{\partial \lambda^\ell V q_\ell}{\partial V}\right) = \lambda^\ell q_\ell + Vq_\ell \left(\ell \lambda^{\ell-1} \frac{\partial \lambda}{\partial V}\right)$$

since q_ℓ is independent of V [Eq. (0.32)], and

$$P = kT \sum \lambda^\ell q_\ell + kT \lambda V q_\ell \ell \lambda^{\ell-1} \left(\frac{\partial \lambda}{\partial V}\right) - \frac{NkT}{\lambda}\left(\frac{\partial \lambda}{\partial V}\right)$$

$$= kT \frac{\sum N_\ell^*}{V} + kT \left[\frac{\ell V q_\ell \lambda^\ell}{\lambda} - \frac{N}{\lambda}\right] \frac{\partial \lambda}{\partial V}$$

$$= kT \frac{\sum N_\ell^*}{V} + kT \left[\frac{\sum \ell N_\ell^*}{\lambda} - \frac{N}{\lambda}\right] \frac{\partial \lambda}{\partial V}$$

so that the last term vanishes with the bracketed term being zero. So

$$P = kT\frac{\sum N_\ell^*}{V} = kT\sum \lambda^\ell q_\ell \tag{D.38}$$

since $\sum \ell N_\ell = N$, $\sum \ell\lambda^\ell V q_\ell = N$ or $\sum \ell\lambda^\ell q_\ell = N/V$.
Expanding this expression then gives

$$q_1\lambda + 2q_2\lambda^2 + 3q_3\lambda^3 + \cdots = N/V. \tag{D.39}$$

Recognizing that the most convenient form for the equation of state is the virial expansion in powers of $1/V$, we shall assume that λ can be similarly expressed

$$\lambda = \frac{a_1 N}{V} + \frac{a_2 N^2}{V^2} + \cdots. \tag{D.40}$$

Two terms will be sufficient to indicate the arithmetic route to a virial coefficient expression. Combining Eqs (D.38) and (D.39) produces

$$q_1\frac{a_1{,}N}{V} + q_2\frac{a_2 N^2}{V^2} + 2q_2\frac{a_1^2 N^2}{V^2} + \cdots = N/V$$

$$q_1 a_1 + (q_2 a_2 + 2q_2 a_1^2)N/V + \cdots = 1$$

requiring that $(q_1 a_2 + 2q_2 a_2) = 0$ and $q_1, a_1 = 1$ or

$$a_1 = \frac{1}{q_1}, \qquad a_2 = \frac{-2q_2 a_1^2}{q_1} = \frac{-2q_2}{q_1^2}.$$

Insertion of these values first into Eq. (D.40) and then using that result in Eq. (D.38) produces

$$P = kT\left[\frac{N}{V} - \frac{q_2}{q_1}\frac{N^2}{V^2} + \cdots\right].$$

From Eq. (D.34) we note that

$$q_1 = \left(\frac{2\pi m kT}{n^2}\right)^{3/2}$$

with $B_1 = 1$ and

$$q_2 = \left(\frac{2\pi m kT}{n^2}\right)^3 \frac{B_2}{2}$$

so

$$P = \frac{N\mathscr{k}T}{V}\left[1 - \frac{N}{2V}B_2 + \cdots\right]$$

with

$$B_2 = -\int f_{ij}d\tau_i = -\int_0^{\infty} f(r)4\pi r^3 dr$$

where in the final expression f_{ij} is replaced by $f(r)$ changing to spherical polar coordinates. The upper limit of integration can be taken as infinity since $f(r)$ vanishes when r is greater than the distance over which interactions are effective. The transformation to polar coordinates involves recognition that f_{ij} was simplified notation for $f(r_{ij})$ and that $f(r_{ij})$ is independent of the angles θ_{ij} and ϕ_{ij} so that integration over them produces the factor 4π. The second virial coefficient

$$B_2 = \int_0^{\infty} 4\pi r^2(1 - e^{-u(r_{ij})})dr \tag{D.41}$$

is to be regarded as the contribution to the configuration integral from two body collisions. The third virial coefficient involves three body collisions, etc.

CONTENT SOURCES AND FURTHER READING

1. E. A. Moelwyn-Hughes, *Physical Chemistry*, 2nd ed., Chapter VII. Pergamon Press, London, 1961.
2. J. E. Mayer and M. G. Mayer, *Statistical Mechanics*, Chapter 8. Wiley-Interscience, New York, 1977.
3. G. S. Rushbrooke, *Introduction to Statistical Mechanics*, Chapter XVI, Oxford Press, London, 1949.

APPENDIX E

ENSEMBLES

An observable thermodynamic property reflects some particular time-averaged behavior of the system components. Gibbs' introduction of the statistical ensemble is a device for conceptually avoiding the time-averaging process. The time-averaged value of a single system property is replaced by the instantaneous average property value taken over an immense number of systems each of which is an exact macroscopic replication of the original system. This large collection of systems is the ensemble and the number of ensemble elements must be sufficiently large that every allowed system state can be represented a statistically meaningful number of times in the instantaneous average. Since the ensemble is fictitious, the number of ensemble elements can easily be taken as approaching infinity.

Macroscopic identity between ensemble elements is specified by an appropriate set of thermodynamic variables such as E, V, and N or T, V, and N. Practical experience has reduced the large number of possible ensembles to three, which are maximally effective. These are called microcanonical (E, V, and N constant, i.e., an ensemble of isolated systems), canonical (T, V, and N constant, i.e., an ensemble of closed isothermal systems), and grand canonical (T, V, and μ constant, i.e., an ensemble of open isothermal systems). Our development in the text focused on the microcanonical ensemble. We anticipate the material that follows to remark that the features of the canonical and grand canonical ensembles are such that if any group of ensemble systems were placed in individual isolation, that collection would be a microcanonical ensemble. There is, then, no real difference between the several ensembles except for possible mathematical convenience.

The canonical ensemble is easiest to visualize; we begin there. Consider that a system of fixed V and N (therefore closed) having diathermal boundaries is immersed in a large constant temperature bath and permitted to reach thermal equilibrium having then the bath temperature T. Replicate this circumstance $\mathcal{N}$ times where $\mathcal{N}$ is a very large number. Now place the systems in diathermal contact with each other and isolate this supersystem which is the canonical ensemble in the limit $\mathcal{N} \to \infty$.

The entire ensemble is an isolated supersystem having energy E_Σ, volume $\mathcal{N}V$ and number of molecules $\mathcal{N}N$. Each system can be regarded as being in a constant temperature bath composed of all other systems and is, therefore, a replica of the original system. Each system will have fixed V and N and will be in one of the quantum energy states consistent with the values T, V, and N. We assume that the distribution of ensemble systems among these quantum states is without favor. Let the energy states be $E_1, E_2, \ldots, E_j$ and let their ensemble populations be $\eta_1, \eta_2, \ldots, \eta_j$, respectively, at any moment the ensemble is sampled. Among the very many ensemble distributions all must satisfy

$$\sum \eta_j = \mathcal{N}, \qquad \sum \eta_j E_j = E_\Sigma.$$

For any distribution there are

$$\Omega_{t(\eta)} = \frac{(\sum^j \eta_j)!}{\prod \eta_j!} = \frac{\mathcal{N}!}{\prod \eta_j!} \tag{E.1}$$

ways in which the ensemble members can be permuted within the distribution. Following the method of undetermined multipliers, which has been detailed in Appendix B, we obtain the maximum value for $\Omega_{t(\eta)}$ subject to the given constraints, again under the assumption that the most probable distribution is overwhelmingly most probable.

From

$$\frac{\partial}{\partial \eta_j}[\ln \Omega_{t(\eta)} - \alpha \sum \eta_j - \beta \sum \eta_j E_j] = 0$$

we obtain

$$\eta_j^* = \mathcal{N} e^{-\alpha} e^{-\beta E_j}$$

then

$$\sum \eta_j^* = \mathcal{N} e^{-\alpha} \sum e^{-\beta E_j}$$

or

$$e^{\alpha} = \sum e^{-\beta E_j}$$

and

$$\bar{E} = \frac{\sum E_j e^{-\beta E_j}}{\sum e^{-\beta E_j}}$$

(both from the equations of constraint). Then $p_j^* = \exp(-\beta E_j)\sum \exp(-\beta E_j)$. The E_j are all possible energy states that individual ensemble members can assume subject to the constraints imposed. The E_j values for a given ensemble member can vary only as a result of energy exchange across the diathermal boundaries. We have demonstrated in a prior argument that systems are in thermal equilibrium when they have a common value of T or β when $\beta = 1/kT$. We will retain this association, ultimately appealing to the internal consistency of the thermodynamic results for confirmation. On the basis of the thermal equilibrium that must prevail, we anticipate that many of the energy states will be identical. We define $\mathscr{Z}$

$$\mathscr{Z} = \sum^{j} e^{-E_j/kT} = \sum^{i} \Omega_{V,N} e^{-E_i/kT} \tag{E.2}$$

where $\Omega_{(V,N)}$ is the degeneracy of an energy value. We are interested in the probability distribution of the E_i around $\bar{E}$.

If the fluctuations about the mean value are Gaussian, the dispersion of values can be characterized by the root mean square deviation (see Appendix C on Statistical Considerations)

$$\sigma_E = [\overline{(E - \bar{E})^2}]^{1/2} = [\overline{E^2} - \bar{E}^2]^{1/2}.$$

If we take

$$\bar{E}\sum e^{-E_j/kT} = \sum E_j e^{-E_j/kT}$$

and differentiate with respect to T we have,

$$\sum e^{-E_j/kT}\frac{\partial \bar{E}}{\partial T} + \bar{E}\left(\frac{1}{kT^2}\sum E_j e^{-E_j/kT}\right) = \sum \bar{E}\frac{E_j}{kT}e^{-E_j/kT}$$

with

$$\mathscr{Z}\frac{\partial \bar{E}}{\partial T} + \frac{(\bar{E})(E)}{kT^2}\mathscr{Z} = \frac{\bar{E}^2}{kT^2}\mathscr{Z}$$

$$\frac{\partial \bar{E}}{\partial T} = \frac{\overline{E^2} + \bar{E}^2}{kT^2} = \frac{\sigma_E^2}{kT^2}$$

$$\sigma_E^2 = kT^2 C_V \qquad \text{and} \qquad \frac{\sigma_E}{E} = \frac{(kT^2 C_V)^{1/2}}{E}.$$

Since $C_V = (0)N\mathscr{k}$ and $\bar{E} = (0)\ N\mathscr{k}T$

$$\frac{\sigma_E}{E} = (0)\left[\frac{(\mathscr{k}T)^2(N\mathscr{k})}{N\mathscr{k}T}\right]^{1/2} = (0)\frac{1}{N^{1/2}} = (0)10^{-12} \tag{E.3}$$

and $\sigma_E = (0)10^{-12}\bar{E}$. This is a completely negligible fluctuation. Effectively all systems in the canonical ensemble have energy $\bar{E}$. Any system in the canonical ensemble could be selected at random with the virtual certainty that it will have energy $\bar{E}$ and of course V and N. We now write

$$\mathscr{Z} = \sum \Omega(E, V, N) e^{-\sum N_j \epsilon_j / \mathscr{k}T}$$

where each energy value E_j for the canonical ensemble is represented by $\Omega(E, V, N)$ ensemble members, each of which consist of N molecules $(= \sum N_j)$ distributed among the energy states. We have, however, demonstrated above that effectively all members of the ensemble have the same energy so that the sum reduces to a single term

$$\mathscr{Z} = \Omega(\bar{E}, V, N) e^{-\sum N_j \epsilon_j / \mathscr{k}T} \tag{E.4}$$

where $\Omega(\bar{E}, V, N)$ is the degeneracy for the effectively common value of E for the ensemble at specified T, V, N. It can also be said that Eq. (E.4) represents the maximum valued term in the sum of all such terms and for $\Omega(E, V, N) = N!|\prod N_j!$ we can again, by familiar procedure, obtain those which maximize the term expression

$$\mathscr{Z} = t_{max} = \frac{N!}{\prod N_j!} e^{\sum N_j \epsilon_j / \mathscr{k}T}$$

since maximizing Eq. (E.2) is equivalent to maximizing Eq. (E.1). The procedure produces directly the familiar result

$$\mathscr{Z} = [\sum (e^{-\epsilon_j / \mathscr{k}T})]^N = \mathscr{z}^N \tag{E.5}$$

where $\mathscr{z}$ is the partition function for an isolated system of N particles at constant E and V. The arguments are not different for indistinguishable particles,

$$\Omega(E, V, N) = \frac{\prod g_i^{N_i}}{\prod n_i!}$$

where the subscript notation refers to energy levels rather than energy states. Arithmetic will produce

$$\mathscr{Z} = t_{max} = \frac{\mathscr{z}^N}{N!} \qquad \text{where } \mathscr{z} = \sum g_i e^{-\epsilon_i / \mathscr{k}T}. \tag{E.6}$$

We could have anticipated this result on the basis that all canonical ensemble members have the same energy. The canonical ensemble (T, V, N) does not differ from a microcanonical ensemble (E, V, N) made up by isolating all members of the canonical ensemble.

The grand canonical ensemble is conceived of as a large ($\mathcal{N}$) collection of open systems, each having volume V, which can be permitted to attain thermal equilibrium with a large reservoir at T and, at the same time, exchange particles with this reservoir so that each system arrives at a common value of chemical potential μ. The ensemble can now be isolated. It will have η systems each of volume V, temperature T, and chemical potential μ. Its total energy is E_Σ and the total number of molecules is N_Σ. Any system in the ensemble can have energy E_j consistent with (T, V, N) but N can have any value consistent with T, V, and μ. In any state E_j let there be $\eta_{j,K}$ systems in the ensemble, then

$$\sum \eta_{j,K} = \mathcal{N}$$
$$\sum \eta_{jK} E_j = E_\Sigma$$
$$\sum N_K \eta_{j,K} = N_\Sigma.$$

Each $\eta_{j,K}$ identifies a number of ensemble systems that are in the energy state E_j and which have N_K molecules. Both $\eta_{j,K}$ and N_K can have any values that are consistent with the above constraints. The number of quantum states in the ensemble is given again by the number of permutations of ensemble systems among the available energy states and is the sum of terms

$$\Omega_t = \frac{(\sum^{j,K} \eta_{j,K})!}{\prod_{j,K} (\eta_{j,K})!} = \frac{\mathcal{N}!}{\prod_{j,K} \eta_{j,K}!}. \tag{E.7}$$

The three constraints require three undetermined multipliers when all prior arguments for selecting the maximum term are applied. From

$$\frac{\partial}{\partial \eta_{j,K}} [\ln \Omega_t - \alpha \sum \eta_{j,K} - \beta \sum \eta_{j,K} E_{j,K} - \gamma \sum \eta_{j,K} N_K] = 0$$

we obtain $\eta^*_{j,K} = \mathcal{N} e^{-\alpha} e^{-\beta E_{j,K}} e^{-\gamma N_K}$.

From $\sum \eta_{j,K} = \mathcal{N}$ and $\sum \eta_{j,K} = \mathcal{N} e^{-\alpha} \sum (e^{-\beta E_{j,K}} e^{-\gamma N_K})$ we obtain

$$e^\alpha = \sum e^{-\beta E_{j,K}} e^{-\gamma N_K} = \Xi \tag{E.8}$$

also

$$\frac{\eta^*_{j,K}}{\mathcal{N}} = p_{j,K} = \frac{e^{-\beta E_{j,K}} e^{-\gamma N_K}}{\sum e^{-\beta E_{j,K}} e^{-\gamma N_K}} = \frac{e^{-\beta E_{j,K}} e^{-\gamma N_K}}{\Xi} \tag{E.9}$$

where $\Xi = \sum (e^{-\beta E_{j,K}} e^{-\gamma N_K})$ defines the grand partition function.

When $\beta = 1/kT$ is inserted into the grand partition function expression $\Xi = \sum^{K} \mathscr{Z}(T, V, N)e^{-\gamma N_K}$. We address the identification of γ and the relationship between Ξ and thermodynamic functions as follows. The undetermined multiplier α in any probability function relates to the partition function

$$\text{microcanonical: } \alpha = \ln z/N, \qquad -kT \ln z = \mu, \qquad \alpha = \frac{-\mu}{kT}$$

$$\text{canonical: } \alpha = \ln \mathscr{Z}, \qquad -kT \ln \mathscr{Z} = A, \qquad \alpha = -\frac{A}{kT} \tag{E.10}$$

$$\text{grand canonical: } \alpha = \ln \Xi, \qquad -kT \ln \mathscr{Z} + \gamma N = A + \gamma N.$$

If $\gamma = -\mu/kT$, $N\gamma = -N\mu/kT = -G$, $A - G = PV$ and $\ln \Xi = PV$. In all cases μ/kT is identified with the undetermined multiplier associated with the constraint on N. The undetermined multiplier is often referred to as the absolute activity.

Following argument applied to fluctuations in E and $\bar{E}$ for the canonical ensemble, we write

$$\bar{N} \sum_{K} \mathscr{Z}(N_K V, T)e^{N_K \mu/kT} = \sum N_K \mathscr{Z}(N_K, V, T)e^{N_K \mu/kT}$$

$$(\sum \mathscr{Z} e^{N\mu/kT})\frac{\partial \bar{N}}{\partial \mu} + \bar{N}\left[\sum \frac{N_K}{kT} \mathscr{Z} e^{N_K \mu/kT}\right] = \sum \left(\frac{N_K}{kT}\right)^2 \mathscr{Z} e^{N_K/kT}$$

$$\Xi\left(\frac{\partial \bar{N}}{\partial \mu}\right) + \frac{\bar{N}}{kT}(\Xi)\sum N_K = \frac{N_K^2}{kT}(\Xi)$$

or

$$\overline{N^2} - (\bar{N})^2 = \sigma_N^2 = kT\frac{\partial \bar{N}}{\partial \mu}. \tag{E.11}$$

With μ on the order of kT, $\sigma/\bar{N}$ is order $N^{-1/2}$, which is the same order as relative fluctuations in E. The variation in N from one system in the ensemble to another is negligible: effectively all systems have $N = \bar{N}$.

We would have, again, anticipated this since $d\mu$ is the gradient that forces the transfer of particles between systems just as dT is the gradient that forces the transfer of thermal energy. The canonical ensemble is virtually at constant E and the grand canonical ensemble is virtually at constant N and E. Both are then basically identical to the microcanonical ensemble. Choice of the partition function is a matter of convenience.

CONTENT SOURCE AND FURTHER READING

1. T. L. Hill, *Introduction to Statistical Thermodynamics*, Chapter 1. Addison-Wesley, Reading, MA, 1960.

APPENDIX F

CLASSICAL CONSIDERATIONS

Although it is generally most convenient now to formulate statistical arguments in terms of quantized energy level (or state) populations, we should recognize that Gibbs, Maxwell, Boltzmann, and others, using classical mechanics, had brought our subject to a high level of refinement decades before quantum ideas were introduced. The most obvious contradictions between the two systems are discrete energy states and the uncertainty principle embodied in quantum mechanics vs the energy continuum and the absence of Planck's constant in classical mechanics. Although we have recognized that in some cases, quantum mechanical energy states are so closely spaced (translation and most often rotation) that the energy can be regarded as a continuous variable, we should still be concerned about congruence between the two formulations on the basis of Planck's constant.

In classical mechanics the total energy of N noninteracting, structureless particles is expressed by the Hamiltonian

$$H(\mathbf{p}, \mathbf{q}) = \sum^{3N} \frac{\mathbf{p}_i^2}{2m} + U(q_1, \ldots, q_{3N}) \tag{F.1}$$

where $\mathbf{p}$, $\mathbf{q}$ refer to the 3N component vectors $\mathbf{p}_1, \ldots, \mathbf{p}_{3N}$, $q_1, \ldots, q_{3N}$ and U is the total potential energy of interaction between the N particles. A single particle in the system would be instantaneously characterized by its momentum value in the above sum and by its coordinate specification. The single particle could be represented by a point in the six-dimensional coordinate axial system p_x, p_y, p_z, q_x, q_y, q_z. The coordinate-momentum space is referred to generally as phase space and this particular six-dimensional phase space is called μ (for molecule) space. The N particle system could be represented by N points in μ space. We could also represent the same system by a single point in a 6N-dimensional space. This phase space we can refer to as Γ (for gas) space. The reader will recognize that μ space is convenient for displaying the momentum and coordinate variables in the N particle system of a microcanonical ensemble while Γ space is convenient for displaying the elements of a canonical ensemble.

For a single particle in μ space we write

$$\mathscr{H}(\mathbf{p}, \mathbf{q}) = \frac{p_x^2 + p_y^2 + p_z^2}{2m} = \epsilon.$$

The classical equivalent to the quantum expression

$$z_{\text{quant}} = \sum^{i} e^{-\epsilon_i/kT}$$

is the integral over accessible regions of phase space

$$z_{\text{class}} = \underbrace{\int \cdots \int}_{6} e^{-(p_x^2 + p_y^2 + p_z^2)/2mkT}\, dx\,dy\,dz\,dp_x\,dp_y\,dp_z. \tag{F.2}$$

The momentum values should be permitted to run from $-\infty$ to ∞, recognizing that very low or very high values of p^2 will contribute very little to the integral. For a confined particle the positional coordinates should run over the dimensions of the container. The 6-fold integral should be written in the form $\int_{-\infty}^{\infty}\int_{-\infty}^{\infty}\int_{-\infty}^{\infty}\int_0^a\int_0^b\int_0^c$.

Since dx, dy, dz do not enter the momentum expression, integration over these variables produces $abc = \text{V}$. Also, since the momentum variables are independent, they are separable and (for a single variable)

$$\int_{-\infty}^{\infty} e^{-p_x^2/2mkT}\, dp_x = 2\int_0^{\infty} e^{-p_x^2/2mkT}\, dp_x \tag{F.3}$$

which is of the general form $\int_0^{\infty} e^{-a^2x^2}dx$ with $a = (2mkT)^{-1/2}$ and for which the definite integral is $\pi^{1/2}/2a$. Each of the integrals over a momentum component contributes a factor $(2\pi mkT)^{1/2}$ to the partition function which becomes

$$z_{\text{class}} = (2\pi mkT)^{3/2}\text{V}. \tag{F.4}$$

The quantum development, which, because of close spacing of the energy levels,

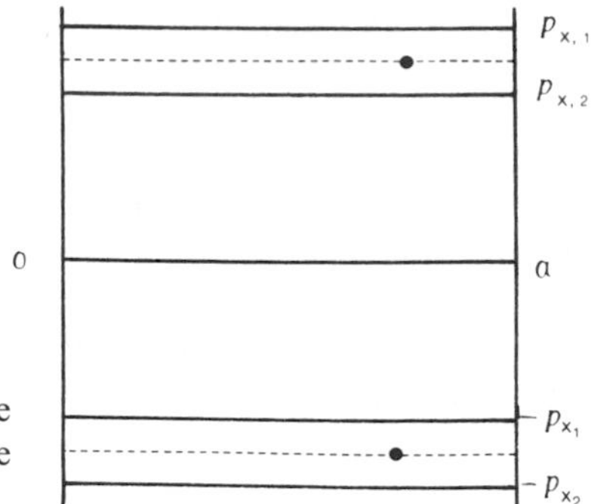

Figure F.1. Illustration of the partition of phase space into volumes h^{-1} for every p, q coordinate pair.

was approximated as an integral, provides

$$z_{\text{quant}} = \left(\frac{2\pi m k \mathrm{T}}{h^2}\right)^{3/2} \mathrm{V}. \tag{F.5}$$

The solution to this problem lies in both discrete energy levels and the uncertainty principle, $\Delta p \Delta q \geqslant h$ for every conjugate variable pair, in quantum mechanical argument. There is no such uncertainty associated with classical mechanical argument and Planck's constant does not formally enter. The contradiction can be resolved if we recognize that subdivisions of phase space smaller than h^3 apparently have no meaning. For a particle moving in one dimension, as in (Figure F.1) the classical, one-dimensional momentum component p_x for a particle constrained to motion $0 < x < a$ can be represented as a point in phase space region defined by $\Delta\epsilon$ for a quantum step where the phase volume $2a[p_{x_2} - p_{x_1}] = \sqrt{2m}(\epsilon_2^{1/2} - \epsilon_1^{1/2})$ is associated with all points $(\cdot)$ $|p_{x_1}| < |p_x| < |p_{x_2}|$. Since $p_x = (2m\epsilon)^{1/2}$ classically and $(2a/h)(2m\epsilon)^{1/2}$ quantally, where p_{x_1} and p_{x_2} are associated with successive energy levels ϵ_1 and ϵ_2

$$[p_{x,2} - p_{x,1}]_{\text{quant}} = \frac{1}{h}[p_{x,2} - p_{x,1}]_{\text{class}}.$$

In three dimensions the lines become constant energy surfaces and the momentum-coordinate area becomes a volume. The imposition of allowed energy surfaces has the effect of partitioning phase space into meaningful volumes of h^{-1} for every coordinate pair p, q. We then recognize the necessity for introducing this factor as an empirical correction to the classical relationships,

$$\sum e^{-\epsilon_i/k\mathrm{T}} = \frac{1}{h^{3\mathrm{N}}} \underset{\substack{\text{phase}\\ \text{space}}}{\int \cdots \int} e^{-\epsilon_i/k\mathrm{T}} dx_1, \ldots, dp_{3\mathrm{N}}. \tag{F.6}$$

The factor $1/h^{3N}$ will clearly have no impact in any of the circumstances in which the partition function logarithm enters as a derivative. When the partition function logarithm enters directly as in S, A, and G or their related functions the factor will appear as $3N \ln h$ but, since in most cases we will be interested only in difference terms involving these functions, it will have no influence on such answers. As before, we should recognize this result as applying to translation, the only mode of motion available to structureless, independent nonlocalized particles. Monatomic gases conform to this model.

Polyatomic and diatomic molecules will have additional modes of motion. Spherical polar coordinates will be most convenient to the classical rotation problem. For the general polyatomic molecule with three principle axes of inertia (01, 02, 03) we will refer these axes to the x, z plane to obtain two angles (θ and ϕ) as indicated in Figure F.2. The third angle ψ describes those orientations of the molecule for fixed θ, ϕ that can be imagined as rotations about the 03 axis. Such rotation is not meaningful in any linear molecule and the complete spatial designation is accomplished by the angles θ and ϕ.

The rotational kinetic energy of the polyatomic nonlinear molecule with moments of inertia I_A, I_B, I_C is

$$\epsilon_{\text{class}} = \frac{1}{2I_A}\left[\sin\phi p_\theta - \frac{\cos\psi}{\sin\theta}(p_\theta - \cos\theta p_\psi)\right] + \frac{1}{2I_B}\left[\frac{\sin\psi}{\sin\theta}(p_\theta - \cos\theta p_\psi) + \cos\psi p_\theta\right]^2 + \frac{1}{2I_C}p_\psi^2. \tag{F.7}$$

If new variables

$$\xi = \sin\psi p_\theta - \frac{\cos\psi}{\sin\theta}(p_\theta - \cos\theta p_\psi)$$

$$\eta = \frac{\sin\psi}{\sin\theta}(p_\phi - \cos\theta p_\psi) - \cos\psi p_\theta \tag{F.8}$$

$$\zeta = p_\psi$$

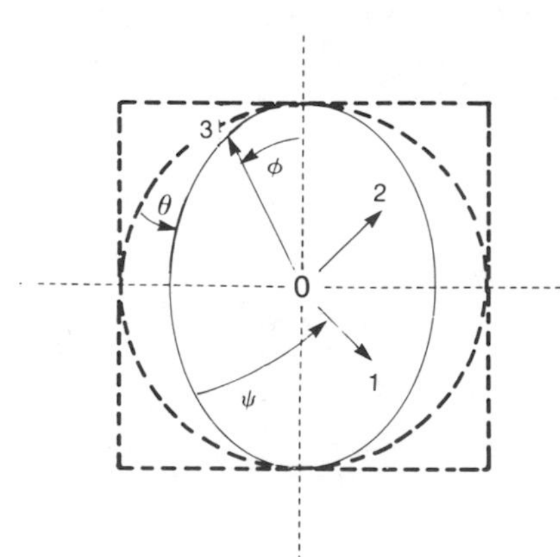

Figure F.2. Identification of Eulerian angles θ, ϕ and ψ for polyatomic molecule rotations.

are introduced, the expression for the partition function is greatly simplified,

$$z_{\mathrm{class,rot}} = \frac{1}{h^3}\int \cdots \int \exp\left(-\frac{\xi^2}{2I_{\mathrm{A}}} + \frac{\eta^2}{2I_{\mathrm{B}}} + \frac{\zeta^2}{2I_{\mathrm{C}}}\right) \Big/ k\mathrm{T} \sin\theta d\xi d\zeta^2 d\theta d\phi d\psi,$$

$dp_\theta dp_\phi dp_\psi$ having transformed into $\sin\theta d\xi d\eta d\zeta$. Each of the squared terms integrates in the same manner as the translational momentum components to, e.g.,

$$\int_{-\infty}^{\infty} \exp(-\xi/2k\mathrm{T}) d\xi = (2\pi I_{\mathrm{A}} k\mathrm{T})^{1/2}.$$

The orientational coordinates considered separately provide

$$\int_0^{2\pi}\int_0^{2\pi}\int_0^{\pi} \sin\theta d\theta d\phi d\psi = (2)(2\pi)(2\pi).$$

All orientations are counted once only when ϕ and ψ run over 0 to 2π while θ runs over 0 to π. The complete partition function is

$$z_{\mathrm{class,rot}} = \frac{1}{h^3\sigma_r}(2\pi k\mathrm{T})^{3/2}(I_{\mathrm{A}} I_{\mathrm{B}} I_{\mathrm{C}})^{1/2} 8\pi^2 \tag{F.10}$$

where the symmetry factor σ_r excludes those regions of phase space in which repeated orientations occur due to symmetry within the molecule. It is the number of indistinguishable structures that occurs in a complete rotation.

The diatomic or linear polyatomic molecule, as indicated above, has a much more simple energy expression

$$\epsilon_{\mathrm{class,rot}} = \frac{1}{2I}(p_\theta^2 + \frac{1}{\sin^2\theta} p_\phi^2) \tag{F.11}$$

where I is the moment of inertia of the molecule about an axis perpendicular to the axis of the molecule and passing through its mass center. The partition function is represented as

$$z_{\mathrm{class,rot}} = \frac{1}{h^2}\int_0^{2\pi}\int_0^{2\pi}\int_0^{\pi} \exp\left[\left(-p_\theta^2 + \frac{1}{\sin\theta}p_\phi^2\right) \Big/ 2Ik\mathrm{T}\right] dp_\theta dp_\phi d\theta d\phi.$$

The limits applied to the integrals follow from arguments developed above. The integration produces

$$z_{\mathrm{class,rot}} = \frac{2\pi}{h^2\sigma_r}[(2\pi I k\mathrm{T})^{1/2}]^2 2 = \frac{8\pi^2 I k\mathrm{T}}{h^2\sigma_r}. \tag{F.12}$$

If the molecule is homonuclear a symmetry factor of 2 is required; otherwise σ_r is 1.

The $3n-6$ (or $3n-5$ if the molecule is linear) internal degrees of freedom of an n atom molecule are taken to be independent vibrational modes and Newton's law of motion for a particle in a parabolic field is applied to each mode considered as a one-dimensional oscillator. In momentum–spatial coordinates the energy is

$$\epsilon_{\text{vib}} = \frac{1}{2m} p_x^2 + \frac{m}{2}(2\pi v^2)x^2 \tag{F.13}$$

where the first squared term is the kinetic energy and the second squared term is the potential energy. The partition function is then, introducing the factor h^{-1} for the single momentum–coordinate pair,

$$z_{\text{class}} = \frac{1}{h}\int_{-\infty}^{\infty}\int_{-\infty}^{\infty} \exp -(p_x^2/2\pi m k\text{T} + mv^2x^2/2k\text{T})dp_x dx \tag{F.14}$$

which is separable into

$$\frac{1}{h}\int_{0}^{\infty} e^{-p_x^2/2mk\text{T}} dp_x = \left(\frac{2\pi m k\text{T}}{h}\right)^{1/2}$$

and

$$\int_{-\infty}^{\infty} e^{-mv^2x^2/2k\text{T}} dx = \frac{2\pi(k\text{T})^{1/2}}{m^{1/2}2\pi v} = \frac{(k\text{T})^{1/2}}{m^{1/2}v}$$

so that

$$z_{\text{class}} = k\text{T}/hv.$$

For a polyatomic nonlinear gas phase molecule these results summarize as

$$z_{\text{class}} = \left[\frac{V}{h^3}(2\pi m k\text{T})\right]\left[\frac{\pi^{1/2}}{h^3\sigma_r}(8\pi^2 k\text{T})^{3/2} I_A I_B I_C\right]\left[\pi^{v_i}\frac{k\text{T}}{hv_i}\right] \tag{F.15}$$

where the final term is the product of $3n-6$ or $3n-5$ vibrational terms. To obtain the energy we utilize $\epsilon = k\text{T}^2\partial \ln z/\partial\text{T}$ obtaining

$$\epsilon_{\text{class}} = \tfrac{3}{2}k\text{T} + \tfrac{3}{2}k\text{T} + (3n-6)k\text{T} \tag{F.16}$$

and for the heat capacity

$$(\partial\epsilon/\partial\text{T})_V = \tfrac{3}{2}k + \tfrac{3}{2}k + (3n-6)k. \tag{F.17}$$

For a linear molecule the classical heat capacity is

$$\left(\frac{\partial \epsilon}{\partial \mathrm{T}}\right)_{\mathrm{V}} = \tfrac{3}{2}k + k + (3n-5)k \qquad \text{(F.18)}$$

reflecting the 2 degrees of rotational freedom and the $3n-5$ degrees of vibrational freedom. For each of the squared terms that appeared in the Hamiltonian, there appears an additive term of $\frac{1}{2}k$ in the heat capacity. This demonstration (not a proof) is known as the principle of equipartition of energy. Classically the energy required to effect a change in temperature of one degree is $\frac{1}{2}\mathrm{N}k$ per mode of independent motion.

In the quantum sense this will be reflected for motions in which the successive energy levels are closely spaced with respect to $k\mathrm{T}$, i.e., if in $\sum e^{-\epsilon_i/k\mathrm{T}}$ the difference $\epsilon_{i+1} - \epsilon_i$ is small with respect to $k\mathrm{T}$, the sum of individual terms can be regarded as a continuous function. For translation this is always true, for rotation it is true except for very low temperature, and for vibration it is almost never true. In this latter case there is effectively no vibration except that associated with the ground state at low temperature and the increasing heat capacity contribution with increasing temperature is given by

$$\frac{\partial^2}{\partial \mathrm{T}^2}\left[\prod \frac{e - h\nu/2k\mathrm{T}}{1 - e^{-h\nu/k\mathrm{T}}}\right]. \qquad \text{(F.19)}$$

The heat capacity associated with vibration does not reach its classical value except at very high temperature (order 10^3) for other than very light molecules.

For many particle systems where each particle is independent of the presence or absence of other particles, the product terms for the partition function associated with the various single particle modes of motion as set down in Eqn. (F.15) is expanded to the total modes of motion of N particles. This leads to, in each case, z^{N} for the system partition function. Internal modes of motion, rotation and, vibration are taken to be uninfluenced by the presence or absence of other particles regardless of particle–particle interaction. These modes of motion are also independent of the statistical feature of localization vs nonlocalization. In effect they are always localized. Translation is taken to be nonlocalized in the gas phase and the division by N! introduced to compensate, $z_{\mathrm{t,system}} = z_{\mathrm{t}}^{\mathrm{N}}/\mathrm{N}!$. Also the presence of particle–particle interactions, which increase with increasing particle density, will be reflected through the volume dependence of the partition function.

Where the Hamiltonian for the translational kinetic energy in an independent particle system is

$$H(\mathbf{p}, \mathbf{q}) = \frac{1}{2m}\sum^{3\mathrm{N}} p_i^2$$

and in an N interacting particle system

$$H(\mathbf{p},\mathbf{q}) = \frac{1}{2m}\sum^{3N} p_i^2 + Q(q_1,\ldots,q_{3N}) \tag{F.20}$$

where $Qq_1,\ldots,q_{3N}$ represents all possible interactions and

$$\mathscr{H}(\mathbf{p},\mathbf{q}) = \frac{1}{2m}\sum^{3N} p_i^2 + \sum_{i>j}^{N} u(r_{ij}) \tag{F.21}$$

where $u(r_{ij})$ represents pairwise interactions, the partition function becomes

$$z = \frac{1}{N!}\frac{1}{h^{3N}}\prod^{3N}(2\pi m k T)^{1/2}\left[\int_N\cdots\int e^{-u(r_{ij})/kT}\,d\tau_1,\ldots,d\tau_N\right] \tag{F.22}$$

where $d\tau_1,\ldots,d\tau_N$ are volume elements $d\tau_1 = dq_1 dq_2 dq_3$, $d\tau_2 = d\tau_4 d\tau_5 d\tau_6,\ldots$. The bracketed term is referred to as the configurational integral.

CONTENT SOURCES AND FURTHER READING

1. J. E. Mayer and M. G. Mayer, *Statistical Mechanics*, Chapter 7. Wiley-Interscience, New York, 1977.
2. E. A. Moelwyn–Hughes, *Physical Chemistry*, Chapter IV. Pergamon Press, London, 1961.

INDEX